Climate Change Management

Series Editor

Walter Leal Filho, International Climate Change Information and Research Programme, Hamburg University of Applied Sciences, Hamburg, Germany

The aim of this book series is to provide an authoritative source of information on climate change management, with an emphasis on projects, case studies and practical initiatives – all of which may help to address a problem with a global scope, but the impacts of which are mostly local. As the world actively seeks ways to cope with the effects of climate change and global warming, such as floods, droughts, rising sea levels and landscape changes, there is a vital need for reliable information and data to support the efforts pursued by local governments, NGOs and other organizations to address the problems associated with climate change. This series welcomes monographs and contributed volumes written for an academic and professional audience, as well as peer-reviewed conference proceedings. Relevant topics include but are not limited to water conservation, disaster prevention and management, and agriculture, as well as regional studies and documentation of trends. Thanks to its interdisciplinary focus, the series aims to concretely contribute to a better understanding of the state-of-the-art of climate change adaptation, and of the tools with which it can be implemented on the ground.

Notes on the quality assurance and peer review of this publication

Prior to publication, the quality of the works published in this series is double blind reviewed by external referees appointed by the editor. The referees are not aware of the author's name when performing the review; the referees' names are not disclosed.

Rubee Singh · Walter Leal Filho
Editors

Climate Neutrality Through Smart Eco-Innovation and Environmental Sustainability

Springer

Editors
Rubee Singh
Institute of Business Management
GLA University
Mathura, Uttar Pradesh, India

Walter Leal Filho
Faculty of Life Sciences
Hamburg University of Applied Sciences
Hamburg, Germany

ISSN 1610-2002 ISSN 1610-2010 (electronic)
Climate Change Management
ISBN 978-3-031-83249-9 ISBN 978-3-031-83250-5 (eBook)
https://doi.org/10.1007/978-3-031-83250-5

This Springer imprint is published by the registered company Springer Nature Switzerland AG
The registered company address is: Gewerbestrasse 11, 6330 Cham, Switzerland

Preface

Climate neutrality represents a transformative approach to combating climate change, centered on balancing greenhouse gas emissions with their removal from the atmosphere, ultimately aiming for net-zero emissions. Simultaneously, the concept of a sustainable ecosystem ensures the responsible management and conservation of natural resources, enabling ecosystems to thrive while meeting the needs of current and future generations. Together, these frameworks provide a pathway to mitigate the impacts of global warming, fostering resilience, adaptability, and sustainability. This book, *Climate Neutrality and Sustainable Ecosystems: Practical Approaches to Addressing Global Warming*, explores these critical themes, offering a comprehensive understanding of the strategies, challenges, and opportunities inherent in addressing climate change.

The book is organized into four parts where Part I lays the groundwork by discussing the essential strategies and perspectives needed to achieve climate neutrality. Chapter 1 begins with an exploration of sustainable production and consumption practices, underscoring their role as vital pathways to resilience in the face of climate change. Chapter 2 highlights the critical roles of private, public, and individual actors in fostering climate adaptation. By examining case studies and best practices, this chapter demonstrates how collaborative efforts across sectors can enhance societal readiness for climate challenges. Chapter 3 delves into the emerging field of climate informatics, illustrating how the interplay of data politics and technology shapes policies and sustainability efforts. Chapter 4 provides an in-depth analysis of the environmental impacts of ice cap reduction, glacial melt, and sea-level rise, while discussing adaptive strategies to mitigate these effects. Chapter 5 in this part offers practical insights into navigating carbon neutrality by addressing environmental challenges through policy integration and technological interventions.

Part II focuses on the transformative potential of innovation in advancing sustainability objectives. Chapter 6 investigates how small and medium enterprises (SMEs) can play a pivotal role in achieving climate neutrality through the adoption of smart eco-innovations. By highlighting the scalability and adaptability of such innovations, this chapter underscores their significance in reducing emissions and fostering economic resilience. Chapter 7 shifts the discussion to artificial intelligence that how

AI technologies can revolutionize sustainability practices by enhancing decision-making, optimizing resource use, and driving innovation. Chapter 8 takes a regional approach by examining eco-innovative strategies for building climate-neutral food systems in the Middle East and North Africa. It discusses how localized solutions can address global challenges, particularly in regions vulnerable to climate impacts. Chapter 9 highlights the importance of voluntary education and awareness campaigns as catalysts for adopting eco-innovations. Chapter 10 concludes this part with a comparative study of human factors influencing sustainable technology adoption in developing countries.

Part III of the book showcases the potential of emerging technologies in advancing climate goals. Chapter 11 presents a bibliometric analysis of alternative energy technologies, evaluating their policies and strategies for promoting a sustainable future. By synthesizing global trends and research, this chapter provides a comprehensive overview of the energy sector's transformation. Chapter 12 examines institutional responses to environmental policies, focusing on how smart ecosystems integrate policy frameworks and technological advancements. Chapter 13 explores the opportunities and challenges associated with emerging green technologies, offering future pathways for achieving environmental sustainability. This chapter underscores the need for continuous innovation and investment to address evolving climate challenges. Chapter 14 concludes this part with editorial insights into the transformation of transport, energy, and communication systems in South Asia.

Part IV provides a critical perspective on the complexities and ethical considerations in climate action. Chapter 15 explores the behavioral dimensions of sustainability by analyzing how behavioral economics influences climate neutrality initiatives. It emphasizes the need for understanding human behavior to drive effective policies and practices. Chapter 16 evaluates the impacts of climate policy changes in Australia, weighing their advantages and limitations. Chapter 17 outlines the causes and impacts of climate change while proposing actionable strategies for achieving climate neutrality. Chapter 18 critically examines the unintended consequences of climate neutrality and smart eco-innovation in both developed and developing nations. By addressing the "dark side" of innovation, this chapter calls for ethical and equitable approaches to sustainability. Chapter 19 discusses the importance of linking global climate finance initiatives with local impacts.

This book serves as a valuable resource for a wide range of stakeholders, including policymakers, researchers, educators, industry professionals, and students. By integrating theoretical insights with practical examples and case studies, it equips readers with the knowledge and tools necessary to drive meaningful climate action. As the world grapples with the challenges of climate change, this book offers not just a roadmap for achieving climate neutrality but also a vision for a sustainable and resilient future. Moving forward, addressing global warming will require a collective commitment to innovation, collaboration, and transformative change. This book aims to inspire and guide these efforts, contributing to a shared goal of preserving our planet for generations to come.

Book has divided in four parts in which there is total 19 chapters from different countries such as India, Sri Lanka, Ghana, Oman, Nigeria, Maldives, UK, USA,

Australia, and Dubai, this book is the result of a collective effort by scholars and experts from various fields, bringing together diverse perspectives and in-depth research. We extend our heartfelt gratitude to all contributors and reviewers whose efforts have made this publication possible. Special thanks go to Springer Nature for their support and dedication to advancing scholarly work in environmental sustainability. In conclusion, *Climate Neutrality Through Smart Eco-Innovation and Environmental Sustainability* aims to inspire and equip readers with the knowledge and tools necessary to drive sustainable world and environmental stewardship. It is our hope that this book serves as a valuable resource, guiding the global community towards a Climate neutral future and fostering a sustainable and resilient innovative ecosystem.

Mathura, India Rubee Singh
Hamburg, Germany Walter Leal Filho

Acknowledgements

We would like to express our gratitude to all those who have contributed to the creation of this book. Firstly, we thank the researchers and experts who have shared their knowledge and insights with us, providing valuable information on the current state of *Climate Neutrality Through Smart Eco-Innovation and Environmental Sustainability*. We also extend our appreciation to the organizations and institutions that have supported our work, including government agencies, non-governmental organizations, and academic institutions.

Finally, we would like to acknowledge the invaluable contributions of our colleagues and editors, who have worked tirelessly to ensure that this book is informative, engaging, and accessible to a wide range of readers. We hope that this book will inspire further research and action towards a Climate Neutral Policies and Smart Eco-Innovation System to promote Environmental Sustainability as per United Nations agenda. We would like to express our sincere gratitude to the experts and stakeholders who contributed their valuable insights and perspectives to this book. Their contributions have been instrumental in shaping our understanding of the challenges and opportunities involved in achieving climate neutrality to addressing global warming with use of eco-innovation and environmental sustainability at globally. We hope that this book will serve as a useful resource for policymakers, practitioners, and researchers working towards a zero-carbon future for the continent.

We are immensely thankful to *Book Series-Editors, Prof. Walter Leal Filho, Editorial Director/Publishing Editor, Dr. Christian Witschel* and the entire team at Springer Nature for bringing out this publication in a very professional and timely manner.

Rubee Singh
Walter Leal Filho

About This Book

Climate Neutrality and Sustainable Ecosystems: Practical Approaches to Addressing Global Warming is an insightful exploration of how climate neutrality and sustainability can be achieved through interdisciplinary strategies, innovation, and collective action. The book delves into the core concepts of balancing greenhouse gas emissions and conserving ecosystems, presenting a comprehensive framework to combat global warming and its associated challenges.

Divided into four parts, Part I of the book lays the foundation by discussing pathways to resilience through sustainable production, climate informatics, and adaptation strategies. It examines the impacts of ice cap reduction and sea-level rise, offering adaptive measures. It examines the contributions of private, public, and individual actors in promoting climate adaptation and resilience, stressing the importance of collaboration across sectors. Part II explores the role of small and medium enterprises (SMEs), artificial intelligence, and eco-innovative solutions in addressing climate challenges, alongside examining human factors and educational awareness as catalysts for change. Part III gives an in-depth analysis of how technological advancements can shape sustainable futures. It showcases emerging technologies like alternative energy, green innovations, and institutional responses that support climate-neutral systems, with specific insights into regional adaptations in South Asia. Part IV critically evaluates the behavioral, ethical, and financial dimensions of climate policies, addressing the unintended consequences and complexities of achieving sustainability. Linking global climate finance with local impacts, it highlights how natural capital can bridge international efforts with community-level sustainability. This book serves as a critical resource for policymakers, researchers, and professionals, providing actionable insights and strategies. It not only highlights the challenges of global warming but also inspires a vision of innovation and resilience to build a sustainable and climate-neutral future.

Contents

Editors and Contributors

About the Editors

Dr. Rubee Singh (Ph.D.) is an Assistant Professor of Human Resource at Institute of Business Management, GLA University India. She is a Post-Doctoral Researcher (D. Litt.) at Department of Management Studies, Kumaun University, Nainital, India. She holds Ph.D. in Management from Noida International University India, an MBA in HR from Dr. A. P. J. Abdul Kalam Technical University India. She has more than 10 years of experience in research and teaching. Apart from teaching and research, she has handled many administrative positions in premier institutes affiliated with A.K.T.U. She has conducted a number of FDPs, Conferences, Workshops, and HR Conclaves in the area of Sustainability, CSR, AI, HR, Human Factors, Women Empowerment and Human rights. She is a United Nations Volunteer, and a contributing author for Voices of Youth (UNICEF). She is an honorary life member of Women Economic Forum and National Secretary of IATM (Youth Wing) International Anti-Terrorism Movement India. Dr. Singh is a Managing Editor of IJARSH India. She is serving on the Editorial Board of more than 30 international journals. She serves as the Advisory Board Member of the *Heliyon Journal*, Elsevier, Research in Globalisation, and reviewer of many reputed journals like *European Journal of Innovation Management* (EJIM), *British Food Journal*, *Sustainability*, *Social Responsibility Journal* etc.

She is life member of International Association of Engineers, Thailand, IMRTC USA, ISRD UK etc. She is a bestselling author with significant experience as an excellent Editor, Eminent researcher, great academician, Global Ambassador (Children's) and Social Activist. She has presented many papers at national and international conferences, published 20 academic books including Springer, Elsevier, De Gruyter Nova Science Publishers, and more than 75+ research papers in SCI, ABDC-A, Scopus and refereed/UGC index journals. Her areas of research are sustainability, Digital HRM, Sustainable HRM, Circular Economy, Digital CE, Net Zero Emissions, Climate Change, Industry 4.0, Artificial Intelligence, Blockchain Technology and Corporate Social Responsibility. Being a thought leader, her books are honored by being added to the library, Ohio Christian University, University of California, Corporacion Universitaria Unisabanta, UMass Chan Medical School, Tunix Community College, USA, Aston University UK, Dow University of Health Sciences, Pakistan, University of Port Harcourt, Nigeria. Her book titled *Training and Development* was added to the MBA syllabus of "The Institute of Aeronautical Engineering" (Jawaharlal Nehru Technological University, Hyderabad). Dr. Singh has been honored with many prestigious awards, which include Most Influential Woman Leaders in 2022, consecutive Best Researchers Award of the Year-2022–2023 and 2023–2024 by GLA University, India, Child Ambassador Award of The Year 2022, Iconic Faculty Award 2019, International Young Researchers Award-2021, Young Scientist Award-2020, and Nari Ratan Award-2023, India Book of Records-2020 and Asia Book of Records Holder for 2021. The list of awards and appreciation is long, but Dr. Singh has always prioritized her work over recognition, making her unique and commendable in her field.

Prof. Walter Leal Filho (D. Litt., Ph.D.) holds the Chairs of Climate Change Management at the Hamburg University of Applied Sciences (Germany), and Environment and Technology at Manchester Metropolitan University (UK). He directs the Research and Transfer Centre "Sustainability Development and Climate Change Management". His main research interests are in the fields of sustainable development and climate change, also including aspects of climate change and health. Professor Leal has been author, co-author, editor or co-editor in many publications on matters related to environment, sustainable development and climate change issues. He has in excess of 700 publications to his credit, which includes 193 books, 102 book chapters and over 430 papers. His field experience in over 70 countries, involves missions undertaken on behalf of various international organizations (e.g., European Union, World Bank, OECD, UNESCO, UNEP) and attendance to specialized events in North America, Latin America/Caribbean, Europe, Africa, Asia and Pacific Region, and the Middle East. Among the many awards he has received, it can be mentioned the 1999 "Leading Editor of the Year Award" and the 2000 "Editor of the Year Award", by MCB University Press, the "Eco-Citizen Award" (2001) in Rio de Janeiro and the Aurelio Giuseppe Prize awarded by the Italian Council of Ministers for his environmental work. Moreover, he was awarded the "Golden Page Award" (April 2003) in London for his editorial work and the North Sea Star Award (2006) by the EU-Interreg Secretariat for his project work in the North Sea Region. In the field of sustainable development, Prof. Walter Leal is the founder of the European School of Sustainability Science and Research and of the Inter-University Sustainable Development Research Programme (IUSDRP). He holds a record of publications on sustainability issues, some of which are available here: https://www.haw-hamburg.de/en/ftz-nk/programmes/iusdrp/ (please scroll down and click on "sharing"). He is also editor of the *World Sustainability* Series with Springer. In addition, he is the founding editor of the *International Journal of Sustainability in Higher Education* (IJSHE) launched in the year 2000. IJSHE is the only journal which fully focuses on sustainability in a higher education context. Professor Walter Leal is also Editor-in-Chief of the Encyclopedia of the UN Sustainable Development Goals currently being prepared. With 17 volumes, one devoted to each SDG, more than 1700 chapters and in excess of 2300 authors, this is the largest editorial project on matters related to sustainable development, ever undertaken.

Contributors

Dawoud Almohammad Department of Business Administration, Faculty of Economics and Administrative Sciences, Artuklu University, Mardin, Turkey

A. Arun Kumar ICFAI Law School, ICFAI Foundation for Higher Education (Deemed to Be University), Hyderabad, Telangana, India

Prince Yeboah Asare Department of Business and Social Sciences Education, University of Cape Coast, Cape Coast, Ghana

Raed Awashreh A'Sharqiyah University, Ibra, Oman

Orhan Balcı Department of Business Administration, Faculty of Economics and Administrative Sciences, Karatekin University, Çankırı, Turkey

Gowri B. Chanal ICFAI Law School, ICFAI Foundation for Higher Education (Deemed to Be University), Hyderabad, Telangana, India

Narayanage Jayantha Dewasiri Department of Accountancy and Finance, Faculty of Management Studies, Sabaragamuwa University of Sri Lanka, Belihuloya, Sri Lanka;
Qasim Ibrahim School of Business, Villa College, Malé, Maldives

Hiranya Dissanayake Department of Accountancy, Wayamba University of Sri Lanka, Kuliyapitiya, Sri Lanka

Jacinta Dsilva SEE Research Center, SEE Institute, The Sustainable City, Dubai, UAE

Walter Leal Filho Faculty of Life Sciences, Humburg University of Applied Sciences, Humburg, Germany

Damith Gangodawilage Qasim Ibrahim School of Business, Villa College, Malé, Maldives

Vikas Garg Symbiosis Institute of Business Management, Noida, Symbiosis International (Deemed University), Pune, India

M. C. R. Jayalath Department of Commerce and Financial Management, University of Kelaniya, Kelaniya, Sri Lanka

P. G. S. Amila Jayarathne De Montfort University, Leicester, UK

Solomon Jeresa University of Huddersfield, Huddersfield, UK

Amit Joshi Department of Management Studies, Kumaun University, Nainital, Uttarakhand, India

Som Aditya Juyal Himalayan School of Management Studies, Swami Rama Himalayan University, Dehradun, India

Vinay Kandpal Department of Management Studies, Graphic Era Deemed to Be University Dehradun, Uttarakhand, India

Raghuveer Katragadda School of Management, V.R. Siddhartha Engineering College, Vijayawada, India

Pooja Kaushik Department of Management Studies, Galgotias College of Engineering and Technology, Greater Noida, Uttar Pradesh, India

Shahbaz Khan Department of Industrial Engineering, University of Tabuk, Tabuk, Kingdom of Saudi Arabia

Cumali Kılıç Department of Business Administration, Faculty of Economics and Administrative Sciences, Artuklu University, Mardin, Turkey

Kishore Kumar School of Management, Bennett University, Greater Noida, India

Lucy Lohiya Amity Institute of Public Policy, Amity University, Noida, India

Ali Najeeb Qasim Ibrahim School of Business, Villa College, Maldives

Donkor Nawaah Department of Finance and Accounting, Woxsen University, Telangana, India

Michael Oyedele Oyenuga School of Business, Woxsen University, Telangana, India

Wei Qian UniSA Business, University of South Australia, Adelaide, Australia

Michael Ayikwei Quarshie Department of Entrepreneurship and Business Sciences, University of Energy and Natural Resources, Sunyani, Ghana

Seema Rani Department of Commerce, Kurukshetra University, Haryana, India

Mananage Shanika Hansini Rathnasiri Department of Marketing Management, Faculty of Management Studies, Sabaragamuwa University of Sri Lanka, Belihuloya, Sri Lanka

Wasswa Shafik School of Digital Science, Universiti Brunei Darussalam, Gadong, Brunei;
Dig Connectivity Research Laboratory (DCRLab), Kampala, Uganda

Ankita Sharma School of Management, IILM University, Greater Noida, India

Akshay Singh School of Management Studies, Indra Gandhi National Open University, New Delhi, India

Amit Kumar Singh Department of Commerce, Delhi School of Economics, New Delhi, India

Rubee Singh Institute of Business Management, GLA University, Mathura, Uttar Pradesh, India;
Department of Management Studies, Kumaun University, Nainital, Uttarakhand, India

T. Sowmyya Department of Forensic Science, University of College of Science, Osmania University, Hyderabad, Telangana, India

Rakesh Suryadevara ICFAI Law School, ICFAI Foundation for Higher Education (Deemed to Be University), Hyderabad, Telangana, India

Lalit Mohan Tewari Department of Botany, DSB Campus, Kumaun University, Nainital, Uttarakhand, India

Apurvaa Trivedi Himalayan School of Management Studies, Swami Rama Himalayan University, Dehradun, India

Valentin Penchev Vasilev Higher School of Security and Economics, Plovdiv, Bulgaria

Lishanthi Wijewardene Business Communication Unit, University of Sri Jayewardenepura, Nugegoda, Sri Lanka

Ming Xue UniSA Business, University of South Australia, Adelaide, Australia

Part I
Climate Neutrality and Sustainable Transition

Chapter 1
Advancing Climate Neutrality Through Sustainable Production and Consumption: Pathways to a Resilient Future

Narayanage Jayantha Dewasiri and Rubee Singh

Abstract This chapter discusses climate neutrality in the context of sustainable production and sustainable consumption as crucial approaches to combating global warming and achieving ecological harmony. Using the Systematic Literature Review (SLR) approach and proceeding from the Population, Intervention, Context, and Outcome (PICO) framework, the study integrates global views and regional findings to identify potential barriers, enablers, and trends toward sustainability. The study sample consisted of thirty-nine seminal studies. Specific recommendations state the shortcomings that have been holding the development back, including the financial and technological factors affecting the sustainability of production, cultural and cost considerations regarding consumption, and the disparities between the global and regional trends toward climate goals. Technological and policy change and governance approaches are suggested to be the four critical game-change levers for progress toward climate neutrality. Moreover, cases always highlight excellence from global adversaries and successes derived from mass actions that are cardinal to South Asia's advancement. The implications for the chapter are policy practitioners, organisations, and civil society by offering the roadmap that should guide stakeholder sustainability integration, advance the capabilities of new technologies, and build global collaboration. This research feeds into the existing literature on climate change by offering a rational and research-backed solution to the problem of climate change.

N. J. Dewasiri (✉)
Department of Accountancy and Finance, Faculty of Management Studies, Sabaragamuwa University of Sri Lanka, Belihuloya, Sri Lanka
e-mail: jayantha@mgt.sab.ac.lk

Qasim Ibrahim School of Business, Villa College, Malé, Maldives

R. Singh
Institute of Business Management, GLA University, Mathura, India

Department of Management Studies, Kumaun University, Nainital, Uttrakhand, India

R. Singh and W. L. Filho (eds.), *Climate Neutrality Through Smart Eco-Innovation and Environmental Sustainability*, Climate Change Management,
https://doi.org/10.1007/978-3-031-83250-5_1

Keywords Climate neutrality · Sustainable production · Sustainable consumption · Systematic Literature Review (SLR) · Policy and governance

Introduction

Due to dramatic climatic changes across the globe, achieving climate neutrality is now an important world agenda. Climate neutrality controls the added mass of greenhouse gases in the atmosphere without accumulating additional global warming pollutants. This idea is crucial for reducing greenhouse gas emissions and contributing to scenario two of the Paris Agreement, limiting global warming to below two °C above pre-industrial levels (Dewasiri et al., 2024). In addition to being a quantitative target, climate neutrality illustrates a transition towards the 'better life' in which production and consumption are in harmony with ecological reality without compromising economic and social prosperity (Mobarakeh & Kienberger, 2022).

Sustainable production and consumption form the core of this shift mechanism. These two correlated elements are crucial for solving the issues of environmental deterioration (Azapagic et al., 2016). Sustainable production is when a company gets involved to minimise the negative impacts on the environment and improve the efficient use of resources while aiming at long-term economic gains. On the other hand, sustainable consumption reconstructs consumers' and society's demand and culture towards production for environmental and responsible consumption habits. Combined, they create the basis of the circular flow of items, minimal energy consumption, and efficient waste management. The effectiveness of these practices in reaching climate neutrality depends on their integration into the primary economic systems.

This chapter discusses the multiple relationships between climate neutrality, sustainable production, and consumption to provide insights into their importance for creating a sustainable world. At the end of the chapter, the authors will describe the concepts for change and best practices for closing the implementation gap by outlining these barriers and opportunities in the policy domains. This discussion covers theoretical considerations, global and regional challenges, and examples of industries and communities worldwide. Hence, the chapter aims to contribute positively to the climate change conversation by analysing how positive change can be entered inside the intersections of the discussed themes.

Methodology

This chapter's methodology starts with literature that employs the systematic literature review (SLR). The study also follows the PRISMA guidelines as applied by Vijerathne et al. (2024), Bandara et al. (2024), Chathuranga et al. (2024), Mishra et al. (2024), Karunarathna et al. (2024), and Rathnasiri et al. (2024). These tools helped maintain a sufficient degree of control and formalised approach to studying climate neutrality, sustainable production, and sustainable consumption topics.

The SLR process began by identifying research questions that align with the chapter's objectives: the impediments, enablers, and routes toward sustainability. An extensive search was performed in these databases using scientific search queries with the following keywords: climate neutrality, sustainable production, sustainable consumption, policy measures, and green technologies. Thus, to enrich the range of sources and get additional value for the result, other sources of information, such as police reports, white papers, and industry cases, were also reviewed.

PRISMA framework helped to be accountable and unambiguous during the inclusion and exclusion of the studies. First, the titles and abstracts of all the studies they identified were studied to select relevant ones, and then the full text was scrutinised. The following criteria were used to exclude any studies from the analysis. First, the researchers did not use data from 2015 and below, nor did they use literature reviews, which were non-empirical and unrelated to the research questions. The last dataset included only peer-reviewed articles published during the last 15 years and contained current trends and innovations in the policies and reports. Accordingly, thirty-nine seminal papers were considered as the sample of this study.

Application of the PICO framework was used to delimit the actual review. The population criterion targeted stakeholders involved in activities related to climate neutrality and key players in businesses, individuals, and governments. The intervention criterion involved analysis of measures targeted at sustainability, such as green technology initiatives, eco-labelling, and renewable power. The context was global, regional, and sectoral, and to draw attention to regional diversity and opportunities, we selected South Asia. The Outcome assessed best practices, policies, and innovations impacting and progressing toward climate objectives.

The selective observation was conducted, and results were classified into themes like sustainable production challenges, consumer behaviour issues, technologies used, and policies imposed. Measures and findings based on quantitative activities included the numbers and effects of the interventions, while qualitative work enriched the situation and theme exploration-based study findings.

The methodological solidity guaranteed the emergence of challenges and opportunities and the generation of practical suggestions for policymakers, enterprises, and civil society. This methodology contributes significantly towards enriching the existing scholarship and informing some of the important policy imperatives for climate neutrality or corporate sustainability by locating theoretical and practical gaps.

Conceptual Review

Climate neutrality has been in the discourse as a concept firmly grounded on other conceptual models that form the backbone of global, regional, and local endeavours. This section identifies the foundation for climate neutrality, analyses sustainable production principles, and assesses the drivers for sustainable consumption. All these elements complement each other, providing a comprehensive perspective on the complexity of strategies needed to tackle the issue of combating climate change.

Climate Neutrality Frameworks

Climate neutrality frameworks are the fundamental principle of international measures to combat climate change. The Intergovernmental Panel on Climate Change (IPCC) stands at the core, providing the relevant science and options for emissions cuts (Beck & Mahony, 2018). The IPCC also states that net-zero emissions by mid-century are required to limit global warming to 1.5 °C above pre-industrial levels. This goal was also incorporated in the Paris Agreement, ensuring that the countries present and sign Nationally Determined Contributions (NDCs) and long-term strategies to attain the state of zero carbon (Dewasiri et al., 2024). Both frameworks highlight extensive system changes across energy systems, industries, or lifestyles.

Also, the United Nations Framework Convention on Climate Change (UNFCC) has been established to enhance and promote the implementation of technology transfer under the Convention (Protocol, 1997); the Green Climate Fund (GCF) helps facilitate international cooperation and funding for developing countries working to achieve climate neutrality (Afful-Koomson, 2015). All these world views emphasise that reaching climate neutrality is not just a techno-scientific question but an issue of socially responsible actions, fair policy, and international solidarity.

Principles of Sustainable Production

Sustainable production is central to emissions mitigation and efficient resource practices as sustainably desired (Alayón et al., 2017). One is the circular economy, which means constant processing of usable items without using essentially new raw materials. The circular economy model is opposed to the existing linear economy model of ‘take-make-dispose’ and thus could work alongside the customers’ need for climate change neutrality.

Another important principle is eco-efficiency, which means being more productive using fewer resources, including energy and water, and involving fewer inputs while generating more value. This disruptive counter includes renewable energy systems, green manufacturing, and low-carbon materials and technologies (Nwokolo et al., 2023a). In addition, resource optimisation is friendly to the economy and the environment since resource management is achieved through better profitability, marginal cost, and optimisation techniques like life cycle analysis, total energy auditing, and innovative supply chain management.

Sustainable Consumption Patterns

Socially responsible consumption involves individual transformation, financial motivation, and a policy environment to achieve responsible decisions (O'Rourke & Lollo, 2015). Consumption is perhaps the most important area that must be transformed to lower demand for these carbon-intensive goods and support green products. Learning about these changes and raising consciousness, campaigns, and social norms have a role in inducing such changes. For example, labels or certifications indicate to the buyer that the products are environmentally and socially appropriate.

More encouragement is provided through generalised economic incentives, such as incentives for green product uptake, putting a price on carbon, and tax credits. Governments have the most responsibility in this area as they pass legislation and rules on sustainable practices like the non-use of plastic bags or setting a minimum efficient regulatory measure for energy consumption in any product. Furthermore, the emerging sharing economy trend, primarily through consumption promoters such as car rental and other utility products, also depicts an efficient consumption model.

In culmination, the triple bottom line concept and sustainable development align directly with the core precepts of the transition to climate neutrality. These frameworks and principles together provide the theoretical and conceptual underpinnings for the transition to climate neutrality. Global strategies and approaches to innovative production and consumption patterns can help society work toward a sustainable and somewhat developed future.

Findings and Discussion

Key Challenges

Climate neutrality, sustainable production, and consumption are critical strategies for an environmentally sustainable future, but achieving these goals is difficult (Guntuka et al., 2024). These challenges are the economic, social, and structural barriers that endeavour to find solutions that are globally just possible. This section explores overcoming these barriers to sustainable production, the challenges and dilemmas of changing the buying public, and overcoming global and regional asymmetry in the climate change equation.

Barriers to Sustainable Production

Economic, technological, and regulatory factors impede sustainable and efficient production practices (Joshi et al., 2024). Economically, one of the most significant drawbacks to green technology and environmentally sound structures is the substantial start-up cost that remains out of reach for small and medium enterprises (SMEs) and most industries in developing countries. This financial burden, coupled with poor access to green financing, less government support, and increased cost of utilising environment-friendly products and technologies, poses a significant disincentive to businesses when planning to change towards sustainability.

Technologically, the slow rate of development and the integrated introduction of sophisticated options are the following challenges (Liu et al., 2021). For example, adopting renewable energy technologies, energy-efficient machines, and waste recycling systems is expensive and sometimes geographically unavailable to most firms. However, there is limited technical know-how and a skilled workforce to support and adopt sustainable production processes.

From a business regulation perspective, there are no strict environmental policies, and the enforcement measures are often unpredictable, allowing organisations to continue their unhealthy practices. Policy measures sometimes do not support sustainability objectives; they do not encourage or require changes in production to low carbon intensity.

Challenges in Shifting Consumer Behavior

Consumers themselves are key actors in promoting a sustainable consumption pattern, although making radical changes to advance this shift remains challenging because of culturally, economically, and informationally influenced barriers (White et al., 2019). The consumption pattern depends highly on culture, and most cultures harbour a convenience and consumption-oriented mentality. This includes reliance on one-time use of products, such as disposables and unrecyclable plastics, and consumption of cheap and trending garments since consumers have been trained to acquire products that meet these characteristics irrespective of the adverse environmental effects. To eliminate these norms, top-down and society-wide campaigns for changing mindsets take time and effort—it is a slow process on an international level.

The second challenge is the cost of implementing the recommendations. Many sustainable products and services remain expensive and beyond the reach of a sizeable population in the first and developing worlds. This seems to be an undesirable thought process because consumers generally are willing to pay less for goods and services regardless of the environmental effects if there are no subsidies or financial rewards for going green.

The breakthrough is, however, limited by informational barriers, which the fact that most consumers can exemplify is ignorant of the existence of eco-friendly products or the sustainable choices available in the market. Most consumers have inadequate information about the effects of their buying behaviour on the environment or the availability of green products in the market. This is further exacerbated by false claims and corporate environmental, social, and governance (ESG) efforts—that reduce confidence and discourage consumers from engaging more in research.

Global and Regional Disparities

Over time, reducing greenhouse gas emissions and encouraging sustainable behaviours should be implemented worldwide. At the same time, there are significant variations worldwide on the implementation level of climate neutrality. This means that developed nations have more financial strength, technology, and better policies to advance in the cause of climate change (Nwokolo et al., 2023c). On the other hand, developed and least developed countries have some limitations in funding, often dated infrastructure, and many other competing projects, including poverty eradication and economic transformation. It leads to a situation where the developed world can record significant cuts in emissions while the developing world lags in digesting these changes.

Further, developed countries, as the sources of historic GHG emissions, share significantly more responsibility for the problem. In contrast, its adverse effects are experienced mainly by vulnerable areas in the developing world. This sharpens conflict in international relations when developing nations call for more resources for financing, technology, and capacity-building.

However, climate policy ambition also varies across regions, which worsens the situation. Although some nations have an attractive policy of reaching climate neutrality by the middle of the century, others fall behind because of political opposition, their economy's complete reliance on the hydrocarbon sector, or ongoing geopolitical tensions. They mock international efforts to fight climate change as one people.

Therefore, achieving climate neutrality through an effective, sustainable production and consumption process is complicated (Zhao et al., 2022). This process encounters several complex and intertwined difficulties. Addressing these barriers requires a multi-pronged approach that includes funding and technology to meet actual needs, policy, and consumer awareness to support current needs. The improved management of these challenges comprehensively will ease the shift towards sustainable practices with appropriate benefits distribution across space.

Pathways to Climate Neutrality

The goal of reaching 'climate neutrality' should be achieved by following an intricate plan that includes a set of designs intended to ensure the environmentally sound condition of the earth, its economic stability, and the well-being of society. This section examines the significant pathways to meeting the challenges described earlier: innovations in sustainable production, effective consumption promotion, policy interventions, and governance. These pathways, therefore, provide a blueprint for decarbonisation consistent with achieving global sustainability goals.

Innovations in Sustainable Production

Technology remains a central component in driving and changing production systems climate-neutral. The best solutions include green technologies, renewable energy integration and waste disposal management. Green technologies refer to innovations that help practitioners and companies achieve less adverse effects on the natural environment or better resource use. Technologies like carbon capture and storage (CCS), energy efficiency in production, biodegradable product manufacturing, etc., are changing industries by allowing them to reduce their emissions while threatening the competitiveness of their services and products (Bui et al., 2018). For instance, implementing industry 4.0 technologies such as the Internet of Things (IoT) and artificial intelligence (AI) makes it possible to monitor and optimise the production process in real-time and with the least resource consumption and pollution.

Solar, wind, and hydropower are imperative in dematerialising energy-intensive industries to conform with the global decarbonisation agenda. Changing to renewable energy sources reduces greenhouse gas emissions and strengthens energy security and the economy. People have come to embrace green energy by incorporating solar-powered plants and wind power into manufacturing industries. In addition, improvements in storage technologies such as battery and utility-scale storage provide the framework and dynamics to support renewable energy integration.

Sustainable production strategies concerning reduction measures perform a central role by focusing on circular economy principles (Khan et al., 2024). Promising methods like upcycling, industrial symbiosis, and closed material loops guarantee that wastes produced in one production process are immediately fed into other processes. For instance, the fashion and construction industries are more likely to incorporate recycled products as a source of raw material than extraneous industrial waste, which has become a source of raw material in these industries.

Promoting Sustainable Consumption

Changing consumption towards sustainable consumption is possible only through awareness-raising campaigns, mechanisms, incentives, and innovative financing concepts (Rubik et al., 2009). Awareness campaigns control consumer education and the effects of their choices. In their bid to encourage citizens to embrace sustainable living, governments, companies, and non-governmental organisations (NGOs) use media, social networks, and community campaigns. Popular slogans that currently exist include "Buy Less," "Think Before You Buy," and "Reduce, Reuse, Recycle." These slogans have cropped up worldwide to encourage people to embrace responsible consumerism.

Eco-labelling offers the consumer quantitative and qualitative information on the sustainability of products and services. Energy Star, Fair Trade, and Forest Stewardship Council (FSC) certifications provide necessary information to help consumers make correct choices (Moore et al., 2012). Clear differentiating allows consumers to trust brands and pressures companies to become environmentally friendly within the supply chain.

Another major enabler is green financing, which seeks to propose invitational incentives to consumers and businesses to make green investments. Incentives include low-rate green loans, subsidies for durable, efficient gadgets and devices, and tax credits for renewable energy use to reduce cost constraints and encourage more people to support sustainable practices. Also, the emergence of green bonds and impact investment points to the emerging trend for products that fit within sustainability objectives.

Policy Interventions and Governance

The prerequisite for climate neutrality is an enabling environment resulting from sound policies and governance structures. Governments are highly involved in regulating sectors, streamlining current policies, offering incentives, and encouraging international efforts. It is unbearable to overlook the role of regulations in specifying responsibility and implementing sustainable practices. Instruments such as carbon prices, emissions trading schemes, and renewable targets pressure firms to act on decarbonisation (Hafner et al., 2022). Besides, specialists, such as the prohibition of plastic products and minimum energy performance standards for appliances, ensure sustainable production and consumption on a large scale.

It is proposed that incentives continue the process even faster because they eliminate some costs while rewarding environmentally friendly behaviour. Governments can afford to encourage green technologies through tax concessions for adopting them, rebates for installing renewable energy sources, and grants for researching sustainable technologies. For example, feed-in tariffs for electricity from clean energy sources and rebates for purchasing electric vehicles are recognised to influence the adoption of the targeted technology.

The climate change issue is international, thus requiring international cooperation to solve the problems. A current example is the Paris Accord and other collaborations, such as the United Nations Sustainable Development Goals (UNSDGs), that work simultaneously, guaranteeing that efforts to reach climate neutrality are consistent and reciprocal (Tiwari, 2022). Knowledge-sharing platforms and technology transfer mechanisms such as the Clean Development Mechanism (CDM) allow third-world nations to attain a higher level of sustainable development, eradicating global inequalities in combating climate change (Sutter & Parreño, 2007; Kumar et al., 2024).

Therefore, technological advancement in sustainable production, efforts on consumerism, and reasonable policy measures are all coherent ideas that allow climate neutrality. Through effective use of technologies, creating consumers' interest and awareness about climate change, and developing mechanisms and structures for governance, society will be able to solve climate change issues. These pathways call for concerted effort and long-term endeavour towards achieving a future in which the protection of the environment and economic development go hand in hand.

Case Studies and Best Practices

Analysing the examples and cases suggests that climate neutrality can be reached by improving sustainable production and consumption patterns. This section looks at influential models in production, consumer-led climate change, and South Asian sustainable development prospects and hurdles.

Successful Models in Production

The industry focuses on various economically sustainable, environmentally sensitive production strategies. A good illustration is Patagonia, a leading company in the apparel industry and a staunch environmentalist company. Recycle content includes post-consumer plastics and organic cotton, making the company reduce its impact on the natural environment. For their part, Patagonia is also an advocate of repair and reuse programmes that compel consumers to lengthen the wearability of the products they have bought, thus promoting a circular economy.

Scholars borrow the automotive sector to provide another example of disruptive innovation in Tesla, the manufacturer of electric vehicles (EVs). Tesla created gigafactories to help generate renewable energy sources, producing sustainable products with less wastage (Cooke, 2020). In the same way, Tesla provides battery storage systems and renewable power generation to improve sustainable transport and assist close to broad clean energy solutions.

In the food and beverage industry, Unilever confirmed its sustainable living plan with goals to decrease greenhouse gas emissions, waste and water consumption in its production processes and supply chains (Lawrence et al., 2019). Another area is the commitment of the company to improve and support regenerative agriculture practices and smallholder farmers that can promote admirable soil and biological farm diversity without compromising financial resource productivity.

Consumer-Led Climate Action

Civil society organisations and community-based approaches are central to changing demand patterns and combating climate change (Nwokolo et al., 2023b). One vivid example is the Zero Waste practice, which encourages people and communities to minimise waste generation through attitude modifications, including using no disposable plastics, home composting, and defending package-free businesses (Zaman, 2015). It has received much international attention, and places like San Francisco have achieved a waste diversion rate of more than 80%, encouraged by the community and supportive municipal policies.

The other brilliant social innovation that seeks to address the problem of plastic pollution is the Plastic Bank. This social business pays individuals from poor backgrounds to segregate plastic waste for recycling and, in return, provides them with some necessities or cash (Katz, 2019). This model regards reducing plastic pollution, generating employment and income for the people, and making societies more resilient.

Youth-led movements such as Fridays for Future, spearheaded by Greta Thunberg, focus on climate change. These global climate strikes and advocacy movements have put pressure on governments and corporations, proving the ability of consumers' voices to push for change.

South Asia's Potential and Progress

South Asian region, comprising one-quarter of the world's population, has been characterised by remarkable achievements, significant development challenges, and opportunities for improving sustainable development. Thus, it is possible to outline the following factors as valuable experiences of the region in combating climate change: The region under consideration demonstrates a unique socio-economic and environmental context that makes its experience especially valuable for analysing climate change.

Countries like India and Bangladesh are now considered among the world's leaders in renewable energy strategy and climate change management. India wants to become a hub for solar energy by targeting the nation's novel generation of 280 GW of solar energy by the end of 2030 under the National Solar Mission (Dawn et al., 2016). Similarly, Bangladesh's Solar Home Systems (SHS) has extended clean energy to millions of houses in the rural region, enhancing lives and minimising reliance on fossil fuels (Hellqvist & Heubaum, 2023).

Sustainability campaigns such as the Sri Lanka Traditional Rice Cultivation Program show that the region is determined to push sustainability as a cultural and economic necessity. As the program adopts indigenous farming practices, using inputs of organic nature and conserving water, the program formulates the prospect of applying traditional practices with modern-day sustainability concepts.

Nonetheless, South Asia has faced challenges in striving for social, economic, political, and technological growth, such as high poverty index rate, high growth and expansion of urban centres and susceptibility to effects of natural disasters influenced by climate changes. Solutions to these problems are seen in regional cooperation within the sphere of environmental management, climate adaptation, and sustainable development, as with the South Asia Co-operative Environment Programme (SACEP).

There is no one-size-fits-all approach to climate neutrality and sustainability. From industries postulating new production standards to bottom-up initiatives changing consumption practices, these cases emphasise the values of innovation, communities, and regions. South Asia's experiences stress that climate interventions should be locus-specific and meet international climate targets. All these success stories are examples of how to scale up interventions worldwide, equally solidifying our collaborative mission of creating a better environment for our planet.

Future Directions and Opportunities

In light of this changing paradigm, it is increasingly important to design new approaches that enable climate neutrality to be embedded within economics, technologies and cooperation systems. This section envisages the possible directions of the further advancement of climate neutrality integration into economic development, the application of cutting-edge technologies to build efficient and sustainable systems, and the increasing focus on the international level to design innovative and systemic cross-industry solutions.

Integrating Climate Neutrality into Economic Planning

Climate neutrality can only be obtained by integrating its principles into economic decision-making. Sustainability is not an afterthought but an intrinsic part of government and business strategies and policies.

Fiscal policies can adopt green economics, where strategies are implemented to achieve sustainable objectives. Carbon pricing—carbon taxes and cap-and-trade—encourages industries to cut emissions and allows governments to create revenue from sales that could be used towards renewables, public transportation, and other climate measures. For instance, Sweden has used carbon taxes effectively to reduce GHG emissions while enhancing growth.

Climate neutrality in economic planning is another important component of sustainable infrastructure investment. Positive governmental intervention in economic development and environmental sustainability can be achieved by allocating finances towards low-carbon transport, energy-efficient buildings, and water infrastructure.

Incorporating sustainability measures within the company reporting language will improve business accountability and transparency. Environmental, social, and governance (ESG) factors have been developed to assess how a company affects the world (Singh et al., 2024). Corporate sustainability investing is gaining broad acceptance in the market, making sustainability an intrinsic part of operations to secure funding (Nwokolo et al., 2024).

Role of Emerging Technologies

Global technology trends, such as AI, blockchain, and IoTs, present a remarkable chance to optimise sustainable systems. AI can better use resources, manage energy, and make a foretold estimate of the environmental effects. For example, by utilising AI-driven predictive analytics, it is possible to match the supply and demand of energy, decreasing the consumption of fossil energies at peak times. Similarly, new innovative technologies in precision agriculture with the application of artificial intelligence allow farmers to take more efficient approaches to water use, seeds, fertiliser, chemicals, and yields, improving environmental damage.

The opportunity to use Blockchain technology is very high because it can bring much-needed transformative direction to achieving sustainable corporate goals and projects. Due to blockchain's high level of transparency, supply chain history concerning the sourcing and production of products can be made visible to the correct authorities. For instance, blockchain can ensure consumers make good choices regarding ethically sourced materials and compel companies to stick to good practices.

IoT changes how resources are monitored and managed by linking gadgets and controlling systems. For example, sensors in the IoT smart grid can determine the right time to convey electricity, detect areas of wastage, and correct anomalies. Under IoT, waste disposal can be made easier by tracking waste in real time with enhanced recycling incorporation and minimal landfill use.

As demonstrated herein, including these technologies in sustainability strategies enables the various stakeholders to overcome the previous impediments to real and sustainable change at scale.

Scaling up Global Collaboration

Since climate change is fundamentally integrative regarding space, time, and the socio-sphere, effective management requires strong global cooperation at the regional, sectoral, and stakeholder levels.

International treaties like the Paris Agreement provide a theoretical background for developing climate action. However, the effectiveness of such agreements lies more in their practical operation and, more so, effective compliance. Countries must be ready to increase their commitment levels through enhanced NDCs and ensure capacity support for developing countries to achieve their goals.

Cross-sectoral and intergovernmental collaboration between ministries, companies, and communities is also essential. A good example of this approach is the United Nations Global Compact, which takes the form of a commitment with businesses worldwide to support sustainable business practices. When carried out effectively, it can be a perfect model of how various organisations can bundle together their specialism, capital, and creativity to solve problems that cannot be fixed by any single actor in the system.

Developing nations' cooperation shows how southern countries can adopt and implement best practices that suit their provinces and states' socio-economic fabric. For example, in recent years, successful collaboration between India and Bangladesh in the renewable energy sector is a sign that regional cooperation will help advance the climate neutrality agenda.

Lastly, education for climate action is needed to create loud and aware citizens worldwide. Projects that focus on global exchange and culture, such as world fellowship conferences, global youth services, and climate change discussions, help foster people and groups and enable them to enhance climate involvement.

The paths expounded in this section stressed the transition possibilities of climate neutrality in the economic structure, the opportunities for using new technologies, and the global partnership. Through such approaches, stakeholders can navigate emergent issues and shape a desirable, inclusive, and durable agenda. As the world moves on this decisive stage, cooperation and creativity will be critical to achieving this vision of a climate-neutral world.

Conclusions

Summary of Key Insights

This chapter has established climate neutrality, sustainable production, and consumption as topical issues closely examined in the internal and global economic platforms, complementing this chapter's framework. The discourse started with the ability to oxidise climate neutrality and its relevance in controlling global heating/upsetting the natural order. The following content was devoted to the theoretical background, emphasising the approach used, for example, in the Paris Agreement and the key principles of circular economy and eco-efficiency.

Looking at the reality and realising the attempts for climate neutrality, the main problems and obstacles in promoting sustainable production and consumption were discussed. Nevertheless, there are approaches to getting to net zero emissions—technology advancement, policy, and synergy brought about by collaboration are fundamental solutions (Bhushan & Singh, 2024). Real-life examples and success stories ranging from corporate management of sustainable manufacturing to consumer uprising for climate activist purchasing show the many-splendored character of this issue.

The chapter also discussed prospects and possibilities, stressed that climate neutrality should be incorporated into economic strategies, and discussed the benefits of using new technologies and expanding cross-country collaboration. These insights offer a framework for the sustainable development of communities, regions, and the world.

Call to Action

Climate Neutrality involves the striving of not only governments, companies, and civil society but every single person and requires tireless action now. The government should move to implement sustainability in both economic and regulatory climate policies regarding developmental goals. This entails practising carbon measures, encouraging green technology, and linking organisations worldwide to share knowledge and resources.

Companies must implement sustainability as a core production enabler, efficiently using renewable energy sources and integrating ESG factors into management strategies. Strategic alliances and commitments in research and development (R&D) will play a critical role in sourcing innovative fundamentals to eliminate technological hitches.

Promoting climate neutrality is also people's responsibility; civil society has an important role here. Consuming sustainably, campaigning for changes in policies and legislation, and investing in environmentally friendly companies are ways that people can force change at the system level. There is a need to increase awareness and education to transform communities into productive units with meaningful resources.

Lastly, decarbonisation is existential for climate change when realised as the quest for climate neutrality as an opportunity to build a fair, defensive, and prosperous future. The time is now, and the voices of humans are rallying together and making this vision statement a mission to be accomplished.

References

Afful-Koomson, T. (2015). The green climate fund in Africa: What should be different? *Climate and Development, 7*(4), 367–379.

Alayón, C., Säfsten, K., & Johansson, G. (2017). Conceptual sustainable production principles in practice: Do they reflect what companies do? *Journal of Cleaner Production, 141*, 693–701.

Azapagic, A., Stamford, L., Youds, L., & Barteczko-Hibbert, C. (2016). Towards sustainable production and consumption: A novel decision-support framework integrating economic, environmental and social sustainability (DESIRES). *Computers & Chemical Engineering, 91*, 93–103.

Bandara, C., Rathnasiri, M. S. H., Dewasiri, N. J., Chathuranga, B., & Singh, N. (2024). Greening the economy for a sustainable future: Opportunities and perceptions. In *Greening our economy for a sustainable future* (pp. 197–213). https://doi.org/10.1016/B978-0-443-23603-7.00016-9

Beck, S., & Mahony, M. (2018). The IPCC and the new map of science and politics. *Wiley Interdisciplinary Reviews: Climate Change, 9*(6), e547.

Bhushan, M., & Singh, R. (2024). Talent management facilitates net-zero transition through employee green behavior. In R. Singh, S. Khan, A. Kumar, S. Luthra, & H. Chokshi (Eds.), *Net zero economy, corporate social responsibility and sustainable value creation. CSR, sustainability, ethics & governance*. Springer. https://doi.org/10.1007/978-3-031-55779-8_7

Bui, M., Adjiman, C. S., Bardow, A., Anthony, E. J., Boston, A., Brown, S., Fennell, P. S., Fuss, S., Galindo, A., Hackett, L. A., & Hallett, J. P. (2018). Carbon capture and storage (CCS): The way forward. *Energy & Environmental Science, 11*(5), 1062–1176.

Chathuranga, B., Rathnasiri, M. S. H., Dewasiri, N. J., Jayarathne, P., Bandara, C., & Bakari, M. A. (2024). Direct and indirect effect on industries and services (A). In *Greening our economy for a sustainable future* (pp. 215–226). https://doi.org/10.1016/B978-0-443-23603-7.00017-0

Cooke, P. (2020). Gigafactory logistics in space and time: Tesla's fourth gigafactory and its rivals. *Sustainability, 12*(5), 2044.

Dawn, S., Tiwari, P. K., Goswami, A. K., & Mishra, M. K. (2016). Recent developments of solar energy in India: Perspectives, strategies and future goals. *Renewable and Sustainable Energy Reviews, 62*, 215–235.

Dewasiri, N. J., Gnanapala, W. K. A. C., Rathnasiri, M. S. H., Yadav, M., Tewari, L. M., & Mishra, U. (2024). The way forward for net zero: Aligning Nationally Determined Contributions (NDCs) and holistic sustainable development in South Asia. In R. Singh & D. Crowther (Eds.), *Transition towards a sustainable future. Approaches to global sustainability, markets, and governance*. Springer. https://doi.org/10.1007/978-981-97-5756-5_5

Guntuka, L., Mukandwal, P. S., Aktas, E., & Paluvadi, V. S. K. (2024). From carbon-neutral to climate-neutral supply chains: A multidisciplinary review and research agenda. *The International Journal of Logistics Management, 35*(3), 916–942.

Hafner, S., Speich, M., Bischofberger, P., & Ulli-Beer, S. (2022). Governing industry decarbonisation: Policy implications from a firm perspective. *Journal of Cleaner Production, 375*, 133884.

Hellqvist, L., & Heubaum, H. (2023). Setting the sun on off-grid solar?: Policy lessons from the Bangladesh Solar Home Systems (SHS) programme. *Climate Policy, 23*(1), 88–95.

Joshi, A., Singh, R., Kumar, V., & Rani, S. (2024). Net zero strategy: Exploring opportunities, confronting challenges for developing countries. In R. Singh & D. Crowther (Eds.), *Transition towards a sustainable future. Approaches to global sustainability, markets, and governance*. Springer, Singapore. https://doi.org/10.1007/978-981-97-5756-5_11

Karunarathna, K. S. S. N., Dewasiri, N. J., Singh, R., & Rathnasiri, M. S. H. (2024). What does artificial intelligence–powered ChatGPT bring to academia? A review. In R. Singh, S. Khan, A. Kumar, & V. Kumar (Eds.), *Artificial intelligence enabled management: An emerging economy perspective* (pp. 3–13). De Gruyter. https://doi.org/10.1515/9783111172408

Katz, D. (2019). Plastic bank: Launching social plastic® revolution. Field actions science reports. *The Journal of Field Actions, 19*, 96–99.

Khan, S., Singh, R., Alnahas, J., Abbate, S., & Centobelli, P. (2024). Navigating the smart circular economy: A framework for manufacturing firms. *Journal of Cleaner Production, 480*, 144007. https://doi.org/10.1016/j.jclepro.2024.144007

Kumar, K., Singh, R., Sharma, A., & Lehri, B. P. S. (2024). Climate change adaptation strategies for achieving net-zero economy. In R. Singh & D. Crowther (Eds.), *Transition towards a sustainable future. Approaches to global sustainability, markets, and governance*. Springer. https://doi.org/10.1007/978-981-97-5756-5_8

Lawrence, J., Rasche, A., & Kenny, K. (2019). Sustainability as opportunity: Unilever's sustainable living plan. In G. G. Lenssen & N. C. Smith (Eds.), *Managing sustainable business*. Springer. https://doi.org/10.1007/978-94-024-1144-7_21

Liu, Y., Wood, L. C., Venkatesh, V. G., Zhang, A., & Farooque, M. (2021). Barriers to sustainable food consumption and production in China: A fuzzy DEMATEL analysis from a circular economy perspective. *Sustainable Production and Consumption, 28*, 1114–1129.

Mishra, U., Rathnasiri, M. S. H., & Dewasiri, N. J. (2024). Net-zero economy and energy sustainability: A Nepalese perspective. In R. Singh, S. Khan, A. Kumar, S. Luthra, & H. Chokshi (Eds.), *Net zero economy, corporate social responsibility and sustainable value creation. CSR, sustainability, ethics & governance*. Springer. https://doi.org/10.1007/978-3-031-55779-8_9

Mobarakeh, M. R., & Kienberger, T. (2022). Climate neutrality strategies for energy-intensive industries: An Austrian case study. *Cleaner Engineering and Technology, 10*, 100545.

Moore, S. E., Cubbage, F., & Eicheldinger, C. (2012). Impacts of forest stewardship council (FSC) and sustainable forestry initiative (SFI) forest certification in North America. *Journal of Forestry, 110*(2), 79–88.

Nwokolo, S. C., Singh, R., Khan, S., Kumar, A., & Luthra, S. (2023a). Remedies to the challenges of renewable energy deployment in Africa. In *Africa's path to net-zero. CSR, sustainability, ethics & governance*. Springer. https://doi.org/10.1007/978-3-031-44514-9_3

Nwokolo, S. C., Singh, R., Khan, S., Kumar, A., Luthra, S. (2023b). Africa's Awakening to climate action. In: *Africa's path to net-zero. CSR, sustainability, ethics & governance*. Springer. https://doi.org/10.1007/978-3-031-44514-9_9

Nwokolo, S. C., Singh, R., Khan, S., Kumar, A., & Luthra, S. (2023c). Impacts of climate change in Africa. In: *Africa's path to net-zero. CSR, sustainability, ethics & governance*. Springer. https://doi.org/10.1007/978-3-031-44514-9_7

Nwokolo, S. C., Singh, R., Khan, S., & Kumar, A. (2024). Threats to the rapidity of sustainability transitions posed by technological changes. In S. C. Nwokolo, R. Singh, S. Khan, A. Kumar (Eds.), *Technological pathways for Africa's net-zero economy* (pp. 39–75). https://doi.org/10.1016/B978-0-443-31486-5.00002-1

O'Rourke, D., & Lollo, N. (2015). Transforming consumption: From decoupling to behaviour change to system changes for sustainable consumption. *Annual Review of Environment and Resources, 40*(1), 233–259.

Protocol, K. (1997). United Nations framework convention on climate change. *Kyoto Protocol, Kyoto, 19*(8), 1–21.

Rathnasiri, M. S. H., Dewasiri, N. J., & Kumar, A. A. (2024). Policy framework and implementation strategies for Sri Lanka's transition to a net-zero economy. In R. Singh, S. Khan, A. Kumar, S. Luthra, & H. Chokshi (Eds.), *Net zero economy, corporate social responsibility and sustainable value creation. CSR, sustainability, ethics & governance*. Springer. https://doi.org/10.1007/978-3-031-55779-8_3

Rubik, F., Scholl, G., Biedenkopf, K., Kalimo, H., Mohaupt, F., Söebech, Ó., Stø, E., Strandbakken, P., & Turnheim, B. (2009). Innovative approaches in European sustainable consumption policies. *Schriftenreihe des Iöw, 192*(09), 1–14.

Singh, R., Gupta, A., & Kandpal, V. (2024). ESG excellence in the digital world. In D. R. Ernesto, S. Gonzalez, V. Kandpal, P. K. Ozili, & P. Chatterjee (Eds.), *Fintech for ESG and the circular economy* (pp. 93–204). Scrivener Publishing LLC. https://doi.org/10.1002/9781394234035.ch10

Sutter, C., & Parreño, J. C. (2007). Does the current Clean Development Mechanism (CDM) deliver its sustainable development claim? An analysis of officially registered CDM projects. *Climatic Change, 84*(1), 75–90.

Tiwari, A. (2022). Climate diplomacy to attain global eco-neutrality. *Advanced Materials Letters, 13*(3), 1697–2203.

Vijerathne, D. T., Wahala, W., Rathnasiri, M. S. H., De Silva, V., & Dewasiri, N. J. (2024). Literature review on sustainability. In *Greening our economy for a sustainable future* (pp. 33–45).https://doi.org/10.1016/B978-0-443-23603-7.00004-2

White, K., Habib, R., & Hardisty, D. J. (2019). How to SHIFT consumer behaviours to be more sustainable: A literature review and guiding framework. *Journal of Marketing, 83*(3), 22–49.

Zaman, A. U. (2015). A comprehensive review of the development of zero waste management: Lessons learned and guidelines. *Journal of Cleaner Production, 91*, 12–25.

Zhao, X., Ma, X., Chen, B., Shang, Y., & Song, M. (2022). Challenges toward carbon neutrality in China: Strategies and countermeasures. *Resources, Conservation and Recycling, 176*, 105959.

Chapter 2
Adaptation and Resilience: Private, Public, and Individual Change Makers

Michael Oyedele Oyenuga and Solomon Jeresa

Abstract Natural climate solutions by private and public as well as individual change agents are needed now more than ever to achieve climate neutrality. This chapter investigates how these sectors are interlinked and contribute to adapting to climate change and dealing with the residual impacts of climate change. The private sector is a crucial innovator that uses technology and financing to push the boundaries of sustainability, while the public sector offers necessary guidance, policies, and investments to support collective action. Social mobilization happens at the grassroots level through individuals who act as change agents, either in how they live their own lives or through advocacy efforts and engagement in their communities. This emphasizes what it means to practice personal responsibility around climate action. Each sector confronts distinct hurdles that may slow advancement, whether due to financial limitations regulatory uncertainties, or social barriers. This chapter demonstrates that there are incredible interdependencies between players in the climate arena, and they need to work together, privately with public and individual actors, not only on policy development but also in making operational their joint strategy for climate neutrality.

Keywords Adaptation · Climate neutrality · Public–private partnership · Resilience

M. O. Oyenuga (✉)
School of Business, Woxsen University, Telangana, India
e-mail: oyenuga.michael@woxsen.edu.in

S. Jeresa
University of Huddersfield, Huddersfield, UK
e-mail: solomon.jeresa@hud.ac.uk

R. Singh and W. L. Filho (eds.), *Climate Neutrality Through Smart Eco-Innovation and Environmental Sustainability*, Climate Change Management,
https://doi.org/10.1007/978-3-031-83250-5_2

Introduction

There is a need to solve climate change problems as it started foreshadowing its aftermath everywhere around the world. While ambitious targets are an integral part of the climate action puzzle for nations today, there is no denying that adaptation and resilience have also continued to emerge as vital components. Adaptation is the measures taken to deal with real or anticipated impacts of climate; and, resilience is the ability of people, communities and systems to absorb and recover from impacts. Climate neutrality, which means reducing net greenhouse gas emissions to zero, is only attainable if all sectors of society from private enterprises and public institutions to every citizen come together but this will take action too. Public–private partnerships (PPPs) represent an emerging solution for catalyzing access between public institutions and private entities to mobilize resources to drive climate neutrality.

Such partnerships capitalize on the advantages of both sectors, combining public supervision and accountability with private sector ingenuity and efficacy. For instance, Khan et al. (2020) argue that consumption-based carbon emissions can only be reduced through public–private partnership investments in China, suggesting the power of partnerships to deliver sustainability. Similarly, Solheim et al. (2022) argue that collaboration across sectors has led to innovative solutions for climate adaptation in Norway which shows how cross-sectoral collaboration has worked in practice, where solution-oriented partnerships work at solving complex environmental problems. Furthermore, the pursuit of climate neutrality gives just as much importance to individual change-makers. Positive personal decisions, actions and activism can make a difference in climate solutions. Informed people not only use energy-efficient products, and buy sustainable ones but also push for policies to lower carbon emissions or they simply become engaged citizens.

As Höhne et al. (2021) write, the net-zero ambition across nations in recent years is driven by an understanding that climate action will need to be inclusive of all actors in society. Given the multi-faceted nature of climate change, this public sector inquiry suggests tailored solutions are required to sufficiently meet the diverse needs and capacities of stakeholders. For public–private collaborations to be effective, they will need to overcome challenges around trust and transparency as well as equity of access to resources. This also means that change makers must find ways to penetrate barriers to engagement and mobilization in their communities. But while the path towards adaptation and resilience is riddled with challenges, there are also opportunities presented in this new reality that we must capitalize on as private, public and individual actors alike. Amidst the ongoing discussions of climate neutrality, this chapter will explore the interplay of adaptation and mitigation, highlighting the role of different actors in facilitating interactions between these two strategies. It will look at the essential linkages we have identified as important for effective responses to climate change coupled with transformative pathways. By understanding the interplay between public–private partnerships and individual actions, we can better appreciate the collective effort required to combat climate change and build a sustainable future for all.

Defining Resilience and Adaptation

Resilience is defined as the ability of something, someone, or a group to anticipate, accommodate, cope, and recover from disruptions or specific dangers. The Resilient Africa Network (RAN) defined resilience broadly as individuals' and systems' ability to prevent, adjust to, recover from, and learn from shocks and stresses in a way that reduces vulnerability and promotes well-being (RAN, 2015).

Adaptation on the other hand simply means the ability of systems to adjust to current or predicted occurrences/events to reduce harm or exploit opportunities. Climate is the process of adjusting to the existing or expected climate and its effects to limit harm or capitalise on positive opportunities (Wannewitz et al., 2023). While resilience is more focused on withstanding and recovering from disruption, adaptation on the other hand is concerned with the ability to cope with disruption over time.

Climate Neutrality and Its Importance

One of the most pressing issues confronting organisations and humanity is how to manage climate change (Kumar et al., 2024) and addressing this issue has become a top priority in building sustainable practices (Nwokolo et al., 2023a, 2023b, 2023c, 2023d). Achieving net-zero emissions by striking an equitable equilibrium between the release of greenhouse gases and removal from the environment is known as climate neutrality (Joshi et al., 2024; UN, 2021). According to Gielen et al., reducing emissions as much as possible through energy optimization, the switch to energy from renewable sources, and sustainable practices can help reach climate neutrality. Importantly, climate neutrality seeks to prevent further global warming by reducing the impact of emissions.

Climate neutrality is significant because it can mitigate the negative consequences of climate change, which endangers ecosystems, economies, and human health. Achieving climate neutrality can also help to lessen these effects by providing more time to adapt and protect vulnerable people. Matos et al. (2022) claim that as governments and enterprises adopt climate-neutral practices, they help to drive innovation in sustainable technologies and renewable energy. These result in a greener economy and more jobs. Therefore, climate neutrality encourages sustainable lifestyles and responsible resource management.

The Role of Change-Makers: Private, Public and Individual Actors in Driving Climate-Neutral Practices

The major contributors to greenhouse gas emissions are businesses. Although, through innovation and resourcefulness, they hold the potential to promote climate neutrality (Mikalai et al., 2024). The change-makers in the private sector, such as corporations influence the market by setting ambitious climate goals. This is done by adopting sustainable production methods and investing in renewable energy. For instance, companies like Apple and Unilever have committed to operating entirely on renewable energy (Apple, 2023; Unilever, 2023). This illustration demonstrates how the business sector may promote environmental stewardship and innovation.

Also, governments (public sector) play an important role in promoting climate neutrality through rules, incentives and public policies. Policymakers can enforce emission reduction targets, invest in green infrastructures, and offer tax breaks to enterprises that use low-carbon technologies (Omri & Ben-Jabeur, 2024). The European Union, for example, has set binding targets to become climate-neutral by 2050. This is part of the European Green Deal, which also lays out a road map for its member states to reduce emissions and switch to renewable energy sources (European Commission, 2023). These strategies not only establish a precedent for aggressive climate goals but also create regulatory standards that incentivise corporations and communities to contribute to climate neutrality. Governments can also encourage climate neutrality by supporting green research and technology. Other means, include, encouraging sustainable urban planning, and informing the public about the importance of reducing emissions. When the government work with private organisations on sustainable projects, its influence grows.

Individuals contribute to climate neutrality by embracing sustainable lifestyle choices and everyday activities. According to Ivanova et al., (2020), reducing waste, taking public transportation, and conserving electricity are ways individuals contribute to the reduction of greenhouse gas emissions. Other means include adopting sustainable products or cutting back on meat intake (Ivanova et al., 2020). Importantly, one person's actions can inspire worldwide action, changing legislation and making large corporations more responsible for their environmental impact, as seen by grassroots projects like Greta Thunberg's Fridays for Future (Science Daily, 2021).

Private Sector as Change Agents

The private sector is arguably the most important change agent. With the effects of climate change becoming ever more apparent, firms are realising that they have to adjust their ways of working and strategy if they are to avoid the threats but also capture the benefits related to a new climate. An increasing number of companies are investing in research and development to develop green technologies/projects that minimize greenhouse gas emissions. Cap-and-trade programs, as Taylor (2012) points out, induce innovation in a way that drives firms to create cleaner technologies that can reduce emissions considerably. It does not only help with the mitigation of climate change but also increases the competitiveness of companies in a fast-changing marketplace.

Funding climate adaptation and resilience requires investment in the private sector. As noted by Morgado and Lasfargues (2017), engaging the private sector will play a significant role in green growth and climate action since development cooperation providers increasingly introduce private sector approaches in climate-related programs. By mobilizing capital, the companies can help fill a gap in financing climate-friendly projects. This is particularly important for developing countries where the sources of public funding may be scanty (Oyenuga & Omale, 2024).

Public–private partnerships are an important mechanism for managing complex climate change issues. According to Agrawala et al. (2011), the public sector may help promote private sector involvement by improving the availability of information and risk management tools so that better choices can be made. Cooperative frameworks enable knowledge, resource, and skill sharing. For instance, the Sendai Framework for Disaster Risk Reduction recommends integrated approaches with the incorporation of the private sector (Leitold et al., 2020).

Private sector players are increasingly realizing the vulnerabilities that climate change imposes on their operations and profitability. Casado-Asensio et al. (2021) argue that businesses, especially MSMEs, are embracing climate resilience. These corporations will be able to withstand climate hazards, hence preventing economic losses due to climate-related hazards when they invest in adaptation measures such as infrastructural upgrading and good supply chain practices. Not only does this proactive approach secure their operations but also contributes to the larger resilience of the community (Barile et al., 2022).

Public Sector as Change Agents

The public sector plays a transformative role in the quest for climate neutrality and resilience. Among the key roles that the public sector plays is pro-climate action policy development and implementation. This consequently sets a regulatory framework that creates an environment of encouragement for sustainable practices in all sectors. For instance, Macassa et al. (2022) weigh the need for incorporating considerations for public health into any adaptation policies, hinting at how coordination of actions in other sectors may boost community resilience. In this regard, effective guidelines by governments should be set up with incentives that drive systemic change among public as well as private actors adopting climate-friendly practices. Public funding should be mobilized in the public sector to fund climate initiatives. Governments can invest in the research sector, provide infrastructure support, and fund community-based adaptation projects. National climate change adaptation policies ought to be meaningfully translated into local agricultural adaptation strategies as observed by Nwokolo et al. (2024) that require financial support and capacity building at the local level. Investing in the climate resilience of the economy will stimulate economic growth while addressing climate change impacts.

The public sector can help to facilitate an alliance by providing an encouraging regulatory setting and cultivating relationships among the private actors and community stakeholders. According to Pauw et al. (2015), "effective engagement with the private sector is a must for scale-up adaptation efforts and ensures that initiatives under climate are sustainable and impactful". The public sector can make actions towards climate easier by providing a space for dialogue and collaboration to be possible. The public sector has the critical role of raising awareness about climate change and engaging citizens in the activities towards the fighting of climate change.

Leiserowitz (2019) writes that the importance of building public and political will over climate action is necessary to bring significant change. The two sources of empowerment for individuals undertaking action for climate resilience are an environmental stewardship culture and community participation. Climate change is known to harshly affect vulnerable groups; hence, measures need to be adapted to appear more inclusive and fair. According to Cintron-Rodriguez et al. (2021), participatory policy design is a very important element in enhancing just climate action as well as shared power and ownership among communities.

Individuals as Change Agents

In the fight against climate change, the individual is becoming an important agent of change by pushing grassroots movements and contributing to broader shifts in society for climate neutrality to be achieved.

Grassroots Movements can be propelled and sustained by passionate actors who want to take action on climate change and the call for sustainability. According to Nardini et al. (2020), social movements can succeed when diverse coalitions of individuals and organizations go into collective actions. Such movements help the person transition from a passive observer into an active participant, imbuing him or her with a sense of agency and responsibility. Grassroots movements have the potential to cause long-term changes in public opinion and policy on climate action through the activation of a community toward common goals.

Momentum change in climate policy as well as expedited action is brought about by the mobilization of the youth, says Han and Ahn (2020). The youth group represented formally in international climate talks, like the UNFCCC, also plays an active role as planners of future climate policy. Their involvement does not only voice their issues but also brings to light traditional dominance by challenging traditional power relations to demand an equal share of inclusiveness in equal climatic solutions. This points out the urgency of involving the young generation to communicate change across generations in meeting the challenges of the climate. Every individual can contribute to climate action through the choices he makes daily through lifestyle changes. From energy reductions, among other things-sustainable transportation and food systems individual activities play a significant role in carbon emissions. According to Tohari (2021), grassroots groups, especially agricultural labourers and landless peasants, use various self-actualization means to voice the call for sustainability within their locality.

This engagement will ensure that climatic policies represent the needs and desires of the communities they affect. Building community resilience is important for communities coming together to address climatic challenges, thus strengthening resilience. Oliveira et al. (2019) point out that participatory planning and grassroots movements form an essential part of resisting hegemony and promoting the needs of the community. The encouragement of collaboration and knowledge sharing among people will increase the adaptability capacity of the communities to the effects of climate change. Grassroots activity assists not only in building resilience but also equips people to own their local environments.

Interconnectedness of Private, Public, and Individual Efforts

Climate neutrality and building resilience to climate change require interconnectedness through private, public, and individual efforts. All sectors play their roles differently but complement each other in addressing the multifaceted challenges presented by climate change, and more effective and sustainable outcomes will be achieved with their collaboration. Public–private partnerships are an important mechanism for ensuring the concurrence of both sectors' strengths in action against climate change. As Leenaars et al. (2013) observe, public health challenges are multidimensional and hence call for intersectoral coordination between both the public and private sectors. It also applies to climate change, whereby pooling of resources, expertise, and innovation from both sectors may drive better adaptation and mitigation measures. For example, according to Baluk et al. (2024), PPP models that could be articulated for the successful promotion of sustainable development can be integratively investigated under a holistic approach to analyzing PPPs. These partnerships could then become a basis to share knowledge, resources, and best practices, and on the way, it would contribute in the long term to making initiatives for climate change mitigation measures effective.

Individuals also shape the direction and pace of climate action through private decision-making and activism. According to Strasser et al. (2021), a closely related factor enabling people to act in concert might lead to significant consequences regarding sustainability. Grassroots movements are usually mobilized and driven by passionate human agency, which may be able to mobilize communities to campaign for policy changes that foster climate resilience. Thus, these individual efforts become not only private but also public ones, multiplying the overall impact of climate action. Building trust and solidarity among stakeholders is fundamental for effective cooperation on climate action. According to Horn and Kerasidou (2020), public trust is established based on the assumption that institutions will be acting in the public interest by carrying out an effective representation of collective interests. Thus, trust forms the basis for building cooperation between the public, private institutions, and people. Building trust can foster effectiveness in public–private partnerships and encourage individuals to actively take part in climate initiatives.

Pooling together each sector's strong attributes in creating innovative solutions geared toward the root causes of climate change, and building community resilience. Transparency and accountability are the bedrock of a successful partnership between the private and public, individual sectors. Ibem et al. (2018) support this notion by stipulating that transparency in the selection and realization process of a project enhances trust and assurances in public–private partnerships. Being open about the goals, processes, and results drives stakeholders to achieve a feeling of accountability that will enforce collaboration and participation.

Challenges Faced by Change Agents

Challenges Faced by the Private Sector in Climate Neutrality Efforts

The private sector faces myriad challenges (Nwokolo et al., 2024) in their quest to minimize respective levels of greenhouse gas emissions while searching for ways to implement sustainability into operations.

The most challenging indirect emissions for companies seeking climate neutrality is how to deal with Scope 3 which involves value chain emissions and those the companies cause their suppliers and customers to emit. According to Seyfried et al. (2023), so far, no company has achieved the scope of achieving climate neutrality at Scope 3, and most do not have proper strategies for handling them in the long run. This multifacetedness makes it difficult to devise comprehensive climate policies. As a consequence, more stakeholders must be involved to achieve relevant reduction results. High investments of money for certain technologies, processes and infrastructures (Apata et al., 2024) are often necessary during the transition to climate neutrality.

Indeed, many SMEs and other companies cannot afford to spend high amounts of money in the sense of spending money on sustainability projects. Casado-Asensio et al. (2021) consider that it is difficult for micro, small, and medium-sized enterprises to know what the long-term financial impact of climate change on profitability and continuity would be. Ambiguity in this regard invests in climate-friendly practices harder, thus slowing further progress toward climate neutrality. The private sector functions within an environment that characterizes the complexity and uncertainty associated with climate policies. It thus presents different jurisdictions of regulation, and firms find it hard to have a uniform climate strategy. Since Odeyemi and Sekiyama (2023) claim that firms may resist aggressive national climate policies and have the fear of losing investment in competitiveness, then such resistance may hinder climate neutrality progress as firms would rather hold on to short-term profitability rather than sustain in the long run. Reducing emissions and achieving climate neutrality requires innovative new technologies. At the same time, technological constraints and knowledge deficits make most companies face difficulties. According to Duffy et al. (2022), there is a call for parallel development of carbon capture infrastructure and industries based on biological science towards supporting net-zero emissions. Companies do not have access to the technologies and expertise necessary to effectuate climate-effective strategies.

The competitive dynamics can act as a source of challenge for businesses that aim to be climate-neutral. Firms may be afraid that consumers will defect to or continue visiting rivals who do not care about the more sustainable set of practices (Oyenuga et al., 2023), and thus firms shy away from investing in climate initiatives to not lose their market share.

Zhang et al. (2022) indicate that competitive pressures can capture the decisions of firms on climate neutrality as organizations trade off the costs of sustainability against potential market disadvantages. The result is a race to the bottom, as companies lose interest in long-term sustainability in the interest of revenues in the short run. With increasing awareness of climate change, stakeholders are becoming more demanding of companies' efforts in environmental accountability. Lastly, cultural and organizational barriers within firms may reduce the adoption of climate-neutral practices. Among the factors that increase the resistance to these changes are resistance to change and lack of leadership commitment and employee engagement. According to Duong et al. (2023), organizations must set priorities in terms of resource allocation building capacity against such barriers and establishing a sustainable culture. If there is no organizational culture that supports activities for achieving climate neutrality; efforts in attaining climate neutrality could slowly fall.

Challenges Faced by the Public Sector in Climate Neutrality Efforts

A major challenge for the public sector is the limitation of financial resources. Most governmental bodies operate under tight budgets that limit investment in climate initiatives. For instance, Carr et al. (2012) observe that the local health departments suffer from inappropriate funding and also minimal political muscle to do climate adaptation planning which may constrain their ability to execute practical climate policy. This implies that the deficit of finance (Oladele et al., 2023) will limit the structure and program development that could be required to achieve climate neutrality. The public sector is faced with a highly bureaucratic legal environment that may create uncertainty in the climate policy. Araos et al. (2015) argue that there is an imperative for precise policy direction and consistent policies in assisting the process of effective adaptation to urban areas.

With this ambiguity, there is the possibility that climate initiatives will not take off until much later (Khan et al., 2024; Nwokolo et al., 2023a, 2023b, 2023c, 2023d). It may also discourage public agencies from making investments as far as long-term sustainability programs are concerned. Furthermore, changed regimes and frequent political leadership promote shifting priorities in government, which does not serve to accelerate a persistent climate agenda in public sectors. Effectual climate action often requires interagency effort and coordination in terms of government levels. Still, interagency coordination remains a challenge due to differences in priorities, bureaucracy, and lack of clear channels of communication. According to Akerlof et al. (2010), public perception regarding climate change often differs considerably across agencies, thereby culminating in fractured approaches towards climate policy.

This weakness may diminish the success of climate action and limit the effective implementation of holistic policies by the public sector. In many instances, knowledge gaps and a lack of capacity building in climate science and adaptation strategies inhibit the effective delivery of public sectors. Mckelvey et al. (2017) point out the barriers to climate change adaptation for environmental health officers, including insufficient guidance and access to information. The evolution of climate science would call for continuous education and support for public sector staff to be well-equipped to counter climate change. Without the proper know-how and skills, public agencies will not be in an appropriate position to make effective climate policies which can better the lives of people. According to Capstick et al. (2014), public perceptions of climate change will change and continue to shift leading to climate fatigue or scepticism. This variability may thus impact the efforts to mobilize community support for action. To encourage further public involvement and garner public support for climate policies, climate change needs to be reframed as a public health issue (Cardwell & Elliott, 2013). Climate change is a "wicked problem" for such a reason as it is very complex and intricately linked to its causes (Pollitt, 2015). This aspect of the problem makes public management research lag on issues regarding climate change.

Challenges Faced by the Individuals in Climate Neutrality Efforts

Individuals are leaders in the pursuit of achieving climatic neutrality; however, there are many issues holding them back from performing effectively. Such barriers range from psychological barriers to socio-economic disincentives that affect an individual's potential to participate in the action on climate change.

Psychological barriers are one of the most crucial barriers to individual action on climate change. Gifford (2011) finds a few "dragons of inaction" such as perceived ineffectiveness, lack of knowledge, and feelings of helplessness. Many people feel their action is little more than a drop in the ocean compared to the giant global issue, which breeds apathy and inaction. This phenomenon is sometimes referred to by the term "diffusion of responsibility," where people feel that others will take responsibility, and breeds this collective inaction. Such psychological barriers will require targeted communication (Oyedele et al., 2023) on the importance of personal contributions to climate action. In many instances, climate change and its impacts do not get people's understanding, thereby hindering individual engagement (Köprülü et al., 2023).

This knowledge gap is expected to limit the extent to which individuals can make better consumption-related choices, appropriate usage of energy, and lifestyle modifications. Public education (Oyedele & Iember, 2021; Singh et al., 2024)) and outreach provide information to encourage people to act meaningfully. Economic factors also play a significant role in the behaviour of individuals concerning climate change. Often, many experience financial constraints that limit their ability to pursue sustainable investments, such as energy-saving appliances or renewable energy. Bhushan and Singh (2024) demonstrate that the stimulus to adapt as an individual and mitigate depends more on economic forces since people care more about short-term financial gains than long-term environmental gain.

Climate action behaviour in the individual also depends on social norms, influence and values (Nwokolo et al., 2023a, 2023b, 2023c, 2023d). However, without these values or norms being normal in a community, an individual would feel de-incentivized to do those things. This social issue might be a source of some barriers to collective action because people fear diverging from the set standards.

The everyday concerns of life may be a source of obstacles for those who want to take action in terms of climate. This can be limited through feelings of powerlessness in the face of bureaucratic red tape, as well as influence by a lack of access to the policy-making process that affects their communities. Such hurdles, however, can be overcome if democratic involvement is strengthened and opportunities for participation in debates concerning climate policy are opened up. It often raises confusion and suspicion among people due to its wide prevalence. Misinformation about climate change questions public understanding of climate science, as well as the need for urgent climate action. According to Maibach et al. (2010), reframing climate change as a public health issue is one important means by which people can be encouraged into further engagement while combating misinformation. Building trust in credible sources of information and promoting scientific literacy can empower those individuals to take informed action on climate change.

Conclusion

In the light of rising challenges of climate change, private, public, and individual actors must engage with each other to achieve climate neutrality as well as develop resilience. Interconnected efforts of different sectors were examined in this chapter to discuss multifaceted roles for every sector in driving climate actions, emphasizing synergies and collaborative frameworks in achieving overall effectiveness.

This makes the private sector a potential change agent in the innovative use of finance, real market dynamics, and sustainable practice, as seen from case studies where evidence reflects an increase in competition for climate action business by compelling them towards emission reduction with competitiveness and long-term sustainability, though challenges are many, like financial constraints, regulatory uncertainties, and even complexity with Scope 3 emissions. With such stubborn obstacles on the way, supporting policies, public–private partnerships, and a commitment to transparency and accountability are needed to overcome them. Again, in climate governance, it is the public sector that bears the more important role of setting the framework for climate action. Therefore, through policy-making and outreach to the community, the government provides an enabling environment for private and personal initiative while the collective commitment to adaptation and resilience will be key in combating climate change and achieving a sustainable, climate-neutral future.

The path to climate neutrality is holistic enough to understand the interacting nature of private, public, and individual endeavours. Encouraging inter-cooperation and, importantly, knowledge sharing will make the collective strength of each sector allow society to develop a robust mechanism for dealing with climate change.

References

Agrawala, S., Carraro, M., Kingsmill, N., Lanzi, E., Mullan, M., & Prudent-Richard, G. (2011). Private sector engagement in adaptation to climate change: Approaches to managing climate risks. *OECD Environment Working Papers*. https://doi.org/10.1787/5kg221jkf1g7-en

Akerlof, K., DeBono, R., Berry, P., Leiserowitz, A., Roser-Renouf, C., Clarke, K., Rogaeva, A., Nisbet, M. C., Weathers, M. R., & Maibach, E. W. (2010). Public perceptions of climate change as a human health risk: Surveys of the United States, Canada and Malta. *International Journal of Environmental Research and Public Health, 7*(6), 2559–2606. https://doi.org/10.3390/ijerph7062559

Apata, S. B., Oyenuga, M. O., Adeoba, M. I., Ugom, M. K., & Abiodun, A. O. (2024) Internet of Things (IoT) Solutions for smart transportation infrastructure and fleet management. *Tuijin Jishu/Journal of Propulsion Technology, 45*(4). https://propulsiontechjournal.com/index.php/journal/article/view/8325

Apple. (2023). Environmental progress report. Apple Inc. 2024 Environmental Progress Report.

Araos, M., Austin, S. E., Berrang-Ford, L., & Ford, J. D. (2015). Public health adaptation to climate change in large cities. *International Journal of Health Services, 46*(1), 53–78. https://doi.org/10.1177/0020731415621458

Asensio, J. C., Kato, T., & Shin, H. (2021). Lessons on engaging with the private sector to strengthen climate resilience in Guatemala, the Philippines and Senegal. *OECD Development Co-operation Working Papers*. https://doi.org/10.1787/09b46b3f-en

Baluk, N., Androshchuk, I., Havadzyn, N., Zhyvko, Z., & Hura, V. (2024). Marketing and sustainable economic development of the energy complex: A case study of the administrative and legal planning of a public-private partnership. *International Journal of Religion, 5*(9), 908–918. https://doi.org/10.61707/eazpbk06

Barile, D., Secundo, G., & Vecchio, P. D. (2022). Food 4.0 for competing during the covid-19 pandemic: Experimenting digitalization in family firms. *European Journal of Innovation Management, 27*(4), 1381–1402. https://doi.org/10.1108/ejim-07-2022-0373

Bhushan & Singh. (2024). Net zero economy, corporate social responsibility and sustainable value creation. In *CSR, sustainability, ethics & governance*. https://doi.org/10.1007/978-3-031-55779-8

Capstick, S., Whitmarsh, L., Poortinga, W., Pidgeon, N., & Upham, P. (2014). International trends in public perceptions of climate change over the past quarter century. *Wiley Interdisciplinary Reviews Climate Change, 6*(1), 35–61. https://doi.org/10.1002/wcc.321

Cardwell, F. S., & Elliott, S. J. (2013). Making the links: Do we connect climate change with health? A qualitative case study from Canada. *BMC Public Health, 13*(1). https://doi.org/10.1186/1471-2458-13-208

Carr, J. L., Sheffield, P. E., & Kinney, P. L. (2012). Preparedness for climate change among local health department officials in New York state. *Journal of Public Health Management and Practice, 18*(2), E24–E32. https://doi.org/10.1097/phh.0b013e31823dea74

Cintron-Rodriguez, I. M., Crim, H. A., Morrison, D. L., Niepold, F., Kretser, J., Spitzer, W., & Bowman, T. (2021). Equitable and empowering participatory policy design strategies to accelerate just climate action. *Journal of Science Policy & Governance, 18*(02). https://doi.org/10.38126/jspg180203

De Oliveira, C., Alberti, C. B., Silva, L. B., & Nodari, G. R. (2019). The people's Bourough plan of action: A counter-project of insurgent citizenship1. *Read Revista Eletrônica De Administração (Porto Alegre), 25*(3), 247–277. https://doi.org/10.1590/1413-2311.274.97304

Duffy, C., Prudhomme, R., Duffy, B., Gibbons, J., Iannetta, P. P. M., O'Donoghue, C., Ryan, M., & Styles, D. (2022). Randomized national land management strategies for net-zero emissions. *Nature Sustainability, 5*(11), 973–980. https://doi.org/10.1038/s41893-022-00946-0

Duong, A. T. B., Yap, T. L., Ngo, V. M., & Quang, H. T. (2023). Critical success factors of food safety management for achieving climate neutrality: A multilevel moderated approach with industry revolution 4.0. *The International Journal of Logistics Management, 35*(3), 865–891. https://doi.org/10.1108/ijlm-03-2023-0112

European Commission. (2023). *Delivering the European green deal*. European Commission. https://commission.europa.eu/strategy-and-policy/priorities-2019-2024/european-green-deal/delivering-european-green-deal_en

Gifford, R. (2011). The dragons of inaction: Psychological barriers that limit climate change mitigation and adaptation. *American Psychologist, 66*(4), 290–302. https://doi.org/10.1037/a0023566

Han, H., & Ahn, S. W. (2020). Youth mobilization to stop global climate change: Narratives and impact. *Sustainability, 12*(10), 4127. https://doi.org/10.3390/su12104127

Höhne, N., Gidden, M. J., Elzen, M. D., Hans, F., Fyson, C., Geiges, A., Jeffery, M. L., Gonzales-Zuñiga, S., Mooldijk, S., Hare, W., & Rogelj, J. (2021). Wave of net zero emission targets opens window to meeting the Paris Agreement. *Nature Climate Change, 11*(10), 820–822. https://doi.org/10.1038/s41558-021-01142-2

Horn, R., & Kerasidou, A. (2020). Sharing whilst caring: Solidarity and public trust in a data-driven healthcare system. *BMC Medical Ethics, 21*(1). https://doi.org/10.1186/s12910-020-00553-8

Ibem, E. O., Onyemaechi, P. C., & Yo-Vaughan, E. A. (2018). Project selection and transparency factors in housing public-private partnerships in Nigeria. *Construction Economics and Building, 18*(2), 15–40. https://doi.org/10.5130/ajceb.v18i2.5771

Ivanova, D., Barrett, J., Wiedenhofer, D., Macura, B., Callaghan, M. W., & Creutzig, F. (2020). Quantifying the potential for climate change mitigation of consumption options. *Environmental Research Letters, 15*(9). https://doi.org/10.1088/1748-9326/ab8589

Joshi, A., Singh, R., Kumar, V., & Rani, S. (2024). Net zero strategy: Exploring opportunities, confronting challenges for developing countries. In R. Singh & D. Crowther (Eds.), Transition towards a sustainable future. Approaches to global sustainability, markets, and governance. Springer. https://doi.org/10.1007/978-981-97-5756-5_11

Khan, Z., Ali, M., Kirikkaleli, D., Wahab, S., & Jiao, Z. (2020). The impact of technological innovation and public-private partnership investment on sustainable environment in China:

Consumption-based carbon emissions analysis. *Sustainable Development, 28*(5), 1317–1330. https://doi.org/10.1002/sd.2086

Khan, S., Singh, R., Alnahas, J., Abbate, S., & Centobelli, P. (2024). Navigating the smart circular economy: A framework for Manufacturing Firms. *Journal of Cleaner Production*, 144007. https://doi.org/10.1016/j.jclepro.2024.144007

Köprülü, F., Hamad, O. A. M., & Kiraz, A. (2023). Perceptions of Libyans towards climate change. *International Journal of Ecosystems and Ecology Science (IJEES), 13*(2), 37–44. https://doi.org/10.31407/ijees13.205

Kumar, K., Singh, R., Sharma, A., & Lehri, B. P. S. (2024). Climate change adaptation strategies for achieving net-zero economy. In R. Singh & D. Crowther (Eds.), *Transition towards a sustainable future. Approaches to global sustainability, markets, and governance.* Springer. https://doi.org/10.1007/978-981-97-5756-5_8

Leenaars, K., Bruggen, M. J. D., & Renders, C. (2013). Determinants of successful public-private partnerships in the context of overweight prevention in Dutch youth. *Preventing Chronic Disease, 10.* https://doi.org/10.5888/pcd10.120317

Leiserowitz, A. (2019). Building public and political will for climate change action. In *Yale University Press eBooks* (pp. 155–162). https://doi.org/10.2307/j.ctvqc6gcq.21

Leitold, R., Diez, J., & Tran, V. (2020). Are we expecting too much from the private sector in flood adaptation? scenario-based field experiments with small- and medium-sized firms in Ho Chi Minh city, Vietnam. *Climatic Change, 163*(1), 359–378. https://doi.org/10.1007/s10584-020-02888-y

Macassa, G., Ribeiro, A. I., Marttila, A., Stål, F., Silva, J. P., Rydback, M., Rashid, M., & Barros, H. (2022). Public health aspects of climate change adaptation in three cities: A qualitative study. *International Journal of Environmental Research and Public Health, 19*(16), 10292. https://doi.org/10.3390/ijerph191610292

Maibach, E. W., Nisbet, M., Baldwin, P., Akerlof, K., & Diao, G. (2010). Reframing climate change as a public health issue: An exploratory study of public reactions. *BMC Public Health, 10*(1). https://doi.org/10.1186/1471-2458-10-299

Matos, S., Viardot, E., Sovacool, B. K., Geels, F. W., & Xiong, Y. (2022). Innovation and climate change: A review and introduction to the special issue. *Technovation, 117*, 102612.

Mikalai, F., Peterson, M. P., Yan, H., Gusev, A., Zhang, L., He, Y., & Yang, S. (2024). Greenhouse gas emissions and reduction strategies for the world's largest greenhouse gas emitters. *Science of the Total Environment, 944*, 173895–173895. https://doi.org/10.1016/j.scitotenv.2024.173895

McKelvey, L., Health, N. B. S. O. H. S. E., & Heacock, H. (2017). Environmental health officers and climate change adaptation in British Columbia. *BCIT Environmental Public Health Journal.* https://doi.org/10.47339/ephj.2017.76

Morgado, N. C., & Lasfargues, B. (2017). Engaging the private sector for green growth and climate action. *OECD Development Co-operation Working Papers.* https://doi.org/10.1787/85b52daf-en

Nardini, G., Rank-Christman, T., Bublitz, M. G., Cross, S. N. N., & Peracchio, L. A. (2020). Together we rise: How social movements succeed. *Journal of Consumer Psychology, 31*(1), 112–145. https://doi.org/10.1002/jcpy.1201

Nwokolo, S. C., Singh, R., Khan, S., Kumar, A., Luthra, S. (2023). Impacts of climate change in Africa. In *Africa's path to net-zero. CSR, sustainability, ethics & governance.* Springer. https://doi.org/10.1007/978-3-031-44514-9_7

Nwokolo, S. C., Singh, R., Khan, S., Kumar, A., & Luthra, S. (2023). Remedies to the challenges of renewable energy deployment in Africa. In *Africa's path to net-zero. CSR, sustainability, ethics & governance.* Springer. https://doi.org/10.1007/978-3-031-44514-9_3

Nwokolo, S. C., Singh, R., Khan, S., & Kumar, A. (2024). Threats to the rapidity of sustainability transitions posed by technological changes. *Technological Pathways for Africa' s Net-Zero Economy*, 39–75. https://doi.org/10.1016/B978-0-443-31486-5.00002-1

Nwokolo, S. C., Singh, R., Khan, S., Kumar, A., & Luthra, S. (2023a). Africa's awakening to climate action. In *CSR, sustainability, ethics & governance* (pp. 299–310). https://doi.org/10.1007/978-3-031-44514-9_9

Nwokolo, S. C., Singh, R., Khan, S., Kumar, A., & Luthra, S. (2023b). Global investment and development in Africa. In *CSR, sustainability, ethics & governance* (pp. 15–58). https://doi.org/10.1007/978-3-031-44514-9_2

Odeyemi, C., & Sekiyama, T. (2023). The roles of four important contexts in Japan's carbon neutrality policy and politics, 1990–2020. *Climate, 11*(12), 233. https://doi.org/10.3390/cli11120233

Oladele, T. O., Oyenuga, M. O., & Adoga, G. J. (2023). Green management practices and organisational efficiency in higher institution. *Budapest International Research and Critics in Linguistics and Education (BirLE) Journal, 6*(3), 276–290. https://doi.org/10.33258/birle.v6i3.7716

Omri, A., & Ben Jabeur, S. (2024). Climate policies and legislation for renewable energy transition: The roles of financial sector and political institutions. *Technological Forecasting and Social Change, 203*, 123347. https://doi.org/10.1016/j.techfore.2024.123347

Oyedele, M. O., Akhaine, M. E., & Omale, S. A. (2023). We need it: How digital marketing tactics influence the purchasing behaviour of Nigerian Millenials. *Konfrontasi Journal: Culture, Economy and Social Changes, 10*(2), 91–105. https://doi.org/10.33258/konfrontasi2.v10i2.275

Oyedele, O. M., & Iember, A. A. (2021, July). Covid-19 and the future of higher education. *IEEE Technology Policy and Ethics, 6*(4), 1–3. https://doi.org/10.1109/ntpe.2021.9778140

Oyenuga, M. O. & Omale, S. A. (2024), Is Africa Jinxed? Exploring the challenges of technology access and adoption in Africa. *African Journal of Economics and Sustainable Development, 7*(4), 142–161. https://doi.org/10.52589/AJESDULN1LRNF

Oyenuga, M. O., Marcus, G. O., & Ahungwa, A. I. (2023). Do consumers care about green marketing practices? Insight from a developing nation. *Budapest International Research and Critics Institute Journal (BIRCI Journal), 6*(3), 1424–1436. https://doi.org/10.33258/birci.v6i3.7665

Pauw, W. P., Klein, R. J. T., Vellinga, P., & Biermann, F. (2015). Private finance for adaptation: Do private realities meet public ambitions? *Climatic Change, 134*(4), 489–503. https://doi.org/10.1007/s10584-015-1539-3

Pollitt, C. (2015). Wickedness will not wait: Climate change and public management research. *Public Money & Management, 35*(3), 181–186. https://doi.org/10.1080/09540962.2015.1027490

Resilient Africa Network. (2015). *The state of African resilience; understanding dimensions of vulnerability and adaptation.*

Science Daily. (2021). Greta Thunberg and Fridays for future, from global myth to local mobilization. https://www.sciencedaily.com/releases/2021/05/210504112519.htm

Seyfried, S., Nagel, L., & Weigold, M. (2023b). Empirical investigation of climate neutrality strategies of companies in industrial production. In *Lecture notes in mechanical engineering* (pp. 999–1007). https://doi.org/10.1007/978-3-031-28839-5_111

Singh, R., Gupta, A., & Kandpal, V. (2024). ESG excellence in the digital world. *Wiley*, 193–204. https://doi.org/10.1002/9781394234035.ch10

Solheim, A., Kalsnes, B., Strout, J., Piciullo, L., Heyerdahl, H., Eidsvig, U., & Lohne, J. (2022). Landslide risk reduction through close partnership between research, industry, and public entities in Norway: Pilots and case studies. *Frontiers in Earth Science, 10*. https://doi.org/10.3389/feart.2022.855506

Strasser, S., Stauber, C., Shrivastava, R., Riley, P., & O'Quin, K. (2021). Collective insights of public-private partnership impacts and sustainability: A qualitative analysis. *PLoS ONE, 16*(7), e0254495. https://doi.org/10.1371/journal.pone.0254495

Taylor, M. R. (2012). Innovation under cap-and-trade programs. *Proceedings of the National Academy of Sciences, 109*(13), 4804–4809. https://doi.org/10.1073/pnas.1113462109

Tohari, A. (2021). From confrontation to collaboration: Peasant negotiations in the arena of decentralization. *Indonesian Journal of Political Studies (IJPS), 1*(2), 172–187. https://doi.org/10.15642/ijps.2021.1.2.172-187

UN. (2021). *A beginner's guide to climate neutrality*. Climate Change. A Beginner's Guide to Climate Neutrality | UNFCCC.

Unilever. (2023). Leading and delivering on sustainability through our Compass commitments.

Wannewitz, M., Petzold, J., & Garschagen, M. (2023). What makes people adapt together? An empirically grounded conceptual model on the enablers and barriers of collective climate change adaptation. *Frontiers in Climate, 5*, 1–10.

Zhang, A., Tay, H. L., Alvi, M. F., Wang, J. X., & Gong, Y. (2022). Carbon neutrality drivers and implications for firm performance and supply chain management. *Business Strategy and the Environment, 32*(4), 1966–1980. https://doi.org/10.1002/bse.3230

Chapter 3
Decoding Data Politics: Leveraging Climate Informatics for Policy, Power, and Sustainability

Mananage Shanika Hansini Rathnasiri, Lishanthi Wijewardene, Damith Gangodawilage, and Lalit Mohan Tewari

Abstract This chapter examines how data is produced, circulated, and used concerning climate politics, particularly climate informatics. This research employs the population, intervention, comparison, and outcome (PICO) approach to explore essential features of climate data politics: stakeholders, interference, contrasts, and consequences. The chapter also uses the Preferred Reporting Items for Systematic Reviews and Meta-Analyses (PRISMA) framework to conduct a systematic literature review and properly synthesise the studies, cases, and empirical data about climate data governance. The final sample consisted of thirty-six seminal studies. Several important issues are brought out, mainly issues of ethics and privacy, data coverage, and issues that stem from the nation's socio-political fabric, making it difficult to repose faith in data. However, it also outlines major prospects like enhancing transparency tools, recognising climate issues as a learning process in policymaking, and sharing climate knowledge worldwide. It presents and discusses examples of one country, city, and company where big data is used and failures of data-driven approaches, especially in carbon markets, disaster response, and the Paris Agreement. When providing equal access to data and incorporating artificial intelligence (AI) and blockchain, policymakers can build a more efficient and cooperative approach to meet

M. S. H. Rathnasiri (✉)
Department of Marketing Management, Faculty of Management Studies, Sabaragamuwa University of Sri Lanka, Belihuloya, Sri Lanka
e-mail: shanika@mgt.sab.ac.lk

L. Wijewardene
Business Communication Unit, University of Sri Jayewardenepura, Nugegoda, Sri Lanka
e-mail: lishanthi@sjp.ac.lk

D. Gangodawilage
Qasim Ibrahim School of Business, Villa College, Malé, Maldives
e-mail: damith.gangodawilage@villacollege.edu.mv

L. M. Tewari
Department of Botany, DSB Campus, Kumaun University, Nainital, Uttarakhand, India

R. Singh and W. L. Filho (eds.), *Climate Neutrality Through Smart Eco-Innovation and Environmental Sustainability*, Climate Change Management,
https://doi.org/10.1007/978-3-031-83250-5_3

the climate change challenges. Hence, climate informatics will only succeed if data practices are coupled with justice, equity, and international cooperation.

Keywords Climate informatics · Data politics · Sustainability · Data governance · Climate action

Introduction

Data politics have become vital in global management, particularly climate change. Data politics involves the gathering, handling, and using data about power relations, control and determinations, and resource distribution (Kitchin, 2014). This study establishes that data is weaponised and central in climate action, making it a site of struggle over access, ownership, and interpretation of climate data. This topic includes data openness, interactivity, and moral concerns about applying third-generation big data analytics to environmental problems. While people and socioeconomic systems try to come to terms with climate change and its effects, it is necessary to take climate data's social and political perspectives to facilitate climate information's natural, warranted, and fair distribution.

At the centre of this discourse, we will identify and explore climate informatics as a relatively new and thriving subfield of data science and artificial intelligence that deals with climate data. Climate informatics links climate science and computation, offering decision-support information for disaster response, carbon accounting, and climate diplomacy. Contemporary governance plays the transformative function needed for designing policies with the help of research findings and increasing accountability and transparency (Khan et al., 2024). However, the problem arises in the unequal distribution of technological resources and biased data usage, increasing worldwide inequality (Hilbert, 2016).

Climate informatics is a relatively young science field that intersects data assimilation and climate science, including machine learning methods (Middel et al., 2022). The use of such a system shall be to apply advanced analytics, big data, and business intelligence on large and difficult-to-analyze sets of data to understand and combat climate change. Climate informatics integrates meteorology, environmental science, computer science, and statistics to effectively handle set problems and understand climate systems and their effects. This convergence makes the scientific evidence much more precise and provides policymakers with the necessary tools to build preventive and responsive strategies for climate risks.

The methods used in climate informatics include efficiency and are among today's technological advances (Krapivin et al., 2015). For example, machine learning techniques are employed in climatic data analysis to learn irregularities and guide the generation of accurate temperature shifts, precipitation, other climatic parameters, and enhanced climatic conditions. In this case, big data analytics involves the analysis of voluminous data sets obtained from satellites, weather stations, and ocean buoys from which average analysis would not be able to pick out some features.

Simulation models assist in simulating the climate in the future, pulling the screen of future climate scenarios, so to speak, under specific emission paths. These tools, when combined, build the foundation for analysing climate variables and interactions of these variables with people.

Climate informatics is a multidisciplinary research field with numerous possible uses ranging from science to policy and mitigation. The most crucial application of high-performing computing is climate prediction, where the computer uses past and current data to draw future trends (Bhattacharyya & Ivanova, 2017). These predictions are useful for governments, businesses, and communities to manage climate variability effects such as droughts, floods, and heat waves. Likewise, climate informatics is critical in disaster response, whereby details of climate informatics shape the early warning and response frameworks. It reduces the impact of disasters caused by the climate by allowing accurate and timely interventions.

Consequently, in the field of carbon accounting, climate informatics can support the measurement of greenhouse gas emissions and the evaluation of the efficiency of the applied reduction strategies. Innovative technologies built on informatics help organisations and countries assess their carbon footprints, which form one of the benchmarks for attaining net-zero goals (Bhushan & Singh, 2024; Kumar et al., 2024). Moreover, climate informatics helps negotiate and monitor international climate agreements and treaties. For instance, the Paris Agreement employs scientifically based climate models and datasets to set up reference points indicating progress in reducing global temperature (Joshi et al., 2024). This approach guarantees that people make noble pledges and pursue goals that have a scientific basis and are realistic.

Climate informatics is a science and policy support tool developed to bridge the gap between climate science and application. Technology's ability to reduce detrimental impacts on climate change reveals why this revolutionary sector needs to be advanced and scaled up. This chapter will, therefore, focus on analysing the complex relationship between data, power, and sustainability within the field of climate informatics. It examines the politics of data as the latter shape climate-related decision-making, from disaster response to climate change negotiations. Understanding the potentials and risks climate informatics poses in this constantly evolving context is the goal of the chapter, which is to identify how best to utilise the concepts in attaining positive climate change outcomes. Lastly, the idea is to draw attention to how complex it is to address climate change. At the same time, simple as it might seem, data must be collected, managed, analysed, shared, protected, and used ethically and sustainably.

Methodology

The researchers of this study used the PRISMA Model, which is specific for systematic reviews and evidence synthesis scholarship. It assists in directing research questions and exploring studies in a particular area. Further, this study uses the PICO framework, a proper method for analysing the focal objects and processes (Frandsen et al., 2020) in data politics and climate informatics.

Population: People subject to or dependent on climate informatics and data governance. Thus, this study's population involves policymakers, governmental agencies, environmental nongovernmental organisations, private organisations, data scientists, and endangered communities due to climate change effects. These stakeholders' manipulation, usage, and control of climate data are essential to data and climate change politics.

Intervention: This methodology incorporates climate knowledge representation, computing, and the ecosystem of activities, technologies, and tools employed to create and utilise climate data. These are satellite-ground images, climate prediction, extensive data analysis, artificial intelligence, and blockchain. Through the present intervention, there is a hope to enhance climate prediction and forecasting, disaster mitigation, record-keeping of carbon footprints, and the efficiency of climate policies and cooperative agreements on an international level.

Comparison: This study's comparison component compares the two contrasting approaches to climate data governance and access. It explores how these approaches work across different political settings (developed and developing world), the extendibility of data and use and misuse, privacy considerations, consent, data justice, and more. If we represent climate action strategies through data, differences must be observed and contrasted with the actuality determining their effectiveness.

Outcome: The present study's research interest is the efficiency, openness, and engagement of data management systems in climate change. Outcomes include increased preparedness for disasters, fair access to carbon markets, the transparency of climate treaties, and attaining sustainability objectives. A frustration of the research is how data politics impacts climate policy formulation and implementation across and within different scales.

Hence, the PICO framework allows for the adequate organisation of the analyses of the topic at hand. It is oriented towards actualities in the real world and questions to be answered.

PRISMA Framework

The PRISMA framework is used systematically in this study to make reviewing and synthesising findings of identified research studies more rigorous and transparent, as highlighted by Vijerathne et al. (2024), Bandara et al. (2024), Chathuranga et al. (2024), Mishra et al. (2024), Karunarathna et al. (2024), and Rathnasiri et al. (2024). The PRISMA guidelines assist in deciding the methodological approach of this chapter and provide the required rigour and accuracy of data analysed.

Protocol and Registration: The cross-sectional approach formalises the protocol adopted when analysing climate data governance, resulting in systematic analysis. This also applies to the choice of stories and case illustrations that would help map the effective use of data in climate action with references to ethics/politics. The protocol of the chapter comprises the critical scrutiny of the available literature, governmental documents, case studies, and key informant interviews regarding the policy/intervention of climate informatics and data politics.

Eligibility Criteria: The students and case examples analysed within this study were chosen based on their concern with climate informatics and its influence on climate management and policy. Sources that meet the following requirements are eligible for admission: Peer-reviewed articles, official climate data reports, international agreements, and industry reports that contain empirical evidence or theoretical analysis in connection with data and climate action. Specifically, works concerning data ethics, data justice, fair/data sharing, and AI and blockchain were considered preferences.

Information Sources: This chapter employs journal articles, government documents, reports from non-government organisations (NGOs), climate databases, and practicum from climate change action organisations. These sources are approved according to their relevance and credibility to the subject matter to achieve diverse viewpoints globally and qualitatively across disciplines.

Study Selection: It was established that the study selection process should involve reviewing articles and reports conforming to specific eligibility criteria. For analysis, only academic papers and case examples that reported on the role of climate informatics in influencing policy and governance or that discussed political and ethical considerations about climate informatics were considered.

Data Extraction: Data was extracted from the chosen studies to identify the significance of climate data in the different settings. Data governance approaches, such as data ownership, access, and usage, were analysed to determine how they affect decision-making in developing policies and international relations on climate change. This data was used to examine current practices and their benefits and flaws, moving to strategies toward future data governance in climate action.

Synthesis and Analysis: The method of analysis used to synthesise findings was a qualitative synthesis where different sources were merged. The synthesis looks at trends and similarities in managing and utilising climate data in various branches and geographical areas. Five thematic areas for analysis are identified data availability and management, ethical issues, emerging technologies on climate informatics, and how power relations frame climate informatics. The understanding I have derived from this synthesis is multifaceted and captures fundamental aspects of climate politics, treating climate data as a sustainable development tool.

The PRISMA framework ensures that the current chapter's methodology is transparent, approachable, inclusive, and informed by the appropriate selection, analysis, and data synthesis. This chapter applies PICO's systematic and specific focus on critical research components to climate informatics and data politics, dwelling on PRISMA's comprehensive framework for conducting systematic reviews. The final sample consisted of thirty-six highly relevant studies published in the recent decade.

Findings and Discussion

The Politics of Climate Data

This study highlights that with the advent of big data, the ownership and control of climate data have become somewhat problematic. Climate information formed with the help of satellites, weather stations, and computational models is generally controlled by government agencies, private companies, and international organisations (Singh et al., 2018). However, this distribution raises critical questions: The immediate concern is who has the right of ownership of this data, and how does the owner regulate its usage? Since private entities own and control most climate data, access is a problem that affects everyone, from researchers to policymakers and marginalised groups (Atik, 2022). This exclusivity is antithetical to the cooperative spirit necessary for fighting climate change, as data must be disseminated and available to everyone.

Worldwide inequalities add another layer to the entire scenario. Developing countries lack the capability, capacity, and facilities for collecting climate data, processing, and analysis; as such, they depend on publications by other countries or international organisations (Kshetri, 2014). Such reliance only creates a vicious circle because such countries have a weak influence on climate regulation policies that impact them. The denial of equal access to information and methods of analysis opaques the voices of weak countries in international talks, including those in the United Nations Framework Convention on Climate Change (UNFCCC) framework (Harrould-Kolieb, 2019). Overcoming these disparities is possible only if international organisations nurture sharing arrangements and provide funding and expertise to enhance local climate data stewardship.

Manipulation and Misuse of Climate Data

The politicisation of climate data is also expressed in manipulating and partially using this information for its purposes. Climatic data became an instrument of manipulation of scientific research, manipulation of public opinion, or manipulation in support of policy aimed at achieving short-term economic performance at the cost of future generations. For example, particular interest parties might selectively seek to rationalise the existence of climate change or exaggerate the financial implications of the measures to combat it (Nwokolo et al., 2023a). The practices hamper prompt international reaction to climate issues and undermine the credibility of climate studies and decision-making.

The examples present the risk of climactic data misinterpretation and implementation biases. The 2009 scandal involving the leaked e-mails of climate scientists organised in Climategate is perhaps the most influential instance of such a kind (Maibach et al., 2012). Despite the scientists' exonerations in subsequent investigations, this led to scepticism and postponement of necessary climate measures.

Likewise, some countries have been labelled as submitting low-emission information to meet international standards but persist with unhealthy practices. Therefore, these cases best illustrate the need for and importance of transparency and accountability in managing climate data.

Such misapplication of climate data shows the need to develop a sound ethical policy regarding collecting, analysing, and reporting this data. Maximising the reliability of the information input with minimal prejudice and making the information universally available to all actors is vital to building trust and coordination in combating climate change. Only by resolving the political aspects of climate data can the global society strive for better solutions to the climate problem.

Challenges and Opportunities in Leveraging Climate Informatics

Challenges

Despite the excellent prospects that climate informatics opens up, several factors minimise its potential and its ability to be used relatively. The first challenges revolve around ethical considerations, including privacy, consent, and surveillance (Tripathi, 2024). Gathering and analysing climate information requires tracking people's actions, like energy or land utilisation, that may sometimes infringe on people's private space or community privacy. The authorities employed unambiguous measures to gather similar data that may end the public's trust and defiance. Furthermore, there is an element of predetermined instrumentalisation of climate informatics to or for surveillance or control, especially in totalitarian governments, which presents some of the most disturbing ethical questions.

The technical constraints also affect the expanded use of climate informatics for analysis (Dwivedi et al., 2022). Sometimes, there is poor coverage and limits, and large areas such as underdeveloped countries and continents make climate modelling and foresight less than perfect. For example, most developing nations still lag in the technological front and cannot afford to put key climate indicators like sensors or satellites in place. In the same way, analytical resources are generally limited, particularly about the extensive and intricate climate data sets that are still commonplace in today's business environment. These limitations play out, particularly the worst for developing nations, pushing the gap in climate informatics leverage any further.

Last but not least, socio-political factors remain the most influential in preventing the use of data in climate decisions (Van Kerkhoff & Pilbeam, 2017). Misleading information, prior misuse of climate data, and cultural and ideological biases can cause suspicion of climate data and its analysis. Elected officials often ignore research findings when unfavourable or inconvenient to their goals, thus making nothing or rolling back progress. Accommodating people's fear and scepticism entails openness and participation of other people, especially those outside the climate informatics field, in managing climate data.

Opportunities

However, for climate informatics, these are the opportunities to transform the world in response to the climate challenge. The other significant potentiality of climate action is to encourage transparency and accountability across various activities (Nwokolo et al., 2023b). Such technological tools can make policy-makers and implementers aware of concrete and tangible yardsticks in observing the level of emissions, deforestation threshold, and other litmus that would inform interventional actions to be taken. For example, real-time monitoring systems for industrial emissions will confirm legislative requirements for environmental issues and increase corporate governance.

Climate informatics serves as a tool to enhance adaptive and evidence-based policymaking since it offers descriptive and prescriptive information. The highly predictive models help policymakers understand the future climate scenario to develop an anticipatory and robust policy framework. They can also define risk mitigation and adaptation measures as less costly; thus, resources will be used optimally. In disaster management, for example, early warning systems based on climate informatics have already been shown to help prevent the loss of lives and property.

Besides, the field offers a promising capacity to unite countries in combating climate change through climate intelligence (Nwokolo et al., 2023c). When several countries share their information and assets, fighting for shared problems such as the increasing sea level and storms becomes easier. Such cases as the Global Carbon Project show how collaborative work with data can help nations build a policy and work in the same direction towards emission reduction. They enhance the quality of climate data and address the issues of inequitable distribution and use of data, chiefly for developing countries.

Therefore, there are many opportunities to maximise climate informatics, while the application faces several challenges on the way to fully realising it. The global community will achieve the appropriate climate information required in the quest for climate solutions through proper consideration of ethical issues, funding for technical structures, and acts capable of building trust with member nations.

Case Studies: Data Politics in Action

Case Study 1: Climate Informatics in Carbon Markets

Voluntary markets and those established under compliance regimes depend on climate informatics to assign and track carbon credits. Satellite data, photography, ground measuring devices, and sophisticated modelling instruments are central to measuring emissions cut and enhancing carbon stock. For example, climate informatics can track plant cover in reforestation and measure tons of carbon dioxide sequestered to validate the credit issued. Likewise, data analytics assist in detecting fraudster practices, including double credit sales or fake cuts in emissions (Tatineni & Mustyala, 2024).

However, it is essential to note that increased data use also has drawbacks. If a specific stakeholder exercises ownership and control of data, this leads to power relations between stakeholders (Prell et al., 2008). Owners of the monitoring assets and data repositories frequently dictate access conditions. Consequently, they may leave behind more compact actors or emerging economies if such actors cannot afford to acquire monitoring data or other services such as analysis. Finally, disputing data quality and availability affects the trust in the system or process. For instance, occurrences in which international carbon market participants present differing emissions figures have occasionally been a cause of disagreement. Such obstacles point to the fact that there is an urgent necessity for equally clear and accessible systems of climate data that would apply to all the markets of carbon equally and impartially.

Case Study 2: Climate Data and Disaster Preparedness

This is especially true in disaster-related areas, where climate informatics provides a foundation for early warning systems, decision-making regarding resource utilisation, and recovery process planning. For example, climate informatics has enhanced disaster management in Bangladesh, a country vulnerable to cyclones (Imam et al., 2017). Raw data from satellite imagery and numerical models allows for an early issue of cyclone forecasts, helping authorities alert communities several hours before the hurricane hits. The outcome has been that the number of deaths caused by cyclones over the past decades has significantly decreased.

However, there have also been very significant failures. This results in a phenomenon where data is either wrong or poorly relayed, which can lead to disastrous outcomes. For instance, in the United States, during the tragic occurrence of the Katrina hurricane disaster, differences in the data used for predictions and poor integration of agencies remained a fundamental influence toward slow response to catastrophe and chaos (Chua et al., 2007). However, socio-political factors worsen the challenges, for example, through different access rates given to warning systems or resources. Various groups may not have the means to respond to warnings appropriately, such as people from vulnerable areas who live in low-income areas. These examples demonstrate that climate informatics can be an enabler of success and a source of failure, suggesting that a more inclusive system must be established and followed rigorously.

Case Study 3: Global Climate Agreements and Data Transparency

The Paris Agreement is an achievement in the international climate regime where data sharing and information have a significant function (Sun et al., 2022). In the form of Nationally Determined Contributions (NDCs), countries agreed to communicate their plans to reduce emissions to a level consistent with the temperature goal of the Paris Accord (Dewasiri et al., 2024). Climate informatics provides the means for assessing progress; it allows countries to observe the emission changes, as well as to estimate the effects of policies and make necessary changes if needed. Another platform, which is scientifically based, including, for instance, the Climate Action Tracker, evaluates whether a country is on the correct path to achieve the said goal or not, which can also increase global coordination.

However, the agreement also raises issues in data politics. Variations in the kind or quality of data collected, methods used, and the presentation style of results denote variations that would limit comparability between countries. Because of resource constraints, developing nations may fail to meet the demanding International Financial Reporting Standards (IFRS) standards, which is why climate management inequities persist. Moreover, some countries have complained that others are cheating—either by not reporting emissions accurately or by claiming to have done more than they have (Walters & Martin, 2016). The present international data collection and verification systems should be enhanced to ensure the ratification of the Paris Agreement and make it credible. In addition, the capacity building of less-developed countries is equally important.

These cases provide an excellent example of how data, politics, and climate change mitigation intersect. Thus, despite the enormous potential of climate informatics to contribute to fair and realistic climate solutions, facing related challenges is pivotal. In conclusion, it may be argued that when it comes to data politics, the problems perpetuate conflicts due to opacity, marginalisation, and imperialism; however, when leaders approach and adhere to transparency and create democracy as well as international cooperation, it becomes possible to change from conflict points that stem from data politics to sustainable solutions.

Conclusion and Future Agenda

Data politics and sustainability or climate informatics have become central to contemporary climate change discourse and activism, transforming climate policy and practice (Nwokolo et al., 2024; Singh et al., 2024). This chapter has pointed out that applying climate data is not simply a technical act but a political endeavour involving power relations, resources, and ethics. The central premise of climate informatics is the enhancement of climate forecasts, disaster risk reduction, carbon management, and climate policy within the globe. However, these issues—starting from the ethical aspects of data protection up to and including data imbalance and technical skills inadequacy—must be overcome to enable this potential.

The examples included in the case studies illustrate how data can be helpful in climate change interventions and how it can be misleading if people rely solely on it, as some lessons learnt from carbon trading schemes, disaster risk reduction and even the Paris Agreement reveal. Such examples demonstrate the duality of climate informatics as the means of empowering people and as the area of potential confrontation. As the global economy proceeds into the future, the right data balance for sustainability must be achieved, and prudent consideration of political and ethical factors is required.

Towards Equitable Data Practices

One of the crucial milestones in realising data politics in climate action is equal access to climate data. Most developed countries have better technological advancements in addressing climate issues and access to data than developing nations, limiting their cooperation on climate change initiatives. There is a need to consciously provide resources to support the program's infrastructure and build capacity among stakeholders to implement the program appropriately. There is a need for developing countries to possess equipment and resources as well as undergo capacity building in producing, analysing and applying climate information. They include open-access databases, fostering collaboration among informatics entities and institutions in developed and developing countries, and international funding for technology transfer to guarantee no region is left behind on the climate informatics revolution.

Furthermore, it is similar to how inclusiveness is promoted in nations to the least-seen communities within these nations. Indigenous people, farmers and the urban poor, who are most sensitive to the impacts of climate change, are most of the time left out of climate decision-making. Therefore, building structures for these groups to bring in local knowledge and get data they can use enables them to develop their resilience to climate impacts. Moreover, equitable data practices improve the reliability and significance of climate informatics while being socially just for climate governance.

Integrating AI and Blockchain in Climate Informatics

Artificial intelligence (AI), blockchain, etc., can revolutionise climate informatics. AI's analytical and computational strengths enable it to analyse big and complex data to produce robust predictions and hidden patterns in anticipatory approaches towards climate patterns. For example, artificial neural networks can be used to upgrade climate predictions, increase the accuracy of disasters, and optimise the use of resources in wind power projects. However, using AI in climate informatics is only possible if ethical norms are established to disallow or avoid biased outcomes and decisions.

Newly minted technology, such as the blockchain, has come with the solution of trust and transparency in climate data. Blockchain can reduce the likelihood of data manipulation and improve accountability where data is exchanged, such as carbon trading or emissions reporting, by creating unalterable, distributed records of data transactions. For example, blockchain technology could help end-users of carbon credits to offer the real-time validation of emissions-cut transactions. International climate agreements may also be monitored automatically from outside and implemented through smart contracts that are electronic contracts saved in the blockchain. Embedding these technologies into climate informatics entails cooperation between technologists, policymakers, and environmental actors to create purposeful and diverse solutions.

Building a Global Framework for Climate Data Governance

To develop an international climate data management system, the political and ethical rules governing climate informatics must be defined urgently; such a framework should set a format for data capturing, transfer, and reporting that would allow comparability of information between nations. Absolute openness and responsibility must exist, which implies a check on data credibility and fines or penalties for fudging.

Policymakers and other international organisations, including the United Nations Framework Convention on Climate Change (UNFCCC), must develop and implement this framework. Some key recommendations include enhancing international cooperation and collaborative ventures in data sharing, establishing innovative financing systems to support climate data structures in low-income nations, and establishing an open system where everyone can share and benefit from climate data. Other delivery mechanisms can also be helpful, involving private actors to use their know-how and capital in developing data capabilities while anchoring it in public institutions.

The last characteristics of the global framework are the ethical dimensions specifying the level of privacy, consent, and surveillance. To ensure that climate data is accepted in society, the following benchmarks should be formulated to guide the usage of climate data, and the following structures should be formulated to incorporate human rights principles.

Climate action can move positively toward equitable, transparent, and collaborative approaches if data politics is reframed and changed for the following progressive orientations. Only when advanced technologies are integrated, data is shared fairly, and the governance structures are strong can the global community fully tap into the capabilities of climate informatics to fight for the future of our planet.

Call to Action

For climate informatics to deliver on the change-making promise, ethical, transparent, and equitable processes of managing climate data must be the norm. International policymakers, technology leaders, and all stakeholders must build global climate data frameworks to drive open access, governance, and policies incorporating new and innovative technology such as artificial intelligence and blockchain. These frameworks should respond to technical and political issues, with the need to consider marginalised communities when making decisions.

The prospect of climate informatics is realised in the unified global cooperation to mitigate the climate change issue irrespective of the politicians' and countries' preferences. It suggests that by making equitable practices and technologies our reference point, we can develop a firmer, data-backed framework for sustainable action. In conclusion, climate informatics will be sustainable if stakeholders are ready to shift from strategic self-interest to epistemic community concern. Climate informatics is guided by openness, participation, and fairness in climate data governance.

References

Atik, C. (2022). Towards comprehensive European agricultural data governance: Moving beyond the "data ownership" debate. *IIC-International Review of Intellectual Property and Competition Law, 53*(5), 701–742.

Bandara, C., Rathnasiri, M. S. H., Dewasiri, N. J., Chathuranga, B., & Singh, N. (2024). Greening the economy for a sustainable future: Opportunities and perceptions. In *Greening our economy for a sustainable future* (pp. 197–213). https://doi.org/10.1016/B978-0-443-23603-7.00016-9

Bhattacharyya, S., & Ivanova, D. (2017). Scientific computing and big data analytics: Application in climate science. In S. Mazumder, R. Singh Bhadoria, & G. Deka (Eds.), *Distributed computing in big data analytics. Scalable computing and communications*. Springer. https://doi.org/10.1007/978-3-319-59834-5_6

Bhushan, M., & Singh, R. (2024). Talent management facilitates net-zero transition through employee green behavior. In R. Singh, S. Khan, A. Kumar, S. Luthra, & H. Chokshi (Eds.), *Net zero economy, corporate social responsibility and sustainable value creation. CSR, sustainability, ethics & governance*. Springer. https://doi.org/10.1007/978-3-031-55779-8_7

Chathuranga, B., Rathnasiri, M. S. H., Dewasiri, N. J., Jayarathne, P., Bandara, C., & Bakari, M. A. (2024). Direct and indirect effect on industries and services (A). In *Greening our economy for a sustainable future* (pp. 215–226). https://doi.org/10.1016/B978-0-443-23603-7.00017-0

Chua, A. Y., Kaynak, S., & Foo, S. S. (2007). An analysis of the delayed response to Hurricane Katrina through the lens of knowledge management. *Journal of the American Society for Information Science and Technology, 58*(3), 391–403.

Dewasiri, N. J., Gnanapala, W. K. A. C., Rathnasiri, M. S. H., Yadav, M., Tewari, L. M., & Mishra, U. (2024). The way forward for net zero: Aligning Nationally Determined Contributions (NDCs) and holistic sustainable development in South Asia. In R. Singh & D. Crowther (Eds.), *Transition towards a sustainable future. Approaches to global sustainability, markets, and governance*. Springer. https://doi.org/10.1007/978-981-97-5756-5_5

Dwivedi, Y. K., Hughes, L., Kar, A. K., Baabdullah, A. M., Grover, P., Abbas, R., Andreini, D., Abumoghli, I., Barlette, Y., Bunker, D., & Kruse, L. C. (2022). Climate change and COP26: Are digital technologies and information management part of the problem or the solution? An editorial reflection and call to action. *International Journal of Information Management, 63*, 102456.

Frandsen, T. F., Nielsen, M. F. B., Lindhardt, C. L., & Eriksen, M. B. (2020). Using the full PICO model as a search tool for systematic reviews resulted in lower recall for some PICO elements. *Journal of Clinical Epidemiology, 127*, 69–75.

Harrould-Kolieb, E. R. (2019). (Re) Framing Ocean acidification in the context of the United Nations Framework Convention on Climate Change (UNFCCC) and the Paris Agreement. *Climate Policy, 19*(10), 1225–1238.

Hilbert, M. (2016). Big data for development: A review of promises and challenges. *Development Policy Review, 34*(1), 135–174.

Imam, N., Hossain, M. K., & Saha, T. R. (2017). Potentials and challenges of using ICT for climate change adaptation: A study of vulnerable community in riverine Islands of Bangladesh. In H. Kaur, E. Lechman, A. Marszk (Eds.), *Catalyzing development through ICT adoption*. Springer. https://doi.org/10.1007/978-3-319-56523-1_7

Joshi, A., Singh, R., Kumar, V., & Rani, S. (2024). Net zero strategy: Exploring opportunities, confronting challenges for developing countries. In R. Singh & D. Crowther (Eds.), *Transition towards a sustainable future. Approaches to global sustainability, markets, and governance*. Springer. https://doi.org/10.1007/978-981-97-5756-5_11

Karunarathna, K. S. S. N., Dewasiri, N. J., Singh, R., & Rathnasiri, M. S. H. (2024). What does artificial intelligence–powered ChatGPT bring to academia? A review. In R. Singh, S. Khan, A. Kumar, & Kumar, V. (Eds.), *Artificial intelligence enabled management: An emerging economy perspective* (pp. 3–13). De Gruyter, Berlin. https://doi.org/10.1515/9783111172408

Khan, S., Singh, R., Alnahas, J., Abbate, S., & Centobelli, P. (2024). Navigating the smart circular economy: A framework for manufacturing firms. *Journal of Cleaner Production, 480*, 144007. https://doi.org/10.1016/j.jclepro.2024.144007

Kitchin, R. (2014). *The data revolution: Big data, open data*. SAGE Publications.

Krapivin, V. F., Varotsos, C. A., & Soldatov, V. Y. (2015). New ecoinformatics tools in environmental science (Vol. 903), Springer.

Kshetri, N. (2014). The emerging role of big data in key development issues: Opportunities, challenges, and concerns. *Big Data & Society, 1*(2), 2053951714564227.

Kumar, K., Singh, R., Sharma, A., & Lehri, B. P. S. (2024). Climate change adaptation strategies for achieving net-zero economy. In R. Singh & D. Crowther (Eds.), *Transition towards a sustainable future. Approaches to global sustainability, markets, and governance*. Springer. https://doi.org/10.1007/978-981-97-5756-5_8

Maibach, E., Leiserowitz, A., Cobb, S., Shank, M., Cobb, K. M., & Gulledge, J. (2012). The legacy of climategate: Undermining or revitalising climate science and policy? *Wiley Interdisciplinary Reviews: Climate Change, 3*(3), 289–295. https://doi.org/10.1002/wcc.168

Middel, A., Nazarian, N., Demuzere, M., & Bechtel, B. (2022). Urban climate informatics: An emerging research field. *Frontiers in Environmental Science, 10*, 867434.

Mishra, U., Rathnasiri, M. S. H., & Dewasiri, N. J. (2024). Net-zero economy and energy sustainability: A Nepalese perspective. In R. Singh, S. Khan, A. Kumar, S. Luthra, & H. Chokshi (Eds.), *Net zero economy, corporate social responsibility and sustainable value creation. CSR, sustainability, ethics & governance*. Springer. https://doi.org/10.1007/978-3-031-55779-8_9

Nwokolo, S. C., Singh, R., Khan, S., Kumar, A., & Luthra, S. (2023a). Remedies to the challenges of renewable energy deployment in Africa. In *Africa's path to net-zero. CSR, sustainability, ethics & governance*. Springer. https://doi.org/10.1007/978-3-031-44514-9_3

Nwokolo, S. C., Singh, R., Khan, S., Kumar, A., & Luthra, S. (2023b). Africa's awakening to climate action. In *Africa's path to net-zero. CSR, sustainability, ethics & governance*. Springer. https://doi.org/10.1007/978-3-031-44514-9_9

Nwokolo, S. C., Singh, R., Khan, S., Kumar, A., & Luthra, S. (2023c). Impacts of climate change in Africa. In *Africa's path to net-zero. CSR, sustainability, ethics & governance*. Springer. https://doi.org/10.1007/978-3-031-44514-9_7

Nwokolo, S. C., Singh, R., Khan, S., & Kumar, A. (2024). Threats to the rapidity of sustainability transitions posed by technological changes. In S. C. Nwokolo, R. Singh, S. Khan, & A. Kumar (Eds.). *Technological pathways for Africa's net-zero economy* (pp. 39–75). https://doi.org/10.1016/B978-0-443-31486-5.00002-1

Prell, C., Hubacek, K., Quinn, C., & Reed, M. (2008). 'Who is in the network?' When stakeholders influence data analysis. *Systemic Practice and Action Research, 21*, 443–458.

Rathnasiri, M. S. H., Dewasiri, N. J., Kumar, A. A. (2024). Policy framework and implementation strategies for Sri Lanka's transition to a net-zero economy. In R. Singh, S. Khan, A. Kumar, S. Luthra, & H. Chokshi (Eds.), *Net zero economy, corporate social responsibility and sustainable value creation. CSR, sustainability, ethics & governance*. Springer. https://doi.org/10.1007/978-3-031-55779-8_3

Singh, C., Daron, J., Bazaz, A., Ziervogel, G., Spear, D., Krishnaswamy, J., Zaroug, M., & Kituyi, E. (2018). The utility of weather and climate information for adaptation decision-making: Current uses and prospects in Africa and India. *Climate and Development, 10*(5), 389–405.

Singh, R., Gupta, A., & Kandpal, V. (2024). ESG excellence in the digital world. In D. R. Ernesto, S. Gonzalez, V. Kandpal, P. K. Ozili, & P. Chatterjee (Eds.), *Fintech for ESG and the circular economy* (pp. 93–204). Scrivener Publishing LLC. https://doi.org/10.1002/9781394234035.ch10

Sun, R. S., Gao, X., Deng, L. C., & Wang, C. (2022). Is the Paris Rulebook sufficient for the effective implementation of the Paris agreement? *Advances in Climate Change Research, 13*(4), 600–611.

Tatineni, S., & Mustyala, A. (2024). Enhancing financial security: Data science's role in risk management and fraud detection. *ESP International Journal of Advancements in Computational Technology (ESP-IJACT), 2*(2), 94–105.

Tripathi, G. (2024). *Sustainability and health informatics: A systems approach to address the climate action induced global challenge*. Springer Nature.

Van Kerkhoff, L., & Pilbeam, V. (2017). Understanding socio-cultural dimensions of environmental decision-making: A knowledge governance approach. *Environmental Science & Policy, 73*, 29–37.

Vijerathne, D. T., Wahala, W., Rathnasiri, M. S. H., De Silva, V., & Dewasiri, N. J. (2024). Literature review on sustainability. In *Greening our economy for a sustainable future* (pp. 33–45). https://doi.org/10.1016/B978-0-443-23603-7.00004-2

Walters, R., & Martin, P. (2016). Trade in 'Dirty Air': Carbon crime and the politics of pollution. In *Environmental crime and its victims* (pp. 121–131). Routledge.

Chapter 4
Pathways to Climate Neutrality: Understanding Ice Cap Reduction, Glacial Melt, and Sea Level Change and Adoption

Wasswa Shafik

Abstract Ice caps and glaciers play important roles in the Earth's systems. In this topic, we go through some of the basic physical definitions and classifications of such ice masses to set the scene for the rest of the talk and demonstrate some of the routes through which they are important to Earth's systems. In the simplest physical terms, ice caps and glaciers are defined depending on the way they flow. An ice cap is defined as ice that flows radially from a central or high axial location. Glaciers, on the other hand, are defined as ice that moves along a valley or belt of low slope. Ice masses can have a mixture of flow patterns and can be classified in other ways, for example, by the way they move or the spatial extent of their covers. Polar ice caps and glaciers are called extremely cold regions and provide the only area of land permafrost in the world, therefore becoming an important sensitive layer in the cryosphere. As a small-scale glacier, ice caps broadly distribute in polar areas, seacoasts, the southern edge of the permafrost region in the high-altitude alpine areas of the plateau, and the northern plateau in western China. There are various types of landforms formed by conglomerated ice, including ice caps, shelves, and bergs. The ice caps are distributed in both polar regions and polar land of the world. Some dry valleys in the ice-free area at the eastern edge of Antarctica have glacier widths of less than 2 km, and these are actually ice caps. Most ice caps are located along the coastlines with marine structures.

Keywords Climate change impact · Coastal erosion · Cryosphere dynamics · Glacial melt · Global warming effects · Ice cap reduction · Oceanic expansion · Polar ice decline · Sea level rise

W. Shafik (✉)
School of Digital Science, Universiti Brunei Darussalam, Gadong, Brunei
e-mail: wasswashafik@ieee.org

Dig Connectivity Research Laboratory (DCRLab), Kampala, Uganda

R. Singh and W. L. Filho (eds.), *Climate Neutrality Through Smart Eco-Innovation and Environmental Sustainability*, Climate Change Management,
https://doi.org/10.1007/978-3-031-83250-5_4

Introduction

The warmth caused by an atmospheric concentration of water apour in excess of the atmosphere's equilibrium water apour content is thus expected to also feed into and enhance the strength of hurricanes (Shafik & Azrour, 2024; Thompson, 2021). However, while the massive drops in air pressure in the Atlantic hurricane basin that produce intense winds and drive water onto land and the torrential rainfall produced in such storms both require the release of latent heat, the role that heat plays in contributing to atmospheric stability and, in turn, enhanced storm activity is not discussed in much of the commentary on oceanic evaporation and hurricane enhancement (Medford et al., 2021). Icecap reduction in the Arctic, Greenland, and the Antarctic affects atmospheric stability, not only by increasing the atmospheric water apour concentration over the Arctic and driving increased global rates of polar easterlies and dipole oscillation activity but also by producing an associated strengthening of the Hudson Bay and Hudson Strait low-pressure systems and NAO and AO activity (Spada et al., 2013; Nwokolo et al., 2023a).

Such changes result from, in the case of ice cap melt, a weakening of the inversion formed in the lower troposphere over warm, open ocean water by how heat flows from it to and fuels evaporation by the water, coupled with the forcing of stronger rising motion in the atmosphere over the Arctic from the ocean water, which, in turn, enhances the atmospheric vertical motion responsible for driving the Arctic's efficient flow of evaporative cooling over the Arctic and for strengthening both Arctic and global atmospheric descents from air mass to atmosphere (Larsen et al., 2019; Nwokolo et al., 2023b). Reducing ice caps and enhancing global evaporation rates causes the Arctic to warm laterally more than vertically, a pattern of warming that favors the growth of storms and also weakens the poles' ability to sustain large sea level pressures. These changes, in turn, drive weather patterns that help propagate surplus tropical heat to the poles and make storms more severe. It is apparent that in the past 30 years, periodically with rapid increases in rate, the transfer of ice by calving from marine-based glacier tongues or the floating ice termini of Antarctic ice shelves to the ocean has been increasing (Benetti et al., 2021).

In contrast to the much larger ice sheets, the smaller-scale ice caps and ice fields are pervasive features in many mountainous and near-polar regions of the globe. Although these forms of glaciers are of a scale that might suggest insignificance in the overall context of global climate change and sea level response, it is apparent that they make a considerable contribution to the sustained and accelerated expansion of the cryosphere (Lane et al., 2023; Shafik, 2024c; Larsen et al., 2019). This chapter, by way of introduction, provides a pronunciation of selected fundamental issues concerning the cryosphere and its contribution to ongoing and imminent sea level change and briefly describes the main components of the cryosphere and the spatial distribution of ice caps and glaciers (Glueder et al., 2022; Nwokolo et al., 2023b). Much attention is given to the ongoing loss of mass from both the Greenland and Antarctic ice sheets. However, in recent years, increased attention has been placed on the much smaller mountain-based glaciers, their significant role, and their potential vulnerability in contributing to sea level rise in the near future. The ice mass stored within individual glaciers is much smaller than that of the ice sheets. Nonetheless, for more than 25% of the Earth's glaciers, predominantly mountainous, we first order calculates that a maximum decrease in glacier height of 2–3 m with consequent freshwater input to the Earth's oceans of approximately 8 mm may occur within the lifetime of the present generation (Lane et al., 2023; Nwokolo et al., 2023c).

Definition and Types of Ice Caps

The terms "ice cap" and "glacier" are often used interchangeably in many popular venues, but these terms are differentiated by both their detailed nature and a matter of scale. When referring to a natural landscape feature, the term "glacier" usually refers to a single-flowing body of ice. An "ice cap," fundamentally speaking, refers to anything that meets the following four criteria: 1. It is round or oval. 2. It wholly resides above the local snowline. 3. It is a large area relative to the surrounding landscape. 4. It covers all landforms: depression fill, hills, mountains, etc. In other words, there is no ice-free topography within the perimeter of the feature (Ou, 2023). The only true ice caps on Earth are those at the poles found at the very peak of our planet.

While polar regions are the ones with the most well-known ice caps, many different types of ice caps form all over the world in either position. In polar climate environments, where all the land is over 60° north or south of the equator, ice can form that either covers an existing mountain or simply pools on the surface or can form only as a flat ice cap. These types are known as alpine glaciers, distribution glaciers, or ice sheetlets. These ice caps can be many different shapes, e.g., conical hills, domes, and a wide variety of others influenced by the landscape beneath the ice (Benetti et al., 2021; Nwokolo et al., 2025). These ice caps receive snow and ice from the lower diurnal zone ice, but when the surrounding heights of the cooling zone ice are greater than the ice cap height, ice spreads out of the ice cap surroundings and onto the flat ice cap. The ice caps in these areas can also be many different shapes and can be found at any latitude. The maximum limitation is found high in the cooling zone elevation where it is below 0 °C. Generally, decreasing the temperature by a degree Celsius allows ice to spread roughly an additional 100 m around the mountain (Glueder et al., 2022).

Generally, in Australia, local government and other sectors need to include climate change in their operations. In doing this, all have a considerable degree of community engagement, as well as focusing on extending the resilience and adaptive capacity of communities (Nwokolo et al., 2025). All focus on the triple bottom line, integrating environmental, economic, and social dimensions. These are all values that are useful in creating effective policies and governance responses for cryosphere change, although the full range of consequences of cryospheric change is not yet fully understood, and values may require adjustment and emphasis may shift. Engaging local communities, building resilience, and education are important policy components. Some suggest that all form the basis of an effective response (Lyså et al., 2021).

Role of Ice Caps in Climate Regulation

Icecap change has serious global implications for human well-being. Ice caps govern the heat balance of the planet by reflecting incoming sunlight; indeed, they are the highest albedo regions on Earth. The latent heat needed to melt ice also acts as a temporary buffer for the climate system, reducing the degree of warming for a given imbalance. Thus, the rapid melting of ice caps is a cause for alarm, as replacing them with open ocean or darker land would create positive feedback to melt other ice regions further (Lane et al., 2023). Salinity plays another significant role; the addition of large volumes of meltwater to the ocean surface can help stabilize the global conveyor belt, whose northward return flow keeps parts of western Europe relatively warm. In addition, freshwater mixing with warm regional currents is directly involved in the development of severe weather systems such as Atlantic hurricanes. On a slightly longer time scale, the erosion of ice caps is expected to cause shifts in jet stream patterns that could threaten agriculture in these regions. Extreme weather events are the clearest threats, namely to regions with high coastal population densities (Ou, 2023). Nearly half the world's population lives in coastal areas, and it is a sobering fact that just seven meters of sea level rise would threaten the lives of 10% of the world's population. While the occurrence of such an event is unlikely within the next century, large changes in the rate of future melting over the next decades could force mass evacuations from low-lying areas. Large changes in the melting rates of vulnerable polar ice sheets, as are presently being observed, would require rapid and substantial adjustments in infrastructure. There are also environmental repercussions, as human settlers living closely with such ice caps—as with the corals and algae in glacier-fed ecosystems—require a stable, cool, and freshwater source. Past eruptions were associated with glaciation; changes to ice caps can also send signals into the atmosphere (Lyså et al., 2021). For all of these reasons, the consequences of ice cap reductions warrant the close attention of policymakers, planners, and scientists.

Causes of Ice Cap Reduction

There are different natural and human-induced causes of ice cap reduction. These are the following: natural causes based on historical times, which relate to the physics of our environment and are also studied to establish future climates and landscapes on Earth, and the present causes based on our society's actual activities that will surely define future climates. The natural ice cap reduction is attributed to climatic and environmental phenomena and changes that have occurred moderately, averaging these non-continuous events with a time frame from 90 to 120 million years, providing evidence within their range of values of a cool to warm climate with a medium that today hits Earth's averages (Parker et al., 2022). Glacial and Interglacial Periods in the Cenozoic, and temperature fluctuations. The variation of the temperature values marks different possible climate states, from "hot to cold" and vice versa. The range of this temperature is approximately 18 °C, indicating relatively high variability. Detailed records of ice and sea level changes may be found in geological changes such as faults, earthquakes, or tilt anomalies can be slow structural changes acting over millions of years or quick changes taking only a few hours, days, or years to occur (Medford et al., 2021). Presently, due to a variety of human-induced factors, we may find a fast change in our Earth's system. Human Causes/ Global Warming: The increases in greenhouse gas concentrations, especially carbon dioxide and methane in the troposphere, within the crucial range of 280 ppm in 1980 to 383 ppm today, characterize the "natural range." These increases are produced in industrial nations, and the sources of this increase are derived from burning fossil fuels to obtain energy. Some countries are also emerging economies, increasing industrial activities, mining, deforestation, and other land-use changes such as building cities, which convert forests into farming lands to enhance and improve their technologies (How et al., 2021). Feedback loops: Moving further into the twenty-first century, a new idea appears—a "feedback loop" where some regions release more methane than others.

Natural Factors

Climate alterations in the past have led to the varying stability of ice caps and are the result of both natural and human-induced occurrences, with the latter being of ever-growing importance and scale. Over geological time scales, it is apparent that the Earth's climate can linger by the water phase transitions of the polar ice caps as a result of a controlled alteration to the natural insolation workforce. Such transitions occur gradually, though with profound impact (Simonsen et al., 2019). Other examples of these changes include significant emissions of CO_2 from volcanism and methane from ice or vegetative-covered ice or frozen soils. One of the mechanisms of altering the ice cap volume is the changes in tectonic forcings, such as the movements of the structural plates that lead to changes in regional climates, which leads to the rise or decrease of ice cap volume over via the changes in the Earth's climate. Other natural factors that may cause the ice to melt are the inclination of the Earth's axis. A number of variables control the Earth's spin axis inclination, the most significant being the gravitational influence of the Moon and Sun (Patton et al., 2021).

The Moon is, without doubt, the single most significant new variable that pushes the Earth's axis around. These factors all have their seasonal impact, which can result in tilting ice-covered carbonate deposits or changing the angle of incoming solar radiation. The sea level change is under the influence of the melting of ice caps; the interaction occurs in time–space and is non-linear via the timing of the processes. The rise of SLR that has to do with ice volume melting is affected by several external factors, such as the Earth's orbital change via natural insolation changes that govern the climate system over geological time scales (Bamber, 2021). In addition, volcanic emissions can also affect SLR; the eruptions that emit large masses of ice and minerals via the plumes will be washed off by the atmosphere, resulting in significant deposition onto the polar ice sheets within months. This has never been seen before since the era of instrumental measurements; the ice cap volume changes as a result of up to 200 GT/y. The decrease in human pumping of water from the ground since 1972 is causing the land to become more solid to support the water load. Given this natural variability, the current observed sea level rise is unusual and has been attributed to anthropogenic influences (Thompson, 2021). Overall, understanding these natural processes and their cumulative effects on the ice cap is imperative when assessing possible future scenarios of melt and sea level rise and distinguishing between natural and anthropogenic involvement.

Human-Induced Factors

Human-Induced Factors. The year 1750 is usually adopted as the starting point of human-induced factors. Atmospheric CO_2 concentration began to rise exponentially when industrially manufactured emissions of greenhouse gases first substantially increased and has currently reached a level 40% higher than that of 1750. Changes in land use, agricultural, and timberland expansion are non-CO_2 sources that lead to land–atmosphere density differences and can have a further impact on warming. Fossil fuel burning, deforestation, and agricultural practices are activities typical of post-industrial society (Baqi & Abbas, 2021). The primary cause of environmental pollution is the rapid economic growth and large-scale urban development in the past century, which makes factories release a great quantity of polluted water, waste, and air. However, as soon as steps are taken to mitigate environmental pollution, the reduction of greenhouse gases is an additional deliverable. Recently, global warming has attracted substantial attention, mainly because of policy responses in the wake of international agreements. Warming is evident in increasing global land near-surface air temperature and ocean surface temperature. The rate of warming has been much larger over the past 50 years than in the past 150 years (Woods et al., 2018). Additionally, glacial, sea ice cover, and snow cover decrease is evident. Deforestation and changes in land use lead to the evaporation of water from soil, and so far, we have had few policies governing air temperature control. These policies are generally at local, not global, levels. The absence of credible climate policy increases the climate impact significantly. The factors making the vulnerable polar bear species extinct are discussed. The magnitude of the anthropogenic and natural drivers of climate change prior to the industrial era was stated to be low. In 2015, it was estimated with high confidence that the extent of the ice cap reduction and glaciers was due mostly to anthropogenic causes (O'Connor & Whyte, 2023).

Impacts of Glacial Melt

While climate impacts can often be disparate and nonlinear, nearly every fjord system and glacier network on Earth is affected by glacial melt, also known as ablation. Environmentally speaking, glacial melt—and calved, or broken-off, icebergs—lead to reduced light penetration and increased community respiration rates in the surface ocean. There are also socio-economic impacts associated with glacial melt, which is often the primary water source for communities living near glaciers and, in many cases, the primary source of water for irrigation (Larsen et al., 2019). This affects rice, fruit, and other agriculture that the people in these regions may rely on for food and income. Given that livelihoods in these areas are often strongly aligned with local agriculture and fisheries, the eventual health—physical or mental—and economic ramifications are widespread. This has been seen in the Nepal region during droughts when multi-day power outages, when there isn't enough hydroelectric power, for example, as river levels fall and melt rates decline, are associated with significant increases in food prices, particularly around grains and lentils (Glueder et al., 2022).

Environmental Impacts

Ecosystem function and biodiversity in glacierized landscapes are now known to be altered along gradients of glacier change. These responses have included changes to the structure and function of ecosystems along new glacier drainage pathways, shifts in species distributions both within affected landscapes and beyond in the potentially isolated sink habitats, and changes in key ecosystem functions, such as primary production and the supply of organic matter to downstream habitats. The loss of glaciers removes a major regional-scale source of water, leading to altered freshwater inputs to both aquatic and terrestrial habitats that will affect river morphodynamics, freshwater habitat alteration in lakes, and food chains (Lyså et al., 2021). The expected hydrological and environmental impacts will be highly variable among climatic regions, which are underlain by different degrees of socio-economic stress. The presence of glaciers particularly strongly modifies river systems draining mountain catchments, and much of the load transported by these rivers derives from the glaciers. If glacial melt raises the level of fluvial sediment transport, then nearby infrastructure will be impacted. The deposition of these sediments on river beds or in deltas will also impact ecological services. The actual rise in sea levels due to the loss above of mass is occurring all over the world (How et al., 2021). Those lands with land-based ice sheets are most impacted because they hold most of the Earth's ice. A further environmental impact has been shown in the case of alpine regions. As the ice retreats to the mountains, rainfall increases, and as the meltwater carries the sediments, it causes landslides and flooding. This has led to a changing mountain landscape and affects the plant and animal life in the valley and surrounding mountains. These changes are very long-term, on the order of thousands of years, with changes in flora and fauna taking not tens but hundreds of thousands of years (Lane et al., 2023; Larsen et al., 2019).

Socio-economic Impacts

An important aspect of groundwater storage is that, depending on the characteristics and capacity of the storage system, water availability can show increases for a short time until the storage is partly emptied and can then drop dramatically over a longer period. A shrinking or phase change in wetland or ice storage occurs between 1.5 and 2 °C. It also directly impacts ice- or meltwater-fed sub-Saharan regions. At the same time, softly increasing temperatures depress agricultural yields (Glueder et al., 2022). An increasingly over-exploited or polluted environment compromises marine and coastal food resources. Therefore, the availability of food, birth, and income is likely to decrease. This can lead to increased migration. Adaptive planning of socio-economic development around declining glacier use becomes a critical issue. As well as rainfall and glacial melt, small- and medium-sized river basins often provide water for domestic use in low-income countries. These basins are those where dependency ratios above 25% are found for at least one sector. Glaciers also provide water for agriculture, hydropower, industrial, and domestic use in lower parts. The decline of glacier storage directly impacts the water supplies of these users (Lane et al., 2023). Direct effects of accelerated glacial melt are water shortages, impaired agricultural yields, loss of soil material, and mortality. Water scarcity may increase dependence on contaminated sources. Reduced agricultural yields lead to a lack of food. Agricultural industries also die or leave the area, pressing local incomes and forcing downland or housing prices. This vicious cycle results in further outmigration. The decline of the productive glacier water resources indirectly threatens non-dependent economic sectors in some river basins, for example, the fishing industry, through changes to the natural regime (O'Connor & Whyte, 2023). It can accelerate conflicts about resource use and management, e.g., who has access to the remaining water and who will be forced to leave the area, bearing increasing socio-economic costs.

Arctic Ice Cap Dynamics

Over the southern Arctic Ocean, air temperatures at the surface have been rising due to the greenhouse effect, as indicated by observed drainage of cold air from the polar cap and the accompanying poleward shift by as much as 2 K over the last 50 years—a rate of increase larger than for any other region of the globe. The observed changes in sea-land climate indicate at least a partial coupling between long-term atmospheric change at the surface and the ocean and extended ice masses. Some aspects of the climate changes, in particular over the sea ice, are crucial for the ability of surface conditions to maintain albedo, thickness, and volume (Baqi &

Abbas, 2021). It has a few undocumented and unquantified elements, particularly concerning open-water reinforcement of atmosphere–ocean coupling in the early autumn after the sea ice has reached its annual minimum extent. Another concern over the greenhouse effect is significant climate warming in the winter hemisphere, coming from atmospheric dynamics at the surface coupled to the general circulation. For the major land areas adjacent to the Arctic Ocean, surface temperatures in winter respond to atmospheric circulation, leading to frozen surfaces and lowered energy and moisture exchanges with the lower atmosphere. Trends and potential changes in these energy and humidity exchanges in order to sustain existing ice covers and precipitation are fundamental for the future trajectory of the ice masses (Patton et al., 2021; Simonsen et al., 2019). Data sets allowed description of overall changes in thickness and area of the five major polar ice masses, together with plausible expressions of volume-area scaling in terms that are appropriate for melting Arctic sea ice.

Antarctic Ice Sheet Stability

A number of physical processes may cause changes in the distribution and mass of the Antarctic Ice Sheet. Changes in air temperature, snowfall, or penetration depths are the obvious candidates for inducing ice sheet variations. However, today, the best candidates for causing rapid and significant changes are (1) melting in the vicinity of ice shelves and (2) the grounding-line dynamics of their outlet glaciers. Ice shelves form at the outermost portions of ice sheets at places of abrupt thickness changes. They are grounded, often deeply so, and are in contact with the ocean for a considerable part of their frontal lengths (Baqi & Abbas, 2021). They are partially floating and are, therefore, the most susceptible to the eustatic effects of climate change. The more ablated the shelf base, the less buoyant force the shelf will have until, at the point of no contact with the ocean, the ice could simply disintegrate and flow rapidly to the sea. This could prevent cold saline shelf waters from interacting with deeper portions of the Antarctic basin (Larsen et al., 2019). However, the breaking of the shelf could release these waters and allow them to participate in the thermohaline circulation of the deep water of the world's oceans.

Glacial Melt in Mountainous Regions

While specific impacts change from place to place, in general, glacial shrinkage will lead to a seasonal reduction of water flow in mountainous areas and potentially to problems like avalanches and different kinds of erosion due to the destabilization of mountain walls. Around the world, people living on rivers originating from snow and ice-covered mountains are directly dependent on water supplies that will change if the ice melts, potentially destabilizing local and regional water supplies (Lyså et al.,

2021; Ou, 2023). In places like the American West, where the snowmelt creates most of the region's summer water supply, shrinking glaciers very likely will have an economically significant impact and will play some part in the allocation of water rights disputes throughout the Intermountain West. However, shrinking mountain glaciers will also affect global sea levels. Like ice caps, many glaciers terminate in the ocean, and those terminuses will be the direct source of increased sea levels associated with glacial melting (How et al., 2021). Unlike continental ice sheets, the primary source of glacial ice loss due to basal melting is the simple temperature of the water surrounding the glacier. So, in theory, slender mountain glaciers should be more sensitive to temperature change than the thick continental ice sheets. In mountainous areas, sunlight is important for local temperatures, and if the white ice and snow covering the mountains disappear, it will be replaced by darker rock, which will absorb more sunlight and result in higher temperatures that enhance local glacial mass loss (Patton et al., 2021).

Sea Level Rise and Its Consequences

At present, the average rate of sea level rise is 3.3 ± 0.3 mm annually, with calculations for sea level change showing an average sea level rise of 1.8 mm per year between 1961 and 2003 and 3.5 mm per year from 1993 to 2003. The warming of ocean water is the primary reason for global sea level rise. For each rise in temperature, the volume of water grows, and the sea level spreads out. Water temperatures increase when they heat up. Thus, ocean water expands when it warms up. In other words, water expands (Spada et al., 2013). Thermal expansion is the increase in volume when temperature rises. If the effects of thermal expansion were eliminated, the observed sea level rise over the past century would be reduced by approximately one-third. The second fundamental cause of global sea level rise is the loss of Greenland and West Antarctica ice caps and glacial mass around the globe. In 2007, glacial isotopic melting, evaporation, precipitation, and direct precipitation of falling snow are the four primary resources that have filled the ocean with extra water (O'Connor & Whyte, 2023).

Mechanisms of Sea Level Rise

Although the mechanisms governing sea level rise are numerous and diverse, a couple stands out significantly when it comes to the total height of the world's oceans: first and foremost, thermal expansion due to the warming of the oceans and the tremendous volume of water stored in polar ice, mostly the land ice of Greenland and Antarctica. A local acceleration of ice sheet loss primarily drives sea level rise from glacial melt and ice cap reduction. How exactly the sea level is affected, in a quantitative way, strongly depends on the region of the world ocean in which the melting happens (Woods et al., 2018). As the salinity of the ocean, along with the volume of ocean water, is expected to increase due to polar contributions and as fresh meltwater discharge might cause stratification, net effects in, for example, the North Atlantic closing down precipitation input into the Mediterranean might differ from those in parts of the North Pacific where many of the ice sheets melt. Likely, the first process that comes to mind when considering sea level rise is the warming of the oceans. By and large, the expansion of a fluid due to warmth, or its change in thermodynamical energy, is its most pronounced effect (Benetti et al., 2021; Kumar et al., 2024; Larsen et al., 2019). Oceans already have a phenomenal amount of heat stored, and they also have a vast heat capacity, but the rate of change of the world's oceans is such that the dissipation of the additional energy being stored guarantees a certain warming for their liquids. Disregarding for a moment the actual amount of oceanic thermal expansion, we can calculate its physical limit. Melting land ice, on the other hand, adds water to the ocean entirely as volume, as floating materials do not contribute to global mean sea level rise (Lane et al., 2023; Glueder et al., 2022).

Impacts on Coastal Communities

Sea level rise increases the frequency and severity of flooding in low-elevation areas. Rapid onset floods frequently accompany high sea levels during storms, potentially generating greater impacts. The most immediate effects of sea level rise occur in coastal zones (Simonsen et al., 2019; Joshi et al., 2024). Flooding damage to homes, roads, and other urban infrastructures is usually multiplied by erosion and the loss of adjacent coastal landscapes. Economies in the EU, in cities around the world, and many low-income nations depend strongly on coastal tourism (Ou, 2023). Future sea level rise would reduce coastlines and beaches and, in tandem with storm surge probabilities, some coastal recreational facilities and natural sporting amenities such as golf courses are threatened. Sea level rise will decrease the stocks of freshwater by causing saltwater intrusion in lowland peat coastal aquifers, which will affect the livelihoods of many coastal communities that still depend heavily on local groundwater sources (Parker et al., 2022). This may require moving drinking water intakes further uphill, causing social issues for local people. Saltwater is known to be hazardous and harmful to most food crops and agricultural soils. In times of sea level rise, international assistance programs must take these future high-risk dependents into account. It has been widely suggested that IFMs in the metropolitan areas assessed should prioritize the improvement of local adaptive potential in relation to sea level rise (Medford et al., 2021).

Impacts on Coastal Communities and Infrastructure

Coastal populations are the most vulnerable to sea level rise. People in coastal areas are very close to the storm system, which can be quite severe in parts of the world, have deep economic connections to the ocean through fishing, shipping, and extraction industries, or live in a city near an ocean where many businesses and workers are connected to economic activities like finance and tourism, which are rooted in geographic conditions reliant on low-elevation coastal settings (Shafik, 2024a). To make matters worse, people living in these areas usually have limited resources. Communities on the perimeter of the Gulf of Mexico are poor by U.S. standards, and I'm not trying to suggest that people living on the coast in lower Bangladesh are living rich lives there (Parker et al., 2022; Nwokolo et al., 2023b). A few centuries ago, the Dutch tamed the ocean and established an economic powerhouse by erecting an extensive system of dikes. The dikes lasted only an instant when the Spanish queen occupied that district. In the long term, however, Dutch economic enterprise won back this collection of islands from the sea and proceeded to expand it in size. The Dutch today have things like the Maeslantkering (Shafik, 2024b). While economics trumps physics in the long term, in the short term, it is the physics that counts. As their priority, the Dutch finished off the complex array of barriers that protected them from the North Sea to give time for the economics to play out. Economic resilience suggests that such projects should be tailored to the specific needs of the affected community (Larsen et al., 2019). For example, the key to protecting the ecological and economic viability of Galveston Bay, Texas, is not to protect the existing shoreline but to build a set of barrier islands that protect the core body of water that allows for the vibrant port. Residents of New York City, on the other hand, want to protect Manhattan, while the most valuable thing about Miami is that beautiful long ribbon of sand (Lyså et al., 2021; Ou, 2023).

Global and Regional Sea Level Trends

Since 1990, mean global sea level rise has decelerated by 31%, with the current rate being unprecedented in the 20-year long modern record. The highest observed sea level change is found in the Southern Hemisphere, where the observed changes are greater in amplitude and have gone through regionally pronounced multi-decadal variability. The overall mean southern expansion in the period from 2007 to 2012 amounted to 1.41 ± 0.43 mm/year (O'Connor & Whyte, 2023). The observed Southern Hemisphere pattern is found to deviate significantly from the expected warming of a cooling ocean, casting serious doubt on the stability of the huge East Antarctic Ice Sheet. In contrast to the southernmost latitudes and consistent with ocean thermal expansion, the 95% confidence interval of the latitude of the 15 mm/yr contour almost coincides with the latitude band between 27°S and 33°N (Patton et al., 2021; Simonsen et al., 2019). A significant feature of the reported maximum sea level rise is that in the time frame 2007–2012, global decreases were recorded in 5 out of the 11 years in this time frame, decreases that are consistent with globally cooling ocean temperatures (How et al., 2021; Medford et al., 2021). Such a pronounced shift in the global heat balance of the observed magnitude is currently not being reproduced by the models, thus indicating potential and shortcomings in their projection skills (Larsen et al., 2019).

Conclusion

If we continue releasing the amounts of CO_2 we are now, more ice caps and glaciers will melt, raising sea levels at a rate of about 0.6–2.3 m by the year 2300. Because of these glacial and ice loss trends, being informed of the importance of glaciers is becoming necessary. However, it can also inflate the magnitude of it. This is because all of our Earth's systems interact with each other. For example, the ice, snow, and water trends in the Himalayan–Tibetan region of Asia affect Asian heat plumes and African climate systems, which in turn affect Atlantic heat plumes that warm ocean systems. These systems are interconnected by the water cycle and the jet stream. In turn, because the water cycle is affected, the glacial melt leads to problems such as decreased fresh water and other issues caused by a lack of freshwater needed for communities and life systems in this region. It is easier to prepare for a storm if meteorologists are able to predict the coming of a dark cloud over a region. This research allows us to predict a dark and foreboding future if we continue to release greenhouse gases at the rate that we are now. Additionally, because of the interconnectedness of global systems, issues in the glacial regions can affect other spheres and lead to the complex multitude of changes that we see already on Earth. These negative consequences are growing unless greenhouse gas emissions are reduced to net zero rather than doubling down on coal, gas, and oil.

References

Bamber, J. (2021). Land ice. In *Climate change: Observed impacts on planet earth* (3rd edn.). https://doi.org/10.1016/B978-0-12-821575-3.00008-6

Baqi, A., & Abbas, A. (2021). Temporal variations in ice cap of Antarctica and Greenland. *International Journal of Innovations in Science and Technology, 3*(2). https://doi.org/10.33411/ijist/2021030201

Benetti, S., Chiverrell, R. C., Cofaigh, C., Burke, M., Medialdea, A., Small, D., Ballantyne, C., Bateman, M. D., Callard, S. L., Wilson, P., Fabel, D., Clark, C. D., Arosio, R., Bradley, S., Dunlop, P., Ely, J. C., Gales, J., Livingstone, S. J., Moreton, S. G., Weilbach, K., et al. (2021). Exploring controls of the early and stepped deglaciation on the western margin of the British Irish ice sheet. *Journal of Quaternary Science, 36*(5). https://doi.org/10.1002/jqs.3315

Glueder, A., Mix, A. C., Milne, G. A., Reilly, B. T., Clark, J., Jakobsson, M., Mayer, L., Fallon, S. J., Southon, J., Padman, J., Ross, A., Cronin, T., & McKay, J. L. (2022). Calibrated relative sea levels constrain isostatic adjustment and ice history in northwest Greenland. *Quaternary Science Reviews, 293*. https://doi.org/10.1016/j.quascirev.2022.107700

How, P., Messerli, A., Mätzler, E., Santoro, M., Wiesmann, A., Caduff, R., Langley, K., Bojesen, M. H., Paul, F., Kääb, A., & Carrivick, J. L. (2021). Greenland-wide inventory of ice marginal lakes using a multi-method approach. *Scientific Reports, 11*(1). https://doi.org/10.1038/s41598-021-83509-1

Joshi, A., Singh, R., Kumar, V., & Rani, S. (2024). Net zero strategy: Exploring opportunities, confronting challenges for developing countries. In R. Singh & D. Crowther (Eds.), *Transition towards a sustainable future. Approaches to global sustainability, markets, and governance*. Springer. https://doi.org/10.1007/978-981-97-5756-5_11

Kumar, K., Singh, R., Sharma, A., & Lehri, B. P. S. (2024). Climate change adaptation strategies for achieving net-zero economy. In R. Singh & D. Crowther (Eds.), *Transition towards a sustainable future. Approaches to global sustainability, markets, and governance*. Springer. https://doi.org/10.1007/978-981-97-5756-5_8

Lane, T. P., Darvill, C., Rea, B. R., Bentley, M. J., Smith, J. A., Jamieson, S. S. R., Ó Cofaigh, C., & Roberts, D. H. (2023). The geomorphological record of an ice stream to ice shelf transition in Northeast Greenland. *Earth Surface Processes and Landforms, 48*(7). https://doi.org/10.1002/esp.5552

Larsen, N. K., Levy, L. B., Strunk, A., Søndergaard, A. S., Olsen, J., & Lauridsen, T. L. (2019). Local ice caps in Finderup Land, North Greenland, survived the Holocene Thermal Maximum. *Boreas, 48*(3). https://doi.org/10.1111/bor.12384

Lyså, A., Larsen, E. A., Anjar, J., Akçar, N., Ganerød, M., Hiksdal, A., Van Der Lelij, R., & Vockenhuber, C. (2021). The last glaciation of the Arctic volcanic island Jan Mayen. *Boreas, 50*(1). https://doi.org/10.1111/bor.12482

Medford, A. K., Hall, B. L., Lowell, T. V., Kelly, M. A., Levy, L. B., Wilcox, P. S., & Axford, Y. (2021). Holocene glacial history of Renland Ice Cap, East Greenland, reconstructed from lake sediments. *Quaternary Science Reviews, 258*. https://doi.org/10.1016/j.quascirev.2021.106883

Nwokolo, S. C., Singh, R., Khan, S., Kumar, A., & Luthra, S. (2023a). Africa's awakening to climate action. In *Africa's path to net-zero. CSR, sustainability, ethics & governance*. Springer. https://doi.org/10.1007/978-3-031-44514-9_9

Nwokolo, S. C., Singh, R., Khan, S., Kumar, A., & Luthra, S. (2023b). Global investment and development in Africa. In *Africa's path to net-zero. CSR, sustainability, ethics & governance*. Springer. https://doi.org/10.1007/978-3-031-44514-9_2

Nwokolo, S. C., Singh, R., Khan, S., Kumar, A., & Luthra, S. (2023c). Remedies to the challenges of renewable energy deployment in Africa. In *Africa's path to net-zero. CSR, sustainability, ethics & governance*. Springer. https://doi.org/10.1007/978-3-031-44514-9_3

Nwokolo, S. C., Singh, R., Khan, S., & Kumar, A. (2025). Chapter 2—threats to the rapidity of sustainability transitions posed by technological changes. In *Technological pathways for Africa's net-zero economy* (pp. 39–75). https://doi.org/10.1016/B978-0-443-31486-5.00002-1

O'Connor, B., & Whyte, L. (2023). Glacial secrets uncovered: Revealing the modes of survival of metabolically active microbial communities entrapped in polar glacial ice. In *ARPHA conference abstracts* (Vol. 6).https://doi.org/10.3897/aca.6.e108883

Ou, H. W. (2023). A theory of orbital-forced glacial cycles: Resolving Pleistocene puzzles. *Journal of Marine Science and Engineering, 11*(3). https://doi.org/10.3390/jmse11030564

Parker, R. L., Foster, G. L., Gutjahr, M., Wilson, P. A., Littler, K. L., Cooper, M. J., Michalik, A., Milton, J. A., Crocket, K. C., & Bailey, I. (2022). Laurentide Ice Sheet extent over the last 130 thousand years traced by the Pb isotope signature of weathering inputs to the Labrador Sea. *Quaternary Science Reviews, 287*. https://doi.org/10.1016/j.quascirev.2022.107564

Patton, H., Winsborrow, M. C. M., & Esteves, M. (2021). The Eurasian arctic: Glacial landforms from the last glacial maximum. In *European glacial landscapes: Maximum extent of glaciations*. https://doi.org/10.1016/B978-0-12-823498-3.00025-X

Shafik, W. (2024a). Industry revolution 4.0 and beyond: Abilities and perils for sustainable zero net economy development. In *Powering Industry 5.0 and sustainable development through innovation* (pp. 292–316). IGI Global. https://doi.org/10.4018/979-8-3693-3550-5.ch020

Shafik, W. (2024b). Environmental conservation and sustainable development. In *Developing digital inclusion through globalization and digitalization* (pp. 142–176). IGI Global. https://doi.org/10.4018/979-8-3693-4111-7.ch008

Shafik, W. (2024b). Sustainable development goal 14: Explainable AI (XAI) for ocean health. In *Artificial Intelligence and edge computing for sustainable ocean health* (pp. 167–198). Springer. https://doi.org/10.1007/978-3-031-64642-3_8

Shafik, W., & Azrour, M. (2024). Building a greener world: Harnessing the power of IoT and smart devices for sustainable environment. In *Technical and technological solutions towards a sustainable society and circular economy* (pp. 35–58). Springer. https://doi.org/10.1007/978-3-031-56292-1_3

Simonsen, M. F., Baccolo, G., Blunier, T., Borunda, A., Delmonte, B., Frei, R., Goldstein, S., Grinsted, A., Kjær, H. A., Sowers, T., Svensson, A., Vinther, B., Vladimirova, D., Winckler, G., Winstrup, M., & Vallelonga, P. (2019). East Greenland ice core dust record reveals timing of Greenland ice sheet advance and retreat. *Nature Communications, 10*(1). https://doi.org/10.1038/s41467-019-12546-2

Spada, G., Bamber, J. L., & Hurkmans, R. T. W. L. (2013). The gravitationally consistent sea-level fingerprint of future terrestrial ice loss. *Geophysical Research Letters, 40*(3). https://doi.org/10.1029/2012GL053000

Thompson, R. (2021). Croll, feedback mechanisms, climate change and the future. *Earth and Environmental Science Transactions of the Royal Society of Edinburgh, 112*(3–4). https://doi.org/10.1017/S1755691021000153

Woods, J., Donaldson, C., White, R. S., Caudron, C., Brandsdóttir, B., Hudson, T. S., & Ágústsdóttir, T. (2018). Long-period seismicity reveals magma pathways above a laterally propagating dyke during the 2014–15 Bárðarbunga rifting event, Iceland. *Earth and Planetary Science Letters, 490*. https://doi.org/10.1016/j.epsl.2018.03.020

Chapter 5
Navigating Carbon Neutrality: Strategies for Sustainable Future and Environmental Challenges

Rubee Singh, Ankita Sharma, and Kishore Kumar

Abstract Carbon neutrality (CN) is imperative to combat climate change, limit greenhouse gas (GHG) emissions, and achieve a net zero economy (NZE). The significance of carbon neutrality is vital for building a resilient and sustainable world. It improves energy security by limiting reliance on renewable energy sources and promotes equity globally by addressing the negative impact of climate change (CC) and global warming on society. This chapter provides insights into the multifaceted perspectives of carbon neutrality, its adoption at the organization & economies level, and the ESG challenges involved in its implementation. Further, this chapter also evaluates issues like lack of uniformity in carbon offsetting practices and trade-offs between resource utilization and carbon neutrality strategies. By highlighting the challenges in the adoption of carbon neutrality and the intertwined environmental sustainability issues, the present study provides pathways for resilient and impactful carbon neutrality strategies for building a sustainable future.

Keywords Carbon neutrality · Carbon credits · Environmental sustainability · Sustainable development · ESG · Net zero emissions · SDGs

Introduction

With the increasing challenges of the negative impact of global warming and climate change, carbon neutrality is considered a cornerstone to achieving environmental sustainability. The notion of carbon neutrality represents a need for collective human

R. Singh
Institute of Business Management, GLA University, Mathura, India

A. Sharma
School of Management, IILM University, Greater Noida, India

K. Kumar (✉)
School of Management, Bennett University, Greater Noida, India
e-mail: akishore001@gmail.com

R. Singh and W. L. Filho (eds.), *Climate Neutrality Through Smart Eco-Innovation and Environmental Sustainability*, Climate Change Management,
https://doi.org/10.1007/978-3-031-83250-5_5

effort to reduce or offset carbon emissions and achieve a NZE. Many prior studies have reported a negative impact of CC on societies, ecology, and economies. Carbon neutrality not only serves as an environmental goal but also provides a critical pathway for inclusive development. It is a global call involving sustainable innovations, public policy actions, and sustainable practices requiring a concerted effort from society, organizations, governments, and other stakeholders (Belgacem et al., 2023; Joshi et al., 2024; Ulpiani et al., 2023). Carbon neutrality also popularly known as net zero emissions, implies striking a balance between GHG emissions and the amount removed from the atmosphere through carbon various offset mechanisms (Tan & Uprasen, 2021; Wiese et al., 2022). This balance helps in reducing GHG emissions from the atmosphere and combating the negative impact of climate change and global warming (Chau et al., 2023; Zhou et al., 2024). The notion of carbon neutrality is comprised of two primary actions. First, reducing carbon emissions as much as possible through human activity. The second involves capturing or removal of the unavoidable carbon emissions from the atmosphere through carbon offset strategies (Nguyen et al., 2023).

The relevance of carbon neutrality lies in its ability to limit the adverse impact of climate change. Global warming, severe weather incidents, and loss of biodiversity are the direct outcomes of excessive carbon emissions. The concept of carbon neutrality is an intervention that addresses the main cause of CC and global warming which is GHG emissions due to human activities (Trouwloon et al., 2023; Room, 2023; Kumar, 2022). It becomes possible through the reduction of carbon emissions and various offsetting mechanisms. It helps to maintain the global temperature and mitigate the adverse impact of CC. Achieving carbon neutrality needs systematic and scientific interventions across industries, regions, and human behavior. With a shift towards non-traditional sources of energy, the adoption of environmentally friendly technology is an important step in this direction (Mishra et al., 2023a, 2023b; Tang, 2023; Saqib, 2022). Moreover, carbon neutrality focuses on preserving natural carbon-capturing sources that play a critical role in capturing harmful gases from the environment, such as oceans, forest areas, and wetlands (Matemilola & Salami, 2023).

The significance of carbon neutrality moves beyond ecological benefits because it is also inherently linked to inclusive and sustainable development. It is central to the theme of many sustainable development goals (SDGs) and global climate alliances and agreements like the Paris Agreement, 2015 (Ghufran et al., 2024; Klaus et al., 2023). This agreement is undertaken to stop the global rising temperature below 2% degrees celsius above the pre-industrial era while making efforts to limit the increase at 1.5 °C. The targets under the Paris Agreement need a drastic reduction in the carbon emission level. This has also led to net zero emission commitments from many countries across the world through their nationally determined contributions (NDCs). This chapter provides a thorough understanding of the concept of carbon neutrality and various strategies to achieve carbon neutrality. Further, this chapter also underscores the importance of CN for achieving SDGs and global climate policies and alliances to achieve the objective of net zero emissions (Nwokolo et al., 2023c).

The rest of the chapter is as follows. The next section explains the concept of carbon neutrality and the evolution of carbon neutrality. The third section presents various strategies to achieve carbon neutrality and followed by Sect. 5.4 related to CN intrinsic role in achieving SDGs. Section 5.5 discusses environmental challenges in achieving carbon neutrality. The final section provides a concluding remark.

Evolution of the Concept of Carbon Neutrality

The origin of the concept of carbon neutrality lies in the growing attention to the negative impact of GHG emissions and the climate change crisis, which became a widespread phenomenon in the mid-twentieth century. The notion of carbon neutrality has an over the past century from an abstract environment protection thought to two main instrument of global climate strategies. The growing awareness of net zero targets shows widespread recognition and understanding of the global warming issues and climate crisis. The consistent progression of carbon neutrality also reflects the need for innovative sustainable solutions to limit carbon footprints (Green & Reyes, 2023; Singh et al., 2024b). Early discourse on reducing carbon emissions was witnessed in the late 1960s. However, it was in the mid-1990s, that the notion of net zero garnered interest among policymakers and society. Researchers, scholars, and policymakers started propagating environment-friendly strategies to reduce carbon emissions and focus on exploring renewable energy sources (Nwokolo et al., 2023b). The call for carbon offsetting strategies in the form of afforestation and solar power also gained momentum (Li & Zhang, 2024; Zhang et al., 2024).

The term carbon neutrality emerged in the early 2000s due to the rising concerns about climate change and global warming. During this time, the concept of carbon footprints also became the mainstream social, political, and economic issue. Initiatives like carbon trading, carbon credits, and carbon offsets started evolving (Sharma et al., 2024). These initiatives allowed individuals, organizations, and governments to take action to reduce GHG emissions and adopt a broader framework to adopt carbon neutrality as a strategic objective. The concept of carbon neutrality witnessed a drastic prominence with the Kyoto Protocol adopted in 1997. This year was a milestone moment in the global climate initiatives, prescribing carbon emission reduction targets for the countries from the developed world (Mor et al., 2023). The Kyoto Protocol laid down the initial platform for carbon emission reduction efforts in the form of a carbon market and related offset systems. This drastically helped in shaping efforts to promote net zero practices. Although this protocol did not specifically mandate carbon neutrality for the countries. Another important milestone in combating the climate crisis was the Paris Agreement 2015. It is considered the most important global framework to date aimed at reducing carbon footprints and limiting global temperature well below 2 degrees Celsius of the pre-industrial levels. With the implementation of the Paris Agreement, carbon neutrality emerged as an important instrument with many countries around the world announcing commitment to net zero targets in the future (Fuglestvedt et al., 2023). This agreement focused on renewable energy sources, carbon, offset mechanisms, eco-friendly practices, and decarbonization of the economy. Global climate policies and initiatives like the Paris Agreement and the Kyoto Protocol promoted the adoption of carbon neutrality at the organizational and national levels (de Lassus St-Geniès, (2024; Mahabadi, 2023). As the world continues to grapple with severe climate crises and global warming issues, the concept of carbon neutrality serves as a transformative goal requiring consulted actions across industries and countries (Nwokolo et al., 2023a).

Figure 5.1 shows the biggest CO_2-emitting countries in the world. It is evident from the figure that China is the largest carbon emitter country in the world which contribute to 31.49 percent of the total emissions in the world. The US contributes to 13% of the total emission followed by India, which accounts for 8.1% of emissions. It is interesting to note that the top five countries on the list (i.e. China, US, India, Russia, and Japan) account for almost 60% of the total carbon emissions of the world (Statista, 2023).

According to the United Nations, more than 100 countries have committed to net zero targets. Many countries have pledged to become NZE. One of the biggest emitters, America has pledged to become NZE by 2050, China by the year 2016, and India has pledged to become carbon neutral by 2070 (Statista, 2024). However, Finland is well on its pathway to becoming carbon neutral by 2035, another European country, Iceland has split to become net zero before 2040. Further countries like Nepal, Germany, and Sweden have committed to becoming net zero in 2045 (Statista, 2024). Figure 5.2 below shows the net zero commitment target of various countries.

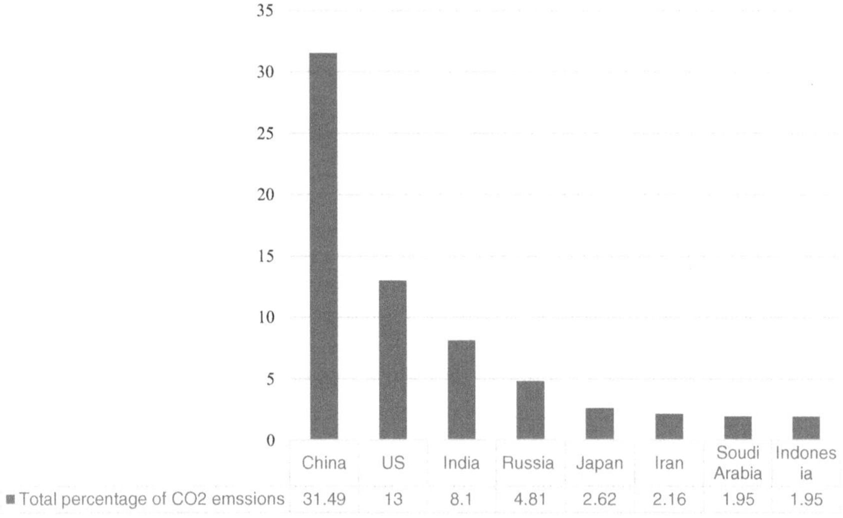

Fig. 5.1 Shows biggest CO_2 emitter countries in the world in 2023 (Statista, 2023)

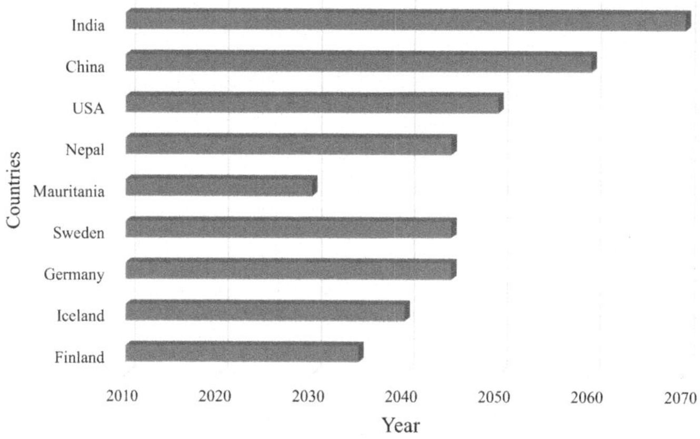

Fig. 5.2 Shows net zero target years of countries (Statista, 2024)

Carbon Neutrality Strategies

Carbon neutrality is a multifaceted approach that takes into consideration, sustainable innovation, offset mechanisms, global policy framework, and individual actions. The following section provides an understanding of the various strategies for achieving carbon neutrality.

Sustainable Innovation

Sustainable innovation plays an important role in reducing carbon footprints and transitioning toward carbon neutrality. Sustainable technological innovations include the adoption of innovative renewable energy sources and energy-efficient technologies (Bekun, 2024; Khurshid et al., 2023). The adoption of innovative renewable sources such as solar energy, and wind energy is central to the idea of decarbonization of energy requirements. Solar energy and wind energy are increasingly channelized to serve the energy requirements (Zastempowski, 2023). The battery technology to store the energy is also witnessing a steep increase in adoption. In the past decade, battery-enabled vehicles have witnessed impetus in the decarbonization efforts as the transportation sector is one of the major carbon emitters. These technologies have a immense potential to reduce the need for traditional energy sources, which are the main source of GHG emissions in the atmosphere.

Energy Efficient Technologies

The improvement in energy-efficient technologies across industries is also an important strategy for achieving carbon neutrality. Smart infrastructure can significantly reduce energy consumption. The development of smart grids also helps in achieving energy efficiency through optimization. A smart grid is a transformative action in energy management (Singh et al., 2024a). The development of advanced energy systems through the leverage of IOT and other digital technologies leads to optimized transmission, minimum losses, and integration of renewable energy sources. Through the advanced systems, real-time monitoring and management also become possible resulting in optimization of energy usage.

Carbon Offset Strategies

Although the reduction in carbon footprints is imperative, there could be still unavoidable GHG emissions that need to be captured or removed. Carbon offsetting strategies involve compensating for carbon emissions by taking actions to capture GHG from the atmosphere. Decarbonization through reforestation and afforestation is one of the most popular and widely adopted strategies of carbon offsetting mechanism (Chen et al., 2022; Zhou et al., 2024). Plantation of forests in the deforested region and the creation of new forest reserves not only help in improving biodiversity, and socio-economic benefits to the local community but also help in capturing harmful gases from the atmosphere. Advanced carbon sequestration strategies have also improved the carbon offsetting mechanism (Li & Zhang, 2024; Mishra et al., 2023a, 2023b). This technology helps in capturing GHG from industrial facilities and storing it underground. These technologies are considered important in the efforts to develop and implement carbon offset mechanisms.

Carbon offset mechanism effectiveness is one of the most important challenges for achieving carbon neutrality. The monitoring, measurement, and disclosure are necessary systems to authenticate the offset outcome and avoid greenwashing (Broderick, 2008; Conte & Kotchen, 2010; Pan et al., 2022). The standard guidelines and frameworks through leverage of advanced technology, such as blockchain can improve accountability and transparency (Prakash et al., 2023; Valiergue & Ehrenstein, 2023). A strong governance structure is also important for driving carbon neutrality. Government bodies and international organizations have an important role in formulating frameworks and guidelines for the widespread adoption of carbon offset mechanisms and environmentally friendly technologies (Dong et al., 2022; Shan et al., 2021).

Carbon Neutrality and SDGs

The United Nations SDGs are an overarching framework to achieve inclusive and sustainable development. It aims to achieve economic growth by balancing environmental protection and social equity (Govindan, 2023). Introduced in 2015, these SDGs are to be achieved by 2030. Carbon neutrality could be one of the most important instruments to attain various sustainable development goals (Dhull et al., 2024). Carbon neutrality aims to reduce emissions and address climate concerns, which helps in achieving several SDGs such as SDG 3, SDG 7, SDG 8, SDG 11, SDG 12, SDG 14, and SDG 15 (Sharma & Kumar, 2022). The transition towards a net zero carbon economy through carbon neutrality will reduce the level of pollution and improve air quality. This will also reduce the health risks associated with pollution and will directly contribute to SDG 3 which is good health and well-being. The usage of clean energy sources and a less polluted environment will prevent a lot of respiratory and other diseases associated with the pollution. Thus, it will improve the health and well-being of society at large (Kumar et al., 2024). Clean energy is concerned with SDG 7. The concept of carbon neutrality focuses on clean and renewable energy sources. Thus, it directly contributes to the SDG 7. Investment in renewable energy alternatives and less dependence on non-renewable sources is important for a resilient and sustainable world. The development of advanced technology to develop sustainable innovation. Renewable energy sources help improve access to clean and reliable energy. Moreover, it also bolsters economic growth and energy security (Ciambra et al., 2023; Ioannou et al., 2023; Yang et al., 2023).

The concept of carbon neutrality also intrinsically caters to SDG 13, which talks about climate action. SDG 13 propagates various measures to tackle the climate crisis and global warming. The concept of carbon neutrality and net zero emission aims to combat climate change and preserve ecology and biodiversity (Tiwari et al., 2025). This proactive approach is crucial in the transition towards net zero and a more sustainable world. This will help to reduce climate risk and protect vulnerable communities from the negative impact of climate change. Large urban cities around the world are primary sources of GHG emissions. Addressing the carbon footprint issue of urban cities could drastically reduce the overall emission level. The concept of carbon neutrality directly caters to SDG 11 which is concerned with developing sustainable cities and communities. Carbon offset projects like the development of sustainable infrastructure, green buildings, sustainable transportation, and afforestation help in building sustainable cities and communities (Agrawal & Kumar, 2024; Yin et al., 2023)). The concept of carbon neutrality also intrinsically contributes to SDG 12, which aims to propagate sustainable consumption and production. Carbon offset to achieve carbon neutrality is an important tool to curb and capture carbon emissions. SDG 14 and SDG 15 are concerned with the protection of life on land as well as life below water respectively. Carbon offset strategies through natural carbon sinks in the form of wetlands, forests, and ocean helps to protect and conserve life on land and water. These carbons offset efforts are directly connect with SDG 14 and SDG 15.

Environmental Challenges in Achieving CN

Although carbon neutrality is a key instrument to address the CC and create a sustainable future the transition towards net zero is fraught with severe environmental challenges. These challenges involve moving beyond traditional energy sources to the development of renewable energy sources and limiting the negative impact of climate mitigation strategies. Understanding these bottlenecks is important to create innovative sustainable solutions for carbon neutrality and the long-term viability of carbon–neutral strategies (Raihan, 2024; Singh et al., 2024a).

One of the most crucial bottlenecks for achieving carbon neutrality is extreme dependence on non-renewable energy alternatives. Although there has been an increasing adoption of renewable energy sources such as solar power hydropower and electric battery energy sources, the rate of adoption is relatively very low among large emitting countries. The shift to cleaner energy sources requires huge investments in clean infrastructure. Furthermore, many countries face extreme financial stringency and other political barriers to reducing excessive dependence on non-renewable sources (Xu et al., 2024; Yang et al., 2023). Thus, striking a balance between the adoption of clean energy sources and the huge cost involved is a critical concern to achieve carbon neutrality. The carbon offset mechanism also has its limitations. For example, afforestation projects require the availability and utilization of a large mass of land which may contradict the other usage of land such as agriculture, housing, and other development activities (Chen et al., 2022; Singh et al., 2024c). Tree plantation strategies in the non-native ecology also involve the risk of loss of biodiversity, leading to severe environmental consequences. Further, carbon-capturing technologies are also in the nascent stage of development. The long-term viability of storing carbon underground from the atmosphere is also uncertain. The authenticity and validity challenges in the carbon offset projects claimed by the organization are also a concern. There have been instances where the carbon offset claims made by the organizations were also questionable.

Another important concern to achieve carbon neutrality is ecological risk and loss of biodiversity. There have been instances where large-scale solar energy projects have been implemented in the vast region with huge investments. However, these projects very often resulted in conflicts with the ecosystem, loss of biodiversity, disruption of wildlife, and displacement of the indigenous communities. The environmental challenges for achieving carbon neutrality are multifaceted and intertwined. These challenges should be addressed through sustainable innovation, inclusive policy, and global effort to transition to net zero and make carbon neutral future.

Conclusion

Carbon neutrality is a transformative goal to combat global warming, protect biodiversity, and achieve inclusive development. It is in the global effort to address the climate crisis and achieve net zero through sustainable innovation, inclusive policy, making, corporate sustainability, and individual behavioural changes. This chapter provides a better understanding of the concept of carbon neutrality and discusses the evolution of carbon neutrality. Further, this chapter also outlines the role of carbon neutrality in achieving several SDGs and environmental challenges associated with the implementation of carbon neutrality. International endeavors like the Kyoto Protocol and the Paris Agreement have resulted in the widespread acceptance of the need to transition towards carbon carbon–neutral world. These agreements focus on aligning economic growth without harming the ecology. Carbon-neutrality strategies significantly contribute to SDG 7 which focuses on clean energy, SDG 11 which focuses on sustainable cities, and SDG 13, which focuses on climate change measures. It will not only protect from environmental risk, but also improve economic resilience, social equity, and inclusivity in society. The implementation of carbon neutrality could be a catalyst for achieving inclusive development and a sustainable future. The journey to carbon neutrality requires concerted effort and commitment from countries, corporations, and individuals. Carbon neutrality has the immense potential to become the most defining story of this century.

References

Agrawal, D., & Kumar, K. (2024). Achieving SDGs through MSMEs: An empirical assessment of environmental consideration initiatives in India. In *Green Transition Impacts on the Economy, Society, and Environment* (pp. 358–376). IGI Global.

Bekun, F. V. (2024). Race to carbon neutrality in South Africa: What role does environmental technological innovation play? *Applied Energy, 354*, 122212.

Belgacem, S. B., Adam, N. A., Khatoon, G., & Pawar, P. S. (2023). Do green finance, low-carbon energy transition, and economic growth help in environmental investment?: Empirical evidence from emerging economies in Asia. *Geological Journal, 58*(9), 3259–3267.

Broderick, J. (2008). Voluntary carbon offsets. In *Sustainable Tourism Futures: Perspectives on Systems, Restructuring and Innovations* (pp. 169–99).

Chau, K. Y., Sadiq, M., & Chien, F. (2023). The role of natural resources and eco-financing in producing renewable energy and carbon neutrality: Evidence from ten Asian countries. *Resources Policy, 85*, 103846.

Chen, L., Msigwa, G., Yang, M., Osman, A. I., Fawzy, S., Rooney, D. W., & Yap, P. S. (2022). Strategies to achieve a carbon neutral society: A review. *Environmental Chemistry Letters, 20*(4), 2277–2310.

Ciambra, A., Stamos, I., & Siragusa, A. (2023). Localizing and monitoring climate neutrality through the sustainable development goals (SDGs) framework: The case of Madrid. *Sustainability, 15*(6), 4819.

Conte, M. N., & Kotchen, M. J. (2010). Explaining the price of voluntary carbon offsets. *Climate Change Economics, 1*(02), 93–111.

de Lassus St-Geniès, G. (2024). The market mechanisms of the Paris agreement and the Kyoto protocol: A legal comparison. In *Research Handbook on the Law of the Paris Agreement* (pp. 240–261). Edward Elgar Publishing.

Dhull, S. B., Rose, P. K., Rani, J., Goksen, G., & Bains, A. (2024). Food waste to hydrochar: A potential approach towards the sustainable development goals, carbon neutrality, and circular economy. *Chemical Engineering Journal*, 151609.

Dong, F., Zhu, J., Li, Y., Chen, Y., Gao, Y., Hu, M., & Sun, J. (2022). How green technology innovation affects carbon emission efficiency: Evidence from developed countries proposing carbon neutrality targets. *Environmental Science and Pollution Research, 29*(24), 35780–35799.

Fuglestvedt, J., Lund, M. T., Kallbekken, S., Samset, B. H., & Lee, D. S. (2023). A "greenhouse gas balance" for aviation in line with the Paris agreement. *Wiley Interdisciplinary Reviews: Climate Change, 14*(5), e839.

Ghufran, M., Aldieri, L., Pyka, A., Ali, S., Bimonte, G., Senatore, L., & Vinci, C. P. (2024). Food security assessment in the light of sustainable development goals: a post-Paris agreement era. *Environment, Development and Sustainability*, 1–29.

Govindan, K. (2023). How digitalization transforms the traditional circular economy to a smart circular economy for achieving SDGs and net zero. *Transportation Research Part E: Logistics and Transportation Review, 177*, 103147.

Green, J. F., & Reyes, R. S. (2023). The history of net zero: Can we move from concepts to practice? *Climate Policy, 23*(7), 901–915.

Ioannou, I., Galán-Martín, Á., Pérez-Ramírez, J., & Guillén-Gosálbez, G. (2023). Trade-offs between Sustainable Development Goals in carbon capture and utilisation. *Energy & Environmental Science, 16*(1), 113–124.

Joshi, A., Singh, R., Kumar, V., & Rani, S. (2024). Net zero strategy: Exploring opportunities, confronting challenges for developing countries. In *Transition Towards a Sustainable Future: Net Zero Policies and Environmental Sustainability* (pp. 233–252). Springer Nature Singapore.

Khurshid, A., Rauf, A., Qayyum, S., Calin, A. C., & Duan, W. (2023). Green innovation and carbon emissions: The role of carbon pricing and environmental policies in attaining sustainable development targets of carbon mitigation—evidence from Central-Eastern Europe. *Environment, Development and Sustainability, 25*(8), 8777–8798.

Klaus, J. P., Nishi, H., Peabody, S. D., & Reichert, C. (2023). CSR activity in response to the Paris agreement exit. *European Financial Management, 29*(3), 667–691.

Kumar, K., Singh, R., Sharma, A., & Lehri, B. P. S. (2024). Climate change adaptation strategies for achieving net-zero economy. In *Transition Towards a Sustainable Future: Net Zero Policies and Environmental Sustainability* (pp. 163–183). Springer Nature Singapore.

Kumar, K. (2022). The uptake of sustainability disclosure practices by environmentally polluting industries in India. In *The Equal Pillars of Sustainability* (pp. 199–224). Emerald Publishing Limited.

Li, L., & Zhang, D. (2024). Forest carbon offset protocols in compliance carbon markets. *Forest Policy and Economics, 165*, 103253.

Mahabadi, D. (2023). Enhancing fairness in the Paris agreement: Lessons from the Montreal and Kyoto protocols and the path ahead. *International Journal of Environment and Sustainable Development, 22*(3), 329–348.

Matemilola, S., & Salami, H. A. (2023). Net zero emission. *Encyclopedia of Sustainable Management* (pp. 2436–2441). Springer International Publishing.

Mishra, R., Singh, R., & Govindan, K. (2023b). Net-zero economy research in the field of supply chain management: A systematic literature review and future research agenda. *The International Journal of Logistics Management, 34*(5), 1352–1397.

Mishra, P., Kumar, K., Fouzdar, A. S., & Singh, A. (2023). Sustainability disclosure and green finance: Driving the transition towards a sustainable future. In *The Sustainable Fintech Revolution: Building a Greener Future for Finance* (pp. 20–36). IGI Global.

Mor, S., Aneja, R., Madan, S., & Ghimire, M. (2023). *Kyoto Protocol and Paris Agreement: Transition from Bindings to Pledges–A Review*. Millennial Asia, 09763996221141546.

Nguyen, T. T., Grote, U., Neubacher, F., Do, M. H., & Paudel, G. P. (2023). Security risks from climate change and environmental degradation: Implications for sustainable land use transformation in the Global South. *Current Opinion in Environmental Sustainability, 63*, 101322.

Nwokolo, S. C., Singh, R., Khan, S., Kumar, A., & Luthra, S. (2023a). Remedies to the challenges of renewable energy deployment in Africa. In *Africa's Path to Net-Zero. CSR, Sustainability, Ethics & Governance*. Springer, Cham.

Nwokolo, S. C., Singh, R., Khan, S., Kumar, A., & Luthra, S. (2023b). Introduction: Africa's net zero transition. In *Africa's Path to Net-Zero: Exploring Scenarios for a Sustainable Energy Transition* (pp. 1–13). Springer Nature Switzerland.

Nwokolo, S. C., Singh, R., Khan, S., Kumar, A., & Luthra, S. (2023c). Africa's Path to Net-Zero. *CSR, Sustainability, Ethics & Governance.*

Pan, C., Shrestha, A., Innes, J. L., Zhou, G., Li, N., Li, J., & Wang, G. (2022). Key challenges and approaches to addressing barriers in forest carbon offset projects. *Journal of Forestry Research, 33*(4), 1109–1122.

Prakash, A., Kumar, K., Khan, W., & Siddiquei, M. I. (2023). Self-induced versus structured corporate social responsibility: The Indian context. *Measuring Sustainability and CSR: From Reporting to Decision-Making* (pp. 167–178). Springer International Publishing.

Raihan, A. (2024). The influence of tourism on the road to achieving carbon neutrality and environmental sustainability in Malaysia: The role of renewable energy. *Sustainability Analytics and Modeling, 4*, 100028.

Romm, J. (2023). *Are Carbon Offsets Unscalable, Unjust, and Unfixable-and a Threat to the Paris Climate Agreement?* University of Pennsylvania, Center for Science, Sustainability, and the Media.

Saqib, N. (2022). Nexus between the renewable and non-renewable energy consumption and carbon footprints: Evidence from Asian emerging economies. *Environmental Science and Pollution Research, 29*(38), 58326–58340.

Shan, S., Genç, S. Y., Kamran, H. W., & Dinca, G. (2021). Role of green technology innovation and renewable energy in carbon neutrality: A sustainable investigation from Turkey. *Journal of Environmental Management, 294*, 113004.

Sharma, H. P., & Kumar, K. (2022). The uptake of environmental management system by small and medium enterprises (SMEs) in India. In *IOP Conference Series: Earth and Environmental Science. 1084*(1), 012015. IOP Publishing.

Sharma, A., Chaudhary, M., Kumar, K., & Daga, S. (2024). Promoting net-zero economy for sustainable development: practice-based view. In *Net Zero Economy, Corporate Social Responsibility and Sustainable Value Creation: Exploring Strategies, Drivers, and Challenges* (pp. 101–116). Springer Nature Switzerland.

Singh, R., Gupta, A., & Kandpal, V. (2024a). ESG excellence in the digital world: Circular economy e-business solutions. *Fintech for ESG and the Circular Economy*, 193–204.

Singh, R., Gupta, A., Bajpai, A., & Kandpal, V. (2024b). Exploring challenges of circular economy. In *Smart Cities and Circular Economy: The Future of Sustainable Urban Development* (Vol. 59).

Singh, R., Kumar, K., & Khan, S. (2024c). A comprehensive view of artificial intelligence (AI)-based technologies for sustainable development goals (SDGs). *Artificial Intelligence Enabled Management: An Emerging Economy Perspective, 183.*

Statista. (2023). https://www.statista.com/statistics/271748/the-largest-emitters-of-co2-in-the-world/

Statista. (2024). Zet zero target year. https://www.statista.com/chart/26053/countries-with-laws-policy-documents-or-timed-pledges-for-carbon-neutrality/

Tan, Y., & Uprasen, U. (2021). Carbon neutrality potential of the ASEAN-5 countries: Implications from asymmetric effects of income inequality on renewable energy consumption. *Journal of Environmental Management, 299*, 113635.

Tang, W., Mai, L., & Li, M. (2023). Green innovation and resource efficiency to meet net-zero emission. *Resources Policy, 86*, 104231.

Tiwari, S., Sharif, A., Sofuoğlu, E., & Nuta, F. (2025). A step toward the attainment of carbon neutrality and SDG-13: Role of financial depth and green technology innovation. *Research in International Business and Finance, 73*, 102631.

Trouwloon, D., Streck, C., Chagas, T., & Martinus, G. (2023). Understanding the use of carbon credits by companies: A review of the defining elements of corporate climate claims. *Global Challenges, 7*(4), 2200158.

Ulpiani, G., Vetters, N., Shtjefni, D., Kakoulaki, G., & Taylor, N. (2023). Let's hear it from the cities: On the role of renewable energy in reaching climate neutrality in urban Europe. *Renewable and Sustainable Energy Reviews, 183*, 113444.

Valiergue, A., & Ehrenstein, V. (2023). Quality offsets? a commentary on the voluntary carbon markets. *Consumption Markets & Culture, 26*(4), 298–310.

Wiese, F., Thema, J., & Cordroch, L. (2022). Strategies for climate neutrality. Lessons from a meta-analysis of German energy scenarios. *Renewable and Sustainable Energy Transition, 2*, 100015.

Xu, P., Adebayo, T. S., Khan, K. A., Özkan, O., & Shukurullaevich, N. K. (2024). United States' 2050 carbon neutrality: Myth or reality? evaluating the impact of high-tech industries and green electricity. *Journal of Cleaner Production, 440*, 140855.

Yang, S., Yang, D., Shi, W., Deng, C., Chen, C., & Feng, S. (2023). Global evaluation of carbon neutrality and peak carbon dioxide emissions: Current challenges and future outlook. *Environmental Science and Pollution Research, 30*(34), 81725–81744.

Yin, C., Zhao, W., Ye, J., Muroki, M., & Pereira, P. (2023). Ecosystem carbon sequestration service supports the Sustainable Development Goals progress. *Journal of Environmental Management, 330*, 117155.

Zastempowski, M. (2023). Analysis and modeling of innovation factors to replace fossil fuels with renewable energy sources-evidence from European Union enterprises. *Renewable and Sustainable Energy Reviews, 178*, 113262.

Zhang, X., Fleskens, L., Huang, Y., & Huang, Y. (2024). Cost, market, and policy constraints on mitigating climate change through afforestation in China. *Environment International, 187*, 108652.

Zhou, Y., Haseeb, M., Batool, M., Bousrih, J., & Wu, J. (2024). Achieving carbon-neutrality goals in Asian emerging economies: Role of investment in clean energy, eco-regulations, and green finance. *Gondwana Research.*

Part II
Sustainability Driving Smart Eco-Innovation

Chapter 6
Climate Neutrality in Small and Medium Enterprises (SMEs): The Role of Smart Eco-Innovation

Gowri B. Chanal, A. Arun Kumar, T. Sowmyya, and Rakesh Suryadevara

Abstract This chapter explores climate neutrality in small and medium enterprises (SMEs), which is imperative for sustainable development and requires the active participation of all economic sectors. SMEs play a critical role in global emissions reduction due to their collective economic influence but face significant challenges in adopting climate-neutral practices. These challenges include limited financial resources, insufficient environmental expertise, and regulatory complexities. Smart eco-innovation emerges as a promising solution, offering cost-effective and scalable technologies that align with SMEs' resource constraints. Leveraging advancements in artificial intelligence, the Internet of Things (IoT), and data analytics, eco-innovation enhances resource efficiency and reduces emissions through strategies such as energy optimization and waste minimization. Beyond environmental benefits, eco-innovative practices can provide SMEs with competitive advantages, including cost savings, improved brand reputation, and access to sustainability-focused markets. However, barriers such as high upfront costs, technical knowledge gaps, and the perception of limited short-term benefits hinder widespread adoption. Policy interventions, including targeted financial incentives, streamlined regulatory frameworks, and capacity-building initiatives, are critical to facilitating the transition of SMEs to climate-neutral operations. The findings underscore the need for tailored solutions that enable SMEs to contribute effectively to global climate goals,

G. B. Chanal · A. Arun Kumar (✉) · R. Suryadevara
ICFAI Law School, ICFAI Foundation for Higher Education (Deemed to Be University), Hyderabad, Telangana, India
e-mail: arunkumar@ifheindia.org

G. B. Chanal
e-mail: drgowribchanal@ifheindia.org

R. Suryadevara
e-mail: rakeshsuryadevara@ifheindia.org

T. Sowmyya
Department of Forensic Science, University of College of Science, Osmania University, Hyderabad, Telangana 500007, India
e-mail: dr.sowmyya@osmania.ac.in

R. Singh and W. L. Filho (eds.), *Climate Neutrality Through Smart Eco-Innovation and Environmental Sustainability*, Climate Change Management,
https://doi.org/10.1007/978-3-031-83250-5_6

highlighting the interconnected benefits of sustainability, resilience, and economic competitiveness within the SME sector.

Keywords Climate neutrality · SMEs · Smart · Eco-innovation · Sustainability

Introduction

Achieving climate neutrality as a pillar of sustainable development has gained international attention in the face of an intensifying climate crisis. A goal enshrined in international agreements like the Paris Agreement, which aims to limit global warming to well below 2 °C above pre-industrial levels, is climate neutrality, commonly defined as the balance between greenhouse gas emissions and their removal from the atmosphere. Every industry, from manufacturing to energy, must lower its carbon footprint for nations to meet these targets. This includes small and medium-sized businesses (SMEs) and big corporations. SMEs are essential to reaching climate goals because they account for more than changes made by SMEs can have a big impact on reducing global emissions because of their combined economic importance. However, the road to climate neutrality for SMEs is fraught with difficulties. In contrast to large corporations with significant financial and human resources, SMEs often have smaller budgets, less access to green financing, and employees who might not have specialized environmental knowledge. Furthermore, implementing sustainability regulations can be expensive and complicated, which makes things even more challenging for smaller companies (Kumar et al., 2024). Therefore, cost-effective strategies that can be adjusted to SMEs' unique constraints are necessary for adopting climate-neutral practices. The full involvement of SMEs in climate neutrality initiatives might remain unattainable in the absence of workable, scalable solutions, endangering nations' capacity to fulfill their climate pledges.

Smart eco-innovation has a role that goes beyond short-term environmental improvements to include wider business advantages. By implementing climate-neutral practices, SMEs can establish themselves as conscientious, progressive companies, increasing their attractiveness to investors, customers, and business associates who care about sustainability. Sustainable supply chains are becoming increasingly important to consumers and large corporations due to growing climate awareness, which provides SMEs with a financial incentive to implement eco-innovative practices. Additionally, several governments and international organizations have started to provide grants, incentives, and support frameworks specially made to help SMEs achieve climate-neutral objectives. These encouraging policies acknowledge that smaller companies with the means to adopt sustainable practices cut emissions and help create a more resilient and competitive economy (Cevat Yaman, 2020).

Many SMEs still encounter implementation challenges even though smart eco-innovation has the potential to promote climate neutrality. Knowledge gaps are the most important of these smaller businesses, and they frequently lack specialized technical or environmental experts who can recognize and apply eco-innovative solutions.

Furthermore, SMEs usually believe that eco-innovation technologies are too costly or out of their price range, particularly if the potential cost savings are not immediately apparent. Because managing sustainability requirements frequently requires resources smaller businesses might not have, regulatory complexity can further discourage SMEs from participating in climate neutrality initiatives. These difficulties highlight the need for easily obtainable, SME-focused solutions that lessen the initial expenses and administrative strains related to eco-innovation. Therefore, policymakers play a crucial role in developing frameworks that promote adoption, such as streamlined regulatory requirements for SMEs, tax breaks, and easily accessible funding sources.

Numerous stakeholders are affected by the research's conclusions. They emphasize to policymakers the significance of supportive frameworks that address SMEs' needs. They highlight eco-innovation's competitive benefits to market leaders, such as reduced costs, enhanced brand recognition, and access to sustainable markets. Additionally, this research offers SMEs themselves a road map for achieving climate neutrality in a way that fits their opportunities and limitations. In addition to helping achieve global climate goals, SMEs can improve their sustainability and resilience in a business environment that is changing quickly by utilizing the power of smart eco-innovation.

The primary global authority on climate science, the Intergovernmental Panel on Climate Change (IPCC), is charged with producing scientific evaluations that assist decision-makers across the globe. The IPCC's reports, which highlight the necessity of reducing emissions immediately to avoid disastrous climate effects, have established the groundwork for our understanding of climate neutrality. Finding a balance between greenhouse gas emissions and removal is necessary for climate neutrality. This entails cutting emissions as much as possible and using carbon sequestration strategies, like afforestation or carbon capture and storage (CCS), to offset any remaining emissions. Limiting global warming to 1.5 °C over pre-industrial levels is essential to preventing severe climate impacts, according to the IPCC's Special Report on Global Warming of 1.5 °C (2018). Reaching net-zero emissions by 2050 and halving global emissions by 2030 are necessary. Current emission trajectories are not enough to meet global climate goals, according to the most recent Sixth Assessment Report (AR6). The IPCC emphasizes how innovation, funding sources, and enabling frameworks can support smaller companies' shift to sustainable practices (IPCC Report, 2024).

The Paris Agreement, adopted in 2015, is a historic global endeavor to address climate change. Its main clauses lay the groundwork for the public and private sectors to work together to achieve climate neutrality. With efforts to keep it to 1.5 °C, the agreement seeks to keep the increase in global temperature to "well below 2 °C" over pre-industrial levels. Nationally Determined Contributions (NDCs), which list countries' goals and plans for reducing emissions, are submitted. Developed nations must give them financial support to help developing nations meet their goals for adaptation and mitigation. Businesses and other non-state actors play a crucial role in accomplishing the goals of the Paris Agreement. This inclusion emphasizes the value of private sector involvement, especially for SMEs, which comprise a sizable portion of the world economy. SMEs are essential for implementing context-specific, locally relevant solutions that help reduce emissions, like low-carbon product innovations, sustainable supply chains, and energy efficiency initiatives. Businesses are encouraged by the Paris Agreement to implement climate-neutral technologies and procedures, such as waste reduction, renewable energy adoption, and sustainable product design. Financial resources are available to assist SMEs in transitioning to climate-neutral operations through national subsidies and international mechanisms like the Green Climate Fund.

The Paris Agreement and the IPCC report emphasize how urgent it is to achieve climate neutrality. This presents both a challenge and an opportunity for SMEs. The shift to a low-carbon economy creates opportunities for innovation, market competitiveness, and long-term sustainability, even though resource limitations and regulatory complexity may impede advancement. Policymakers must develop frameworks and incentives tailored to SMEs to encourage climate-neutral practices. The international community, including SMEs, can cooperate to achieve climate neutrality and lessen the worst effects of climate change by adhering to the IPCC's scientific recommendations and the Paris Agreement's policy directives. Small and medium-sized businesses (SMEs) are essential to reaching the EU's sustainability and climate goals, according to the European Commission (EC). They are essential to the EU's larger climate action plans, such as the Fit for 55 packages and the European Green Deal, because of their combined impact on energy consumption, resource use, and carbon emissions (Nwokolo et al., 2023a, 2023b, 2023c).

The main findings from the European Commission's reports on the economic significance of SMEs and their contribution to the advancement of climate goals are summarized below. With 56.4% of the economy's value-added coming from SMEs, they are the foundation of the EU economy. Significant greenhouse gas emissions are caused by a few SME-dominated industries, including light manufacturing, logistics, and construction. Achieving the EU's climate goals requires these industries to switch to sustainable practices. By adopting green technologies, circular economy principles, and sustainable innovation, SMEs play a crucial role in realizing this vision (Hultman et al., 2012). Funding for research and innovation, especially in environmentally friendly technologies, is one of the funding mechanisms the EC has set up to assist SME transitions. provides guarantees and funding to SMEs investing in sustainable and green projects.

This chapter will highlight successful examples of climate-neutral SMEs and offer insights into the eco-innovative practices that made them possible through an analysis of case studies and industry data. In doing so, this study aims to provide useful suggestions for legislators, corporate executives, and environmentalists trying to hasten the SME sector's shift to climate neutrality.

Historical Emission Reduction Trends and Challenges

Although the historical average annual reduction rate of 1.08% is well below the 3.7% needed to meet the 2030 targets, the EU has shown a commendable commitment to climate action. Increased efforts and systemic changes in all sectors are required to close this gap. Renewable technology adoption has advanced significantly in some industries, such as energy production. However, because green practices and technologies are being adopted slowly, industries like heavy industry, transportation, and agriculture continue to present difficulties. Because of differences in economic capabilities, policy priorities, and resource availability, some member states have made notable progress in lowering emissions, while others are lagging. This discrepancy jeopardizes EU-wide cooperation.

Significant financial obstacles are presented by the high costs of switching to sustainable technologies and renewable energy, especially for SMEs and member states with weaker economies. Geopolitical tensions, including dependence on outside energy sources, hamper efforts to quickly transition from fossil fuels to renewable energy sources. Public support for climate policies may decline due to rising energy and inflation rates, especially if people prioritize immediate economic stability more than long-term environmental objectives. Implementation and enforcement of policies may be delayed by the EU's multi-layered governance structure, particularly when juggling the disparate interests of its 27 member states. Notwithstanding these obstacles, the EU can accelerate its progress toward the 2030 climate neutrality targets.

Reducing emissions in the energy sector can be accelerated by taking advantage of the EU's leadership in renewable energy technologies like solar and wind. Research and development expenditures for energy storage, green hydrogen, and carbon capture technologies can yield scalable ways to cut emissions in industries that are challenging to decarbonize. Financial obstacles for companies and member states can be removed by enlarging programs like the EU Green Deal Investment Plan and attracting private investment. Improving resource efficiency, recycling, and waste reduction, especially for SMEs, can greatly lower emissions while increasing economic resilience.

Government Guidelines on SME Innovation and Climate Strategies by OECD and UNEP

The United Nations Environment Programme (UNEP) and the Organization for Economic Co-operation and Development (OECD) have been leading proponents of incorporating small and medium-sized businesses (SMEs) into international climate policies. Their government guidelines and position papers highlight SMEs' importance in attaining climate neutrality and sustainable development. Additionally, they draw attention to the difficulties that SMEs encounter and offer workable solutions to encourage creativity and adaptability in this industry (Di Wang et al., 2024a, 2024b).

SMEs are the foundation of most economies, making substantial contributions to employment and GDP. However, because of their quantity and resource intensity, they have a significant overall environmental impact. The OECD's position papers advocate for including SMEs in national and international climate action plans, emphasizing the significance of "greening" them to conform to global climate goals. UNEP emphasizes that SMEs are important change agents in accomplishing sustainability goals and are susceptible to climate risks. Their versatility and inventiveness are crucial for advancing low-carbon technologies and green growth. The OECD and UNEP identify obstacles preventing SMEs from implementing climate strategies (Joshi et al., 2024). One of the biggest obstacles is the lack of funding and investment opportunities. SMEs' capacity to invest in green technologies is limited by the perception held by traditional lending institutions that they are high-risk borrowers. Many SMEs do not understand the advantages of sustainability and the technical know-how needed to put eco-innovative solutions into practice. Create green finance tools tailored to SMEs, like lesser interests, tax waivers, and grants, to encourage investment in sustainability. Encourage public–private partnerships to direct funding toward climate projects aimed at SMEs (Baoyi et al., 2024).

Create training courses and knowledge bases to help SMEs better grasp the opportunities and difficulties associated with sustainability. To promote knowledge sharing and access to best practices, promote cross-sector cooperation. Simplify laws to make them easier for SMEs to understand and less onerous. Encourage eco-innovation with grants, prizes, and recognition initiatives. SMEs should be positioned as important stakeholders in national climate policies to guarantee their involvement in reaching net-zero goals. Urge governments to develop roadmaps for sustainability and emissions reduction tailored to SMEs. Encourage collaborations with academic institutions and tech companies to reduce the price of environmentally friendly solutions. Encourage technology-sharing websites to give SMEs more equitable access to green innovations.

According to the OECD and UNEP, SMEs have enormous potential to spur sustainable innovation and growth but face major obstacles when implementing climate strategies. Governments and international organizations can unleash the transformative potential of SMEs in accomplishing global climate goals by addressing financial, technological, and regulatory barriers and promoting focused research and policy support (Hultman et al., 2012).

Methodology

This chapter is grounded entirely in secondary data sources, drawing extensively from reports, published academic papers, organizational publications, and reputable websites. These sources provide comprehensive insights into the challenges and opportunities associated with achieving climate neutrality in small and medium-sized enterprises (SMEs) through the lens of smart eco-innovation. The data encompasses findings from international agreements like the Paris Agreement, reports from the Intergovernmental Panel on Climate Change (IPCC), and analyses by organizations such as the European Commission, United Nations Environment Programme (UNEP), and Organization for Economic Co-operation and Development (OECD). Additionally, it integrates information from sector-specific studies, case analyses, and global sustainability frameworks. By synthesizing this wide range of reliable secondary sources, the chapter offers a well-rounded exploration of the subject, ensuring the validity and relevance of its conclusions for policymakers, business leaders, and environmental stakeholders.

Results & Discussion

Small and Medium Businesses (SMEs) can become climate neutral by increasing their operational efficiency, lowering their carbon footprints, and coordinating with global sustainability objectives thanks largely to eco-innovation. SMEs, which play a major role in economic growth, encounter difficulties in their decarbonization initiatives, including a lack of resources, knowledge, and restricted access to funding.

Through material waste reduction and energy optimization, eco-innovation promotes resource efficiency. In addition to reducing operating expenses, this helps SMEs adhere to more stringent environmental standards. Even though it requires initial organizational and financial investments, implementing cleaner production technologies has demonstrated long-term benefits (Energy, Sustainability, and Society). Market demands, supply chain requirements, and stakeholder pressures all impact SMEs. Many are driven to implement sustainable practices to improve their legitimacy and reputation in their communities. Environmentally friendly labels on eco-innovative products can also give them a competitive edge. Reports emphasize how financial institutions can help SMEs shift to net-zero operations (Hultman et al., 2012).

Customized financing options such as green loans, grants, and recommendations for emissions reporting and sustainable practices are essential to overcome hurdles to implement eco-innovation. Wider adoption is hampered by the high perceived costs and risks of implementing low-carbon initiatives, especially among microenterprises that use less energy. Government programs encouraging SMEs to invest in eco-innovation include regulatory incentives and subsidies for green technologies. In addition to internal incentives like cost savings and ethical commitments, regulatory "push" factors have proven successful in promoting research and development in cleaner technologies.

For small and medium-sized businesses (SMEs), reports from think tanks like the World Resources Institute (WRI) and consulting firms like McKinsey and Accenture offer important insights into the financial effects and return on investment (ROI) of sustainable practices (Kelliher et al., 2020).

According to McKinsey's research, reaching net-zero emissions will require a large increase in investment in low-emissions technologies, underscoring the high capital requirements for the sustainability transition. Rethinking operations, materials, and energy use are all part of SMEs' integration of sustainable practices. Despite these expenses, McKinsey points out possible financial gains, like access to new markets that prefer eco-friendly products and long-term cost savings from energy efficiency (McKinsey, 2024).

Reports from Accenture highlight the rising demand for sustainability from investors and consumers, which can give SMEs implementing eco-innovations a competitive edge. Improved stakeholder trust, employee engagement, and brand perception correlate with sustainable practices. Accenture also emphasizes that to achieve these advantages, a strategic plan to integrate sustainability throughout operations and leadership commitment is essential (Accenture, 2024).

WRI focuses on the strategic and operational changes SMEs must make to meet climate targets. Their research supports the use of tools such as carbon accounting and creative business models to maximize return on investment while reducing ecological impact. They stress that resource optimization and the scalability of renewable energy solutions are essential for adopting sustainability at a reasonable cost (World Resources Institute, 2024).

Although success frequently necessitates overcoming initial financial and operational challenges, SMEs can increase efficiency, access market opportunities, and contribute to global climate goals by implementing sustainable practices. Together, these sources imply that although achieving sustainability may necessitate a large initial outlay of funds, there will be substantial long-term economic and environmental advantages (Klewitz, 2015).

Barriers to Implementation of Climate Neutrality and Sustainable Practices in SMEs

Due to their substantial contribution to global economic activity, small and medium-sized enterprises (SMEs) are essential to reaching climate neutrality. However, they encounter many obstacles in pursuing climate neutrality, eco-innovations, and sustainable practices. These obstacles include knowledge gaps, regulatory difficulties, and financial and technological limitations. To effectively mobilize SMEs as agents of climate action, it is imperative that these barriers be addressed (Gossling et al., 2024).

The high initial cost of putting eco-innovations, renewable energy systems, and energy-efficient technologies into place is one of the biggest obstacles to adopting sustainable practices. Financing these investments is difficult for SMEs since they frequently have limited access to capital and operate on tight margins. McKinsey asserts that significant capital expenditures are necessary to shift to sustainability, particularly in sectors that depend heavily on carbon-intensive processes or have high energy consumption. SMEs usually have a harder time getting grants or loans for sustainable projects than large corporations. In developing nations, where banking systems might not have the resources or desire to provide green financing options, this problem is made worse (OECD, 2021).

Sustainability can result in long-term cost savings and market opportunities, according to World Resources Institute (WRI) studies, but smaller businesses may be discouraged from pursuing these measures due to the initial financial risk. The operational and financial advantages of implementing sustainable practices are unknown to many SME managers and owners. According to studies, some SMEs may not fully comprehend how sustainability relates to profitability and growth, even though they acknowledge the significance of environmental stewardship. Climate-neutral and eco-innovation tactics frequently require expertise in supply chain optimization, carbon accounting, and energy efficiency (Barragan-Hernández and Aguilar-Fernandez, 2024).

This disparity is more noticeable in SMEs engaged in conventional or resource-intensive industries. Due to a lack of SMEs-specific training and education programs, it is challenging for these companies to upskill their employees or implement innovative practices. This problem is worsened by the absence of knowledge-sharing networks, which restricts SMEs' access to successful case studies. For SMEs, navigating environmental regulations can be challenging, particularly in areas with disjointed or unevenly enforced policies (Nayak et al., 2024).

Adopting climate-neutral practices may be delayed because many SMEs lack the resources necessary to understand and abide by complicated regulations. Government incentives for implementing sustainable practices are often inadequate or poorly explained. Although some nations provide tax breaks or subsidies for installing renewable energy or energy efficiency improvements, SMEs may be unable to take advantage of these programs because of administrative barriers or a lack of knowledge. SMEs face uncertainty due to frequent changes to environmental laws and regulations. Businesses might hesitate to invest in long-term sustainability investments if the regulatory environment is unstable (Arun et al., 2024).

Many sustainable technologies like solar panels, energy storage systems, or sophisticated manufacturing equipment remain unaffordable for SMEs. The adoption of these technologies is hampered by the lack of reasonably priced options, especially for smaller businesses with tighter budgets. Due to a lack of cooperation with research institutions or technology providers, SMEs frequently have difficulty accessing cutting-edge eco-innovations. Bigger companies are better equipped to embrace and apply new technologies because of their wider networks and resources. Integrating new technologies into old systems can be difficult for SMEs, especially if their operations depend on antiquated infrastructure. Problems with compatibility may raise expenses and delay the adoption of sustainable practices (Rathnasiri et al., 2024).

Many SMEs are reluctant to alter long-standing procedures, especially in conventional industries where innovation is considered unnecessary or dangerous. Adoption of sustainable practices may be slowed by organizational cultural resistance. SMEs frequently put immediate financial stability ahead of long-term sustainability objectives. Many SMEs rely on outside vendors for goods and services and are a part of wider supply chains. If suppliers do not prioritize sustainability, SMEs may struggle to align their operations with climate neutrality goals.

On the other hand, SMEs that provide goods to bigger companies are forced to adhere to stricter sustainability guidelines. This can encourage action and add to SMEs' operational and financial burdens. Other difficulties SMEs face in developing nations include inadequate financial systems, restricted access to renewable energy, and a lack of infrastructure. The adoption of climate-neutral practices is divided globally due to these discrepancies. Because they depend heavily on resource-intensive processes, some industries, like manufacturing and agriculture, have difficulty cutting emissions. Customized solutions are needed to meet the unique requirements of these industries. Some markets have low demand for sustainable products and consumer awareness (Singh et al., 2024).

SMEs might not see the benefits of implementing eco-innovative practices without obvious market incentives. SMEs frequently work in fiercely competitive markets where price is a key factor. They may be at a competitive disadvantage to less sustainable rivals due to the additional cost of implementing sustainable practices. The ability of SMEs to switch to cleaner energy sources may be hampered by inadequate access to RE infrastructure, like wind and solar energy. This problem is more noticeable in rural or impoverished areas. Adopting sustainable resource management techniques is difficult for SMEs in sectors that produce a lot of waste or rely heavily on water (Kumar et al., 2024). Legislators, financial institutions, and industry participants must work together to overcome these obstacles. It is feasible to unleash SMEs' potential as catalysts for global sustainability and climate action by creating an environment that supports them (Singh et al., 2023).

Comparative Analysis with Other Sectors

Comparing the obstacles, possibilities, and effects across different industries is essential when analyzing the adoption of eco-innovative technologies to comprehend the difficulties SMEs face. Because of economies of scale and regulatory pressures, larger manufacturing companies are frequently the first to adopt eco-innovation. Using energy-efficient equipment and circular economy principles, they can use government subsidies designed for large-scale projects and spread the cost of new technologies over increased production volumes.

For instance, SMEs frequently rely on small, gradual improvements, which results in slower adoption rates, whereas large manufacturing companies may upgrade their entire production line. Because these businesses can directly incorporate sustainability into their product development, capitalizing on consumer demand for greener technologies, return on investment (ROI) is frequently higher. However, the business case for eco-innovation is less strong for SMEs in traditional industries, since they might not feel the consumer-driven need for sustainability (Bhushan et al., 2024).

Obstacles to eco-innovation in agriculture include using non-renewable fertilizers and water-intensive methods. However, innovations like organic farming methods and drip irrigation have demonstrated a significant return on investment over time. Compared to SMEs, green farming practices receive more government support through subsidies. Due to lower economies of scale, SMEs in the food processing industry frequently face higher upfront costs when switching to sustainable sourcing and packaging, with slower returns. Due to consumer demand and regulatory requirements, energy companies are at the forefront of adopting renewable technologies such as wind and solar (Rathnasiri et al., 2024).

Cost–benefit Analysis of Eco-Innovative Technologies

High upfront costs are necessary for eco-innovative technologies like solar panels, energy-efficient HVAC systems, and sustainable production equipment. Due to their inability to negotiate advantageous terms or take advantage of bulk purchasing, SMEs incur disproportionately higher costs. Despite their high initial costs, eco-innovative technologies usually have lower operating costs because they use less energy and produce less waste. Over the course of their lifetime, solar energy systems can save up to 70% on electricity costs (Nwokolo et al., 2023a, 2023b, 2023c). According to studies, eco-innovations, especially in energy-intensive operations, result in significant cost savings over the medium to long term. For instance, switching to LED lighting within five years can reduce electricity costs by 80%. Reducing waste through circular economy methods also lowers raw material costs. Access to new markets is made possible by eco-innovations, especially for companies looking for green-certified suppliers and environmentally conscious consumers (Singh et al., 2022).

Indirectly, these benefits lead to improved financial performance. Energy storage devices, wind turbines, and solar panels are common ways to cut carbon emissions. High upfront installation expenses, ranging from $10,000 to $30,000 for small enterprises. Tax breaks and long-term electricity savings balance out initial expenditures. Power consumption can be decreased by 30 to 50% by retrofitting energy-efficient parts into existing machinery. Retrofits, $5,000 to $15,000. ROI is achieved in two to four years with annual energy savings. Reusable and biodegradable packaging materials satisfy consumer preferences while lessening their negative effects on the environment. More expensive materials than conventional packaging. Enhanced brand recognition and adherence to future packaging laws. The comparative study shows that although SMEs encounter greater obstacles than larger companies when implementing eco-innovative technologies, the advantages frequently exceed the drawbacks in the long run.

Solutions and Enabling Frameworks for SMEs

Targeted solutions and enabling frameworks are needed to integrate smart eco-innovation and transition SMEs to climate neutrality. These steps should promote creativity and cooperation while addressing money, technology, regulations, and cultural obstacles. Low-interest loans and grants should be made available by governments and international organizations, especially SMEs, so they can invest in energy efficiency, renewable energy, and other environmentally innovative technologies. SMEs can receive funding to create sustainable innovations through the Horizon 2020 program of the European Union. The high upfront costs can be partially offset by tax rebates or deductions for installing renewable energy sources or energy-efficient systems. Tax credits are available for investments in green technology in nations like the UK and Germany. By sharing risks and pooling resources, partnerships between public and private organizations can increase access to financing. Partnerships like the Global Environment Facility (GEF) support SME sustainability initiatives in developing nations. SMEs can participate in carbon markets by selling carbon credits to offset emissions or implementing low-carbon practices. This encourages emissions reductions and offers an extra source of income.

Governments and international organizations can support creating and implementing reasonably priced eco-innovative technologies. Small-scale renewable energy system installation is partly financed by World Bank programs such as the Renewable Energy for SMEs initiative. SMEs can access common resources like pilot plants, labs, and mentorship by creating innovation hubs. Through innovation hubs, the European Institute of Innovation and Technology (EIT) assists SMEs in incorporating green technologies. SMEs can access cutting-edge eco-innovations and expertise through partnerships with academic institutions. In Europe, university-led initiatives offer technical assistance to SMEs shifting to sustainable practices (Nwokolo et al., 2023a, 2023b, 2023c).

The compliance burden is lessened by streamlining environmental regulations and offering precise advice suited to SMEs. Specific actions to assist SMEs in shifting to sustainability are part of the European Green Deal. When stable and predictable policies are in place, SMEs are encouraged to invest in eco-innovative solutions over the long term. SMEs' adoption of renewable energy systems in Germany has increased due to consistent feed-in tariffs. Implementing phased sustainability reporting requirements promotes the slow adoption of climate-neutral practices while giving SMEs time to adapt (Singh et al., 2022).

SME suppliers are indirectly impacted by the UK's Net Zero Strategy, which mandates that larger companies reveal their carbon emissions. Governments can encourage sustainable practices by giving SMEs eco-credential preferences in public procurement. The EU's Green Public Procurement program encourages SMEs to implement sustainable solutions. Demand increases, and SMEs are encouraged to adopt eco-innovations when consumers know sustainable products' advantages. Initiatives like "Choose Green" encourage environmentally friendly products, which helps SMEs that sell sustainable goods. Awareness campaigns highlighting sustainability's long-term financial and environmental advantages can change SME leaders' perspectives. The International Chamber of Commerce's (ICC) programs emphasize integrating sustainability into the leadership of SMEs (Klewitz, 2015).

Policy Recommendations

Governments and financial organizations should improve access to green loans, grants, and subsidies designed especially for SMEs. Tax breaks or public–private partnerships can be used to encourage private sector investments in the sustainability of SMEs. Create government-supported initiatives that give SMEs reasonably priced access to digital tools, energy-efficient equipment, and renewable energy systems. Invest in shared infrastructure, such as water recycling systems or grids powered by renewable energy. Launch public awareness campaigns about environmentally friendly goods and services to generate demand for SMEs implementing sustainable practices. Access global funding sources like the Green Climate Fund to assist SME transitions in developing nations. Boost international cooperation to exchange eco-innovation resources, technology, and knowledge. Adapt national SME strategies to the UN and the Paris Agreement's Sustainable Development Goals (SDGs). Create specialized metrics to monitor SMEs' advancements in implementing eco-innovations and reaching climate neutrality.

Conclusion

Due to their substantial contributions to GDP and employment, small and medium-sized businesses (SMEs) are essential to global economies. However, reaching climate neutrality is hampered by their environmental impact. By implementing clever eco-innovations, SMEs can establish themselves as important contributors to sustainable development, promoting resource efficiency and lowering emissions. Despite these advantages, SMEs cannot fully adopt eco-innovative practices due to obstacles like high upfront costs, a lack of technical know-how, and restricted access to financing. The distinct difficulties SMEs encounter are highlighted by the comparative study of eco-innovation across industries, underscoring the necessity of focused solutions. Cost–benefit analyses show that even though eco-innovative technologies require large initial investments, they are profitable in the long run due to increased competitiveness, operational savings, and regulatory compliance. Enabling frameworks that include financial incentives, technological assistance, regulatory policies, and capacity-building programs are essential to overcome these obstacles. SMEs can attain climate neutrality and make a significant contribution to global sustainability objectives, such as those delineated in the Paris Agreement, by tackling these issues comprehensively.

References

Accenture. (2024). Climate transition plan. https://www.accenture.com/content/dam/accenture/final/accenture-com/document/Climate-Transition-Plan.pdf

Arun Kumar, A., Sowmyya, T., Sambrani, S., &Shekhar, S. (2024). Behavioral changes and public engagement in net-zero policies: Issues and challenges. In R. Singh, & D. Crowther (Eds.), *Transition Towards a Sustainable Future. Approaches to Global Sustainability, Markets, and Governance.* Springer. https://doi.org/10.1007/978-981-97-5756-5_14

Barragán-Hernández, A. D., & Aguilar-Fernández, M. (2024). Strategic dimensions of eco-innovation adoption in manufacturing SMEs in Mexico City. *Systems, 12*(7), 246. https://doi.org/10.3390/systems12070246

Bhushan, M., & Singh, R. (2024). Talent management facilitates net-zero transition through employee green behavior. In R. Singh, S. Khan, A. Kumar, S. Luthra, & H. Chokshi (Eds.), *Net Zero Economy, Corporate Social Responsibility and Sustainable Value Creation. CSR, Sustainability, Ethics & Governance.* Springer. https://doi.org/10.1007/978-3-031-55779-8_7

Gössling, S, & Reinhold, S. (2024). Accelerating small and medium sized tourism enterprises' engagement with climate change. *Journal of Sustainable Tourism*, 1–18. https://doi.org/10.1080/09669582.2024.2350659

Hasan, M. J., & Rahman, M. S. (2023). Determinants of eco-innovation initiatives toward sustainability in manufacturing SMEs: Evidence from Bangladesh. *Heliyon, 9*(7), e18102. https://doi.org/10.1016/j.heliyon.2023.e18102

Hultman, J., & Corvellec, H. (2012). The European waste hierarchy: From the sociomateriality of waste to a politics of consumption. *Environment and Planning a: Economy and Space, 44*(10), 2413–2427. https://doi.org/10.1068/a44668

IPCC. (2024). Opening remarks by the IPCC chair–"leading the pathway towards climate neutrality: The role of European NECPs. https://www.ipcc.ch/2023/10/20/ipcc-chair-jim-skea-climate-neutrality-the-role-of-european-necps/#:~:text=Limiting%20warning%20to%201.5%20degrees,see%20a%20global%20emissions%20peak.

Joshi, A., Singh, R., Kumar, V., & Rani, S. (2024). Net zero strategy: Exploring opportunities, confronting challenges for developing countries. In R. Singh, & D. Crowther (Eds.), *Transition Towards a Sustainable Future. Approaches to Global Sustainability, Markets, and Governance.* Springer. https://doi.org/10.1007/978-981-97-5756-5_11

Kelliher, F., Mellett, S., & Harrington, D. (2020). Enablers of green innovation in the micro-firm perspectives from Ireland and Canada. *Journal of Small Business & Entrepreneurship, 34*(1), 74–93. https://doi.org/10.1080/08276331.2020.1789826

Khan, S., Singh, R., Alnahas, J., Abbate, S., & Centobelli, P. (2024). Navigating the smart circular economy: A framework for manufacturing firms. *Journal of Cleaner Production, 144007*. https://doi.org/10.1016/j.jclepro.2024.144007

Klewitz, J. (2015). Grazing, exploring and networking for sustainability-oriented innovations in learning-action networks: An SME perspective. Innovation. *The European Journal of Social Science Research, 30*(4), 476–503. https://doi.org/10.1080/13511610.2015.1070090

Kumar, K., Singh, R., Sharma, A., & Lehri, B. P. S. (2024). Climate change adaptation strategies for achieving net-zero economy. In R. Singh, & D. Crowther (Eds.), *Transition Towards a Sustainable Future. Approaches to Global Sustainability, Markets, and Governance.* Springer. https://doi.org/10.1007/978-981-97-5756-5_8

Liao, A.-T., Pan, C.-L., & Zhiying, W. (2024). Digital transformation and innovation and business ecosystems: A bibliometric analysis for conceptual insights and collaborative practices for ecosystem innovation. *International Journal of Innovation Studies.* https://doi.org/10.1016/j.ijis.2024.04.003

Liu, T., Guan, X., Wang, Z., Qin, T., Sun, R., & Wang, Y. (2024). *Technological Forecasting and Social Change, 209*, 123812. https://doi.org/10.1016/j.techfore.2024.123812

McKinsey. (2024) What is net zero? https://www.mckinsey.com/featured-insights/mckinsey-explainers/what-is-net-zero

Möslinger, M., Ulpiani, G., & Vetters, N. (2023). Circular economy and waste management to empower a climate-neutral urban future. *Journal of Cleaner Production, 421*, 138454, https://doi.org/10.1016/j.jclepro.2023.138454

Nayak, D. V., Arun Kumar, A., Suryadevara, R., & Khan, S. (2024). Financing the transformation to net zero goals: IREDA investment for sustainable initiatives. In R. Singh, & D. Crowther (Eds.), *Transition Towards a Sustainable Future. Approaches to Global Sustainability, Markets, and Governance.* Springer. https://doi.org/10.1007/978-981-97-5756-5_4

Nwokolo, S. C., Singh, R., Khan, S., Kumar, A., & Luthra, S. (2023). Africa's awakening to climate action. In *Africa's Path to Net-Zero. CSR, Sustainability, Ethics & Governance.* Springer. https://doi.org/10.1007/978-3-031-44514-9_9

Nwokolo, S. C., Singh, R., Khan, S., Kumar, A., & Luthra, S. (2023). Global investment and development in Africa. In *Africa's Path to Net-Zero. CSR, Sustainability, Ethics & Governance.* Springer. https://doi.org/10.1007/978-3-031-44514-9_2

Nwokolo, S. C., Singh, R., Khan, S., Kumar, A., & Luthra, S. (2023). Remedies to the challenges of renewable energy deployment in Africa. In *Africa's Path to Net-Zero. CSR, Sustainability, Ethics & Governance.* Springer. https://doi.org/10.1007/978-3-031-44514-9_3.

Nwokolo, S. C., Singh, R., Khan, S., & Kumar, A. (2025). Threats to the rapidity of sustainability transitions posed by technological changes. In *Technological Pathways for Africa's Net-Zero Economy* (pp. 39–75). Elsevier. https://doi.org/10.1016/B978-0-443-31486-5.00002-1

OECD. (2021). *No net zero without SMEs: Exploring the key issues for greening SMEs and green entrepreneurship.* OECD SME and Entrepreneurship Papers, 30. OECD Publishing. https://doi.org/10.1787/bab63915-en

Okolo, V. O., Ohanagorom, M. I., Okocha, E. R., Muoneke, O. B., & Okere, K. I. (2023). Does financing SMEs guarantee inclusive growth and environmental sustainability in the European union? *Heliyon, 9*(4), e15095. https://doi.org/10.1016/j.heliyon.2023.e15095

Pitkänen, K., Antikainen, R., Droste, N., Loiseau, E., Saikku, L., Aissani, L., Hansjürgens, B., Kuikman, P. J., Leskinen, P., & Thomsen, M. (2016).What can be learned from practical cases of green economy?–studies from five European countries. *Journal of Cleaner Production, 139.* https://doi.org/10.1016/j.jclepro.2016.08.071

Rathnasiri, M. S. H., Dewasiri, N. J., & Kumar, A. A. (2024). Policy framework and implementation strategies for Sri Lanka's transition to a net-zero economy. In R. Singh, S. Khan, A. Kumar, S. Luthra, & H. Chokshi (Eds.), *Net Zero Economy, Corporate Social Responsibility and Sustainable Value Creation. CSR, Sustainability, Ethics & Governance.* Springer. https://doi.org/10.1007/978-3-031-55779-8_3

Reike, D., Vermeulen, W. J. V., & Witjes, S. (2018). The circular economy: New or refurbished as CE 3.0?—exploring controversies in the conceptualization of the circular economy through a focus on history and resource value retention options. *Resources, Conservation and Recycling, 135*, 246–264. https://doi.org/10.1016/j.resconrec.2017.08.027

Sang, Y., Xie, M., Bai, X., & Guo, F. (2024). Does natural resource dependence influence the impact of financial technologies on corporate ESG and digital governance in China's listed enterprises? *Resources Policy, 91*, 104948. https://doi.org/10.1016/j.resourpol.2024.104948

Singh, R., Khan, S., & Centobelli, P. (2022). Investigating the interplay between social performance and organisational factors supporting circular economy practices. *Sustainability, 14*(24), 16781. https://doi.org/10.3390/su14241678

Singh, R., Khan, S., & Dsilva, J. (2023). A framework for assessment of critical factor for circular economy practice implementation. *Journal of Modelling in Management, 18*(5), 476–1497. https://doi.org/10.1108/JM2-06-2021-0145

Singh, R., Gupta, A., Bajpai, A., & Kandpal, V. (2024). Exploring challenges of circular economy initiatives for smart cities. In V. Kandpal, E. D. Santibanez-Gonzalez, P. Chatterjee, & M. K. Nallapaneni (Eds.), *Smart Cities and Circular Economy, Emerald Publishing Limited, Leeds* (pp. 59–69). https://doi.org/10.1108/978-1-83797-957-820241005

Taleb, M. A., & Farooque, O. A. (2021). Towards a circular economy for sustainable development: An application of full cost accounting to municipal waste recyclables. *Journal of Cleaner Production, 280*(2), 124047. https://doi.org/10.1016/j.jclepro.2020.124047

Velenturf, A. P. M., & Purnell, P. (2021). Principles for a sustainable circular economy. *Sustainable Production and Consumption, 27*, 1437–1457.

Viglioglia, M., Giovanardi, M., Pollo, R., & Peruccio, P. P. (2021). Smart district and circular economy: The role of ICT solutions in promoting circular cities. *Sustainability, 13*(21), 11732. https://doi.org/10.3390/su132111732

Wang, B., Khan, I., Ge, C., & Naz, H. (2024a). Digital transformation of enterprises promotes green technology innovation–the regulated mediation model. *Technological Forecasting and Social Change, 209*, 123812. https://doi.org/10.1016/j.techfore.2024.123812

Wang, D., Chen, L., & Dong, L. (2024b). A critical review of climate change mitigation policies in the EU based on vertical, horizontal and policy instrument perspectives. *Journal of Cleaner Production, 467*, 142972. https://doi.org/10.1016/j.jclepro.2024.142972

WRI. (2024). Rising to the climate change challenge WRI works with governments, businesses and civil society to stabilize the climate so that people and nature can thrive. https://www.wri.org/climate#:~:text=The%20Climate%20Program%20at%20WRI,to%20a%20zero%2Dcarbon%20economy.

Yaman, C. (2020). Investigation of greenhouse gas emissions and energy recovery potential from municipal solid waste management practices. *Environmental Development, 33*, 100484. https://doi.org/10.1016/j.envdev.2019.100484

Chapter 7
Leveraging Artificial Intelligence for Climate Neutrality: Human-Centric Mitigation Strategies to Combat Global Warming

Akshay Singh, Rubee Singh, and Raghuveer Katragadda

Abstract In this research chapter, we discuss the idea of climate change mitigation as climate change with the use of AI to support human centred initiatives to combat global warming. While dangerous, such further rise in the world temperature and global warming beyond 1.5 pre-industrial levels is unavoidable, and any further rise takes place to the detriment of such sensitive areas as this idea outweighs it. One example it briefly describes is that AI uses such databases to predict climate, how AI can help optimize renewable energy, and how sporadic resources can be directed to reduce carbon footprints. Therefore, the authors propose integrating AI with humans, who present how they approached the main themes theoretically and practically, the focus on decentralization and inclusion and development in global contexts for effective application of AI in different socio economic settings. Globally recognized examples of such issues include algorithmic bias, data constraints, disparity of information and communication technology access. In so saying, the paper asserts the need of international policies, of climatized polices presented through international policies, and joint work between AI specialists, climatologists and policy makes. Future studies must augment AI capacities, normative theories, and favorable policies, but need to convene cross disciplinary research to involve stakeholders constantly, in order that the AI will be a helpful tool in the direction towards climate neutrality.

Keywords AI · Climate change · Mitigation strategies · Human-centered approaches

---

A. Singh
School of Management Studies, Indra Gandhi National Open University, New Delhi, India

R. Singh (✉)
Institute of Business Management, GLA University, Mathura, Uttar Pradesh, India
e-mail: dr.rubeerajput@gmail.com

Department of Management Studies, Kumaun University, Nainital, Uttarakhand, India

R. Katragadda
School of Management, V.R. Siddhartha Engineering College, Vijayawada, India

R. Singh and W. L. Filho (eds.), *Climate Neutrality Through Smart Eco-Innovation and Environmental Sustainability*, Climate Change Management,
https://doi.org/10.1007/978-3-031-83250-5_7

Introduction

Overview of the Current State of Global Warming and Its Impacts

Global warming is slow increase in the temperature of surface of the earth due to different anthropogenic activities like, emission of greenhouse gases such as; carbon dioxide, methane and nitrous oxide among others and so forth. A rate of increase of 1.2 °C in global mean surface temperatures from the pre-industrial revolution has been observed. These include the increase in the intensity and frequency of hot extremes, the sea level rise, the reduction of arctic Sea ice, and change in ecosystems and bio diversity. Most especially in the African continent, the effects of climate change are more severe including frequent occurrences of drought, food scarcity and other social economic problems as studies have mandated Africa to forcefully address the challenge of climate change and its effects (Kaack et al., 2022; Nwokolo et al., 2024).

The IPCC stresses that the most severe impacts of climate change can be prevented only if the global temperature rise is limited to 1.5 °C. Global initiatives are now directed towards attaining a net zero economy, a process that demands many changes and initiatives, especially in developing nations that have their own special needs concerning climate actions because of conditions such as lack of infrastructure and resources (Piyathanavong et al., 2024).

Climate neutrality or carbon neutrality is the situation whereby the total GHG emissions produced by a given entity in a fiscal year equal to zero. This is done by matching the GHGs that are being released to the environment with that which is being absorbed or avoided thereby. Climate neutrality is important in order to stop the climate change and its negative impacts on the planet. The Paris Agreement signed in 2015 calls for the decarbonisation of the global economy by mid-century to hold the increase in global temperature to well below 2 °C above pre-industrial levels and to pursue efforts towards 1.5 °C. Developing countries, in particular, face a dual challenge: Chasing economic development at the same time targeting to achieve net-zero emissions calls for unique policy interventions and multilateral agreements (Khan et al., 2024; Singh et al., 2024).

Human-Induced Factors Contributing to Global Warming

Recent global warming is mainly attributed to human activities. The use of fossil fuels in power generation and vehicles add to the atmosphere a large quantity of CO_2. Deforestation and land use changes, heavy proportions of which are due to agriculture, decrease the Earth's ability to absorb CO_2 which leads to an increase in greenhouse gases. Borrowing from agricultural activities, CH_4 and N_2O are emitted to the atmosphere worsening the effects of global warming. Industrial processes and waste management also produce different GHGs that rein force the greenhouse effect and produce global warming. The effects of these activities are most disastrous in areas that have weak ecosystems for instance the African region where both environmental and economic systems are easily affected by climate change (Nwokolo et al., 2024). These human activities need to be changed to reduce GHG emissions and support sustainable practices for climate change mitigation especially in economies that are resource based (Kaack et al., 2022).

Need for Human-Centered Approaches to Mitigation

Combating climate change requires behavioural, policy, and technological changes that can be directed at people. This entails changing the energy suppliers to renewable sources, improving energy efficiency, encouraging sustainable land management and use, and putting in place policies that encourage the use of low carbon technologies (Singh et al., 2024). In Africa, for instance, the use of renewable energy has been identified to be a major strategy towards combating carbon footprint and energy access deficit, but there are several challenges that beset it (Nwokolo et al., 2023d). Involving people and organisations in climate change action helps make the solutions fair and efficient to implement based on the requirements of various groups in the society (Sovacool et al., 2019).

Emergence of Artificial Intelligence (AI) in Climate Solutions

AI, is the study of making machines with the capability to think, learn, solve problems, and perform tasks that would have otherwise been done by human beings. In the area of climate change work, AI can contribute to the simulation of climate processes, energy management, and increasing the effectiveness of renewable energy solutions. Some of the applications of AI algorithms include; Climate prediction through the use of big data, tracking environmental changes, and formulation of measures to be taken in the face of climate change (Joshi et al., 2024).

Rationale for Integrating AI with Human-Driven Mitigation Strategies

This research paper clinches that incorporating AI into already human-led strategies for climate change mitigation can augment the efficacy and efficiency of these strategies. AI can analyze large and intricate data to support policy making, management of resources and generation of strategies that can help reduce emissions. For instance,

AI can enhance the operation and maintenance of energy networks, establish the availability of renewable energy resources, and enhance the logistics of transport systems with a view of minimizing the carbon prints. Yet, the principles of fairness and justice–including data ownership and privacy as well as algorithmic justice in climate action–have not elicited much concern (Raghav et al., 2024; Vinuesa et al., 2020).

Literature Review

Overview of Climate Neutrality and Global Warming

Climate neutrality or net zero is a state in which a fixed volume of greenhouse gases are emitted into the environment and removed or are sequestered in equal volumes. As the consequences from climate change upon ecosystems, resources, and lives continue to intensify, this idea has become topical in the world (Nwokolo et al., 2023c). Robustness of such a policy has been dictated by the long absence of an effective and efficient climate change mitigation policy, which is aimed at inducing climate neutrality, through a hodgepodge of policies, innovations and sustainable practices in diverse sectors of the economy (Bhushan & Singh, 2024). Over the past century leaders of the world have put forward pledges for cutting emission rates to climate change outcomes and AI has been viewed as a key enabler of improvement of renewable energy and technology for carbon capture (Nwokolo et al., 2023d; Raghav et al., 2024). However, technology alone cannot provide climate neutrality, people-centred approaches are also important. It is important to use AI applications in line with and to support human led efforts to reduce emissions and manage impacts from climate change that are otherwise associated with such efforts (Raghav et al., 2024; Yadava et al., 2022).

Achieving Climate Neutrality and Global Policies

Climate neutrality became an internationally recognized concept after the signing of the Paris Agreement in 2015 that aimed at the global warming of no more than 1.5 °C Pre-industrial levels. Climate Neutrality in this context means the goal of attaining net-zero emissions by the middle of the century, a goal supported by global economic powerhouses including the European Union, the United States of America and China. Stated policies of climate neutrality involve: Decarbonization, which is achieved through promotion of renewable energies, energy efficiency improvement and increase of carbon removals by means of afforestation and carbon capture technologies (Nwokolo et al., 2023b; Zeshan et al., 2024). For instance, The European Green Deal published in 2020, lays out a plan for the EU to achieve climate neutrality by 2050 through putting investments in green innovation and competitive and sustainable agriculture.

However, implementing these ambitious policies is still difficult because dependence on fossil energy sources and the use of renewable energy sources are still limited in many countries including the developing ones. The policy frameworks and the actual practice tend to be rather different in many ways. This is despite the fact that global policy commitments have been on the rise; actual carbon emissions remain on the rise indicating the gap between policy and practice (Rolnick et al., 2022). Other factors which also have an impact on the process include Socio-economic in this case due to the fact that the economy of the two areas is mostly pulled by oil and gas exploration.

Integrating AI with Human-Centric Mitigation Strategies

Codifying technology-based approaches to climate change is essential because AI, in particular, can greatly contribute to achieving climate neutrality, but only if technology is integrated with human factors. AI in achieving climate neutrality: AI tools on their own cannot do the job, and must be guided by people and by certain sets of principles concerning oversight, governance and ethics, including fairness and responsible use. Research also shows that, AI-based interventions have to be equitable in order to solve the climate change which is a challenge in most deprived communities (Tariq, 2024; Vinuesa et al., 2020).

Hence, the social and ethical implication of AI in climate change policies are crucial factors, for example, the high computational needs for AI models that can lead to the generation of carbon footprint is counter to their objective. This challenge points to the importance of energy conscious Artificial Intelligence and green computing. Thus, the present study should extend the discussion of AI applications in climate action by considering cultural and social factors to guarantee that technology is implemented with respect to human-centered mitigation measures. Systems which involve the participation of the AI scientists, regulators and stakeholders in the society are critical in developing problem-focused solutions that are sustainable and applicable on a large scale (Habib et al., 2024; Vinuesa et al., 2020).

Thus, the further climate neutral pathway should be based on the synergy of human-centric and cognitive AI-based approaches to climate change mitigation. The global policies define the action plan, whereas the AI increases productivity and flexibility. Nevertheless, to make the most of AI in climate mitigation, it has to be done in a right manner that is inclusive, culturally sensitive and equitable. This integrated approach suggests that AI should be used as a way to enhance community participation in fighting climate change worldwide and not as their replacement (Habib et al., 2024).

Applications of AI in Climate Mitigation

Existing AI Applications in Environmental Monitoring, Energy Management, and Carbon Reduction

AI has become an influential technology that aids climate change solutions in all industries through environment management, energy efficiency, and carbon footprint. In environmental monitoring, the AI algorithms analyze satellite images and sensor data to assess the impact of deforestation; count the species; and track changes in ecosystems, including in real-time (Yadava et al., 2022). These technologies facilitate monitoring of movement of animals and transfer of relevant information to the managers and conservationists.

In energy management, AI improves the grid management, incorporates the renewable energy and predicts energy consumption which greatly improves efficiency. For example, predictive algorithms regulate consumption of energy in the smart grid to meet the demand as well as to avoid energy wastage (Ahmad et al., 2021; Alder et al., 2024). In the area of carbon reduction, AI is applied in the development of better materials and procedures in industries and reducing carbon impact on industries (Ansari et al., 2024; Tariq, 2024).

Challenges and Gaps in Current Research

Limitations in AI-Driven Climate Mitigation

However, there is a number of constraints that can be revealed regarding the use of AI in climate mitigation. First of all, AI models need big and high-quality datasets with which AI models can operate efficiently, and such a dataset may not be available, or it may be of poor quality in some regions. This issue hampers the precision of the model and hence the viability of AI outputs in various environmental conditions (Piyathanavong et al., 2024; Raghav et al., 2024). Moreover, the training of the AI models is process which requires a large amount of computational resources that consume considerable energy and thus produced carbon footprint, which is, however, a key issue that AI is aimed to prevent (Nordgren, 2022; Sovacool et al., 2019). A second type of bias is algorithmic bias. Climate outcomes based on such AI models may be unfair as they are trained on biased data leading to unfavorable implications for vulnerable populations (Nordgren, 2022). These biases mean AI algorithms must be made more transparent and responsible in order to introduce fairness in the climate change solutions.

Gaps in Integrating AI with Human-Centric Approaches

Another challenge is the combination of AI with humanistic climate change management approaches. First, the collaboration between the AI professionals and the climate scientists is rare, hence, there is no proper ways of sharing knowledge and ideas. This gap can prevent developing AI models that can assist human efforts for mitigation.

The above part also raises issues to do with ethics in the integration of AI. Applying AI in climate mitigation must not be counterproductive to the social justice principles that underpin its use. Nonetheless, there are no systematic approaches that can help to manage the ethical use of AI in climate change settings (Vinuesa et al., 2020). However, the expansion of AI solutions for SDG applications at the international level requires reliable infrastructure and capital, which may not be easily available in developing countries, hence the inequality in the distribution of AI outcomes (Nishant et al., 2020; Nwokolo et al., 2023a, 2023b, 2023c, 2023d).

Conceptual Framework for AI-Driven Climate Neutrality

To effectively utilise AI in attaining climate neutrality, a framework that incorporates usage of AI in climate data, human centered climate change solutions and AI supported policies must be adopted. Modern techniques of data acquisition allow for constant observation of the environment dynamics and immediate reaction (Singh et al., 2022a, 2022b). The human-centric AI applications encourage behavior change through the use of feedback and engage sustainable habits (Joshi et al., 2024). Also, AI enhances the formulation of policies when it comes to data analysis that may be challenging to understand, thus comes up with strategies that are consonant with climate change objectives (Ghaffarian et al., 2023). It also appears that any such implementation of AI might require collaboration with a number of different parties to be optimally useful (Gaur et al., 2023).

There is therefore potential of AI in the area of climate data collection and observation that if well harnessed may help in enhancing the real time knowledge and prediction on climate change as a way of achieving climate change neutrality (Joshi et al., 2024). The availability of climate data is enhanced by AI through the use of satellites and IoT sensor information. Satellites have made it easier to get detailed spatial data in monitoring large scale environmental shifts such as deforestation, glacier melt, and ocean heat and on the other hand IoT sensors collect data on the ground, including heat islands in cities, air quality, and moisture in the soil (Singh et al., 2022a, 2022b). From these sources, AI can assist in giving climate information that can assist in decision making on the sustainability agenda.

These advanced Smart Systems with the help of AI, machine learning, deep learning etc. shall make a real-time monitoring of data by collecting large volumes of data and then categorizing and analyzing the pattern of climate change (Nwokolo et al., 2023a). For instance, neural networks in AI-based climate models predict the signs of climate change and other climatic disturbances (Joshi et al., 2024). These models incorporate historical climate data and real-time information to provide operational climate outlooks and hazards outlooks. Thus, through enhancing climate predictions, AI helps develop a preventive mechanism through which societies and governments prepare for potential climate-related dangers thereby promoting prevention measures (Nwokolo et al., 2024).

Moreover, AI in climate monitoring allows adaptive measures because it provides information on the regional and global levels. This adaptive capacity is crucial for specific human-centred mitigation measures that focus on the groups most affected by climate change (Zhao & Fariñas, 2022). In particular, AI-driven climate monitoring is an important tool for climate neutrality and endowing societies with the tools to tackle climate change proactively using data.

Conceptual Representation Diagram for Leveraging AI for Climate Neutrality

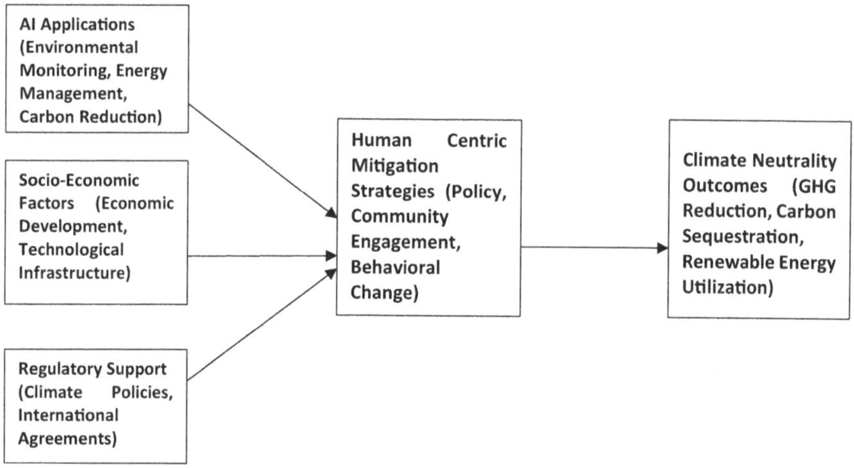

The conceptual variable diagram presented in this paper provides a model for implementing AI in the context of the climate neutrality with Human-Centric Mitigation Strategies in the core of the model. This central element captures the role of policy, community involvement and dynamic change in behaviour towards proper and sustainable climate action. Four elements are linked to this core. AI Applications include environmental monitoring and energy management that are operational tools to decrease carbon footprint and enhance resource utilization. Socio-Economic Factors such as economic development and technological infrastructure affects the environment in which AI operates to make it clear that economic and technological preparedness are critical to the success of AI Climate Change interventions. Climate Neutrality Outcomes, such as Greenhouse Gas (GHG) emission mitigation or Carbon Drawdown, captures the strategic Objective of AI in tackling climate change. Finally, the Role of Regulatory Support as climate policies and/or international agreements provide fundamental the governance and setting which will allow

AI application functioning within the framework of climate change objectives. This structure emphasizes an integrated and people's oriented perspective to realizing AI for a sustainable future.

Human-Centric Mitigation and Behavioral Change: The Role of AI

AI is now crucial for steering the change in the behaviour of people and societies in relation to climate change. Through processing large data sets, AI systems are able to estimate consequences of people's actions on the environment and, consequently, offer targeted measures that encourage pro-environmental behavior. AI creates individualised feedback mechanisms which increases awareness of an individual's carbon footprint, and also provides recommendations on how to reduce carbon footprint; this increases the level of accountability and engagement (Bhushan & Singh, 2024; Kumar et al., 2024). Furthermore, the AI systems assist communities in adopting climate risk visualization to foster community actions and sustainable (Singh et al., 2022a, 2022b).

This can be done since AI has the ability to use real time data to create messages that are most likely to be seen by the various targets. NLP algorithms can analyze the public mood, and thus AI can present the audience with content that focuses on sustainable behavior while also taking into account regional environmental issues (Zhao & Gómez Fariñas, 2022). Also, Virtual personal assistant, and chatbots, can use AI to engage with the user and give them relevant information pertaining to the environment, creating an environmental awareness culture (Gaur et al., 2023). With these strategies, AI not only educates but engages and enables people and societies to be part of climate neutrality actions. In essence, AI provides a pluralistic means of achieving humanistic solutions to mitigation through endorsement of conduct consistent with environmental objectives. When applied properly, AI-generated knowledge can therefore assist in changing people's decisions and social practices and move the world towards sustainability.

Implementation and Operationalization of AI Solutions

A systematic approach, inter-disciplinary collaboration, and stakeholder engagement are needed to bridge the gap between artificial intelligence (AI), as Singh et al. (2022a, 2022b), and Vinuesa et al. (2020) stated. Idea implementation with AI fundamentally asks the question of data quality, domain specific model selection, addressing the bias with data and promoting the operational needs. Involving feedback mechanisms is crucial for a continuous improvement to fight climate challenges. Machine learning, for example, contributes to renewable energy because it improves wind and solar power predictions, simultaneously reducing energy loss and improving grid stability. AI powered platforms for storage and distribution of energy further improve power supply efficiency (Singh et al., 2022a, 2022b). AI helps emissions reduction in industrial applications by tracking, inspecting, and performing predictive maintenance reducing energy and carbon foot prints. In addition, AI driven resource management facilitates water saving measures as well as crop breeding strategies for resilience to climate (Gaur et al., 2023). Ethical and political concerns are important, in addition to technological readiness, to mitigate biased use of AI and foster a climate action sufficiently sustainable for human values.

Risks of AI Misuse or Over-Reliance

Moreover, this has come amid concerns of over or misuse of AI in climate change mitigation. In contrast to humans, AI can process huge quantities of data to get quicker insight than people, but if entrepreneurs over-rely on these systems they may discount risks in climate-related decision making. Such dependency could further erode environmental science and climate policy domain expertise, resulting in decisions that fail to take context sensitivity or uncertainty outside of AI's realm of predictive scope (Rolnick et al., 2022). In fact, these challenges have only been exacerbated by misapplication of AI, like greenwashing. For example, organizations might take the steps toward reducing their carbon footprint, but utilize AI driven carbon footprint calculators to showcase emission reductions even while doing other polluting practices (Singh et al., 2024). It corrodes stakeholder trust and blocks real climate action. In order to solve this problem ethical norms and standards of accountability must guarantee that AI solutions enhance environmental outcomes as opposed to being misrepresented as doing so (Cowls et al., 2021).

Moreover, AI surveillance for environmental monitoring like monitoring the deforestation or emissions as it is very extensive to collect and store data, also has privacy risks. Robust policies and ethical frameworks are needed to balance the level of monitoring effectiveness with individual rights (Nwokolo et al., 2023a, 2023b, 2023c, 2023d; Olawade et al., 2024) as it is to safeguarding the individual privacy. AI applications must put the emphasis of ethics ahead of being misused and over dependence by ethically working with the principle of responsibility, transparency, fairness and human oriented. AI solutions must avoid bias, focus on real environmental gains and be overseen by humans, say researchers and practitioners. But to do so effectively, AI must be used as a tool for ethical stewardship and towards positive contributions to climate neutrality (Olawade et al., 2024).

Potential and Limitations of AI in Achieving Climate Neutrality

AI is a powerful tool that can help to enhance the effectiveness of climate neutrality by working with data, minimizing waste and offering timely inputs on how to reduce carbon footprint. Thanks to AI, energy consumption can be predicted, renewable energy generation may be anticipated, and supply chains may be managed in ways that minimize emissions and wastage of resources (Rolnick et al., 2022). For instance, the AI models help increase the efficiency of wind/solar energies by employing the weather and load forecasting for the energy storage and reduced reliance on fossil energies (Vinuesa et al., 2020). Nevertheless, the challenges of AI in climate action should not be overlooked such as; inadequate data, algorithms' prejudice and high computational needs. These factors can unintentionally prevent sustainable results because they depend on strong data infrastructure and the ability to obtain high-quality, extensive environmental data that can be lacking in some areas (Nwokolo et al., 2023d).

Also, there are still ethical and governance issues that are still an issue of concern to the organization. AI, if not well implemented with consideration of the target groups likely to be impacted by climate change, can only worsen the existing social economic divides (Nwokolo et al., 2023c). To this end, AI frameworks that are clear and non-discriminatory when it comes to the generation of low-carbon outcomes must be developed in order to avoid the adverse effects of AI.

Analysis of AI's Contribution to Climate Goals and Future Support

AI applications have positively impacted climate objectives, primarily in tracking emissions and boosting RE performance. For instance, AI based sensors and machine learning can accurately track and predict emissions of greenhouse gases in industrial settings (Gaur et al., 2023). Having incorporated AI into emissions tracking and predictive maintenance, industries can cut down on emissions and increase energy efficiency (Habib et al., 2024). Future development of AI is expected to enhance these contributions in the future. New application areas of predictive analytics and autonomous energy management can provide opportunities to decouple demand from supply by moving consumption to off peak periods thus easing the burden on the grid and increasing the utilization of renewable energy.

Nevertheless, to accomplish climate objectives, it is not enough to develop technical AI solutions; it is essential to harmonize AI potential with the socio-political context of interaction. AI can also support the improvement of decision-making in climate policy as it can simulate a range of different approaches to climate change mitigation and the consequences of these approaches on different populations (Vinuesa et al., 2020). But, the awareness of AI in achieving climate objectives will only be as useful as the integration of artificial machines with human experience to fight complex and interrelated climate problems.

Insights on Synergizing AI with Human Mitigation Strategies

Overarching human-centered responses for addressing emissions reduction, resilience and adaptation intensify the benefits that can be derived from AI. AI can assist enhance and amplify human abilities mainly in climate modeling and data analysis. For example, the application of AI platforms can allow the planners to create better layouts for cities which will minimize the energy consumption, while incorporating the human understanding of needs of the communities and residents (Rolnick et al., 2022). In farming, AI technology assist farmers in tracking the aspects of soil and water consumption in order to harvest high value per acre while using low resources. Here, the human element is employed in the interpretation of the recommendations made by AI in the domain of agriculture particular to the area (Gaur et al., 2023). One of the most powerful insights here presents the fact that the AI systems cannot work properly without interactions with human inputs and control. Local climate risk analysis, socio-economic factors, ecological, and cultural differences can all be handled better by human expertise and incorporated when incorporating the AI solutions. This, turns makes the use of AI tools for climate action more effective by flexibility to accommodate the environmental and social conditions of the regions.

Academic and Managerial Implications

For academia, it is important to adopt cooperation research covering aspects of both AI and environmental sciences, sociological and ethical research. It could open the way for the creation of the next generation of AI tools, co-designed with climate neutrality and tailored for multiple regions (Ghaffarian et al., 2023). Academics could also pitch in by creating guidelines for the right use of AI within the environmental space while also advocating for an appropriate integration of AI towards achievement of human-oriented climate targets.

For managers, therefore, the result of this study is a message that cultivating AI-readiness needs to start with addressing the issue of sustainable development. In order to pursue both AI applications and sustainability goals, managers need to incorporate AI into sustainability objectives and form teams comprising technical experts, climate science and policy specialists. Collaborating with participants from other disciplines improves the translatable application of AI in relation to specific climatic issues affecting specific industries, such as power, farming, transportation, and production (Vinuesa et al., 2020).

Conclusion

Situating Climate Neutrality within the context of Artificial Intelligence, this research underscores the importance of AI for sustainability and more importantly rolls out Human-Centered Artificial Intelligence as an avenue that would underscore Climate Neutrality. AI can be particularly beneficial in orchestrating diverse climate action frameworks catapulting emissions reduction, renewable energy utilization as well as resource efficiency. Main discoveries suggest that the utilization of AI can provide valuable contribution to the accuracy of renewable energy predictions, the stability of the grid, and efficient usage of resources. AI also applies in real-time monitoring of the emission intensive industries for predictive maintenance and lower energy consumption which in real term influences the emissions directly (Kumari & Pandey, 2022). Using large data sets, AI insights contribute to better adapted and less vulnerable climate initiatives, making it a tool for climate neutrality (Kaack et al., 2022). However, the AI use enshrouds complexity that calls for cross-sector integration and continuous engagement of stakeholders, a result that points to a key need of human centred and ethical framework (Vinuesa et al., 2020).

Future Research Directions

This research chapter recognizes the possibilities of Artificial Intelligence (AI) to advance climate neutrality and highlights Human Centered AI as a must have. Climate action frameworks enabled by the use of AI improve emissions reduction, renewable energy use, and resource efficiency. Vinuesa et al. (2020) noted on how AI can help boost the accuracy of renewable energy forecasts, stabilize power grids, and fine tune resource use. In emission intensive industries, artificial intelligence helps perform online monitoring, for predictive maintenance and decreases energy consumption (Kumari & Pandey, 2022). AI, however, relies on large datasets to deliver insights toward resilient climate initiatives, and therefore constitutes an important tool of sustainability (Kaack et al., 2022). Yet cross sector integration and the ethical frameworks remain critical (Vinuesa et al., 2020).

Future work should also refine machine learning algorithms, making predictions of renewable energy and climate more robust in general, and adaptable across different environments (Habib et al., 2024). One can use improved integration of AI and IoT sensors to make real time data available for resource efficient strategies (Nishant et al., 2020). It would enable behavioral studies to see how AI driving recommendations can enhance energy consumption, at individual and organizational levels, and would help in better adoption strategies (Nwokolo et al., 2023a, 2023b, 2023c, 2023d). Jurisprudential research into the AI in emissions reduction and renewable energy should also consider ethical issues, such as data privacy and algorithmic bias (Rolnick et al., 2022), as well as the conditions under which AI can substitute for, or to some extent complement, human efforts and expertise in technology development, planning, and implementation.

But AI innovations and societal and environmental goals should be aligned and their integration designed by conjunction of climate scientists, AI experts and policy-makers. This research highlights and contributes to the multiple functions that AI can play is to combat climate change and lays out a path for using technology, science, and policy to create sustainable and resilient environments. More progress in these areas will enable the utilization of AI in reaching climate neutrality and the scenario of eradication of global warming.

References

Ahmad, T., Zhang, D., Huang, C., Zhang, H., Dai, N., Song, Y., & Chen, H. (2021). Artificial intelligence in sustainable energy industry: Status Quo, challenges and opportunities. *Journal of Cleaner Production, 289*, 125834. https://doi.org/10.1016/j.jclepro.2021.125834

Alder, N., Ebert, K., Herbrich, R., & Hacker, P. (2024). AI, climate, and transparency: Operationalizing and improving the AI act. *arXiv (Cornell University)*. https://doi.org/10.48550/arxiv.2409.07471

Ansari, F. A., Haque, S. N., Khan, S., & Khan, W. (2024). Modeling the enablers of implementing green HRM practices: An ISM-MICMAC approach. *Corporate Social Responsibility and Environmental Management*. https://doi.org/10.1002/csr.2913

Bhushan, M., & Singh, R. (2024). Talent management facilitates net-zero transition through employee green behavior. In R. Singh, S. Khan, A. Kumar, S. Luthra, & H. Chokshi (Eds.), *Net zero economy, corporate social responsibility and sustainable value creation. CSR, sustainability, ethics & governance*. Cham: Springer. https://doi.org/10.1007/978-3-031-55779-8_7

Cowls, J., Tsamados, A., Taddeo, M., & Floridi, L. (2021). The AI gambit: Leveraging artificial intelligence to combat climate change—opportunities, challenges, and recommendations. *AI & Society, 38*(1), 283–307. https://doi.org/10.1007/s00146-021-01294-x

Gaur, L., Afaq, A., Arora, G. K., & Khan, N. (2023). Artificial intelligence for carbon emissions using system of systems theory. *Ecological Informatics, 76*, 102165. https://doi.org/10.1016/j.ecoinf.2023.102165

Ghaffarian, S., Taghikhah, F. R., & Maier, H. R. (2023). Explainable artificial intelligence in disaster risk management: Achievements and prospective futures. *International Journal of Disaster Risk Reduction, 98*, 104123. https://doi.org/10.1016/j.ijdrr.2023.104123

Habib, M., Habib, A., Albzaie, M., & Farghal, A. (2024). Sustainability benefits of AI-based engineering solutions for infrastructure resilience in arid regions against extreme rainfall events. *Discover Sustainability, 5*(1). https://doi.org/10.1007/s43621-024-00500-2

Joshi, A., Singh, R., Kumar, V., & Rani, S. (2024). Net zero strategy: Exploring opportunities, confronting challenges for developing countries. In *Approaches to global sustainability, markets, and governance* (pp. 233–252). https://doi.org/10.1007/978-981-97-5756-5_11

Kaack, L. H., Donti, P. L., Strubell, E., Kamiya, G., Creutzig, F., & Rolnick, D. (2022). Aligning artificial intelligence with climate change mitigation. *Nature Climate Change, 12*(6), 518–527. https://doi.org/10.1038/s41558-022-01377-7

Khan, S., Singh, R., Alnahas, J., Abbate, S., & Centobelli, P. (2024). Navigating the smart circular economy: A framework for manufacturing firms. *Journal of Cleaner Production*, 144007. https://doi.org/10.1016/j.jclepro.2024.144007

Kumar, K., Singh, R., Sharma, A., & Lehri, B. P. S. (2024). Climate change adaptation strategies for achieving net-zero economy. In *Approaches to global sustainability, markets, and governance* (pp. 163–183). https://doi.org/10.1007/978-981-97-5756-5_8

Kumari, N., & Pandey, S. (2022). Application of artificial intelligence in environmental sustainability and climate change. In *Elsevier eBooks* (pp. 293–316). https://doi.org/10.1016/b978-0-323-99714-0.00018-2

Nishant, R., Kennedy, M., & Corbett, J. (2020). Artificial intelligence for sustainability: Challenges, opportunities, and a research agenda. *International Journal of Information Management, 53*, 102104. https://doi.org/10.1016/j.ijinfomgt.2020.102104

Nordgren, A. (2022). Artificial intelligence and climate change: Ethical issues. *Journal of Information Communication and Ethics in Society, 21*(1), 1–15. https://doi.org/10.1108/jices-11-2021-0106

Nwokolo, S. C., Singh, R., Khan, S., Kumar, A., & Luthra, S. (2023a). Africa's awakening to climate action. In *CSR, sustainability, ethics & governance* (pp. 299–310). https://doi.org/10.1007/978-3-031-44514-9_9

Nwokolo, S. C., Singh, R., Khan, S., Kumar, A., & Luthra, S. (2023b). Global investment and development in Africa. In *CSR, sustainability, ethics & governance* (pp. 15–58). https://doi.org/10.1007/978-3-031-44514-9_2

Nwokolo, S. C., Singh, R., Khan, S., Kumar, A., & Luthra, S. (2023c). Impacts of climate change in Africa. In *CSR, sustainability, ethics & governance* (pp. 237–262). https://doi.org/10.1007/978-3-031-44514-9_7

Nwokolo, S. C., Singh, R., Khan, S., Kumar, A., & Luthra, S. (2023d). Remedies to the challenges of renewable energy deployment in Africa. In *CSR, sustainability, ethics & governance* (pp. 59–74). https://doi.org/10.1007/978-3-031-44514-9_3

Nwokolo, S. C., Singh, R., Khan, S., & Kumar, A. (2024). Threats to the rapidity of sustainability transitions posed by technological changes. In *Elsevier eBooks* (pp. 39–75). https://doi.org/10.1016/b978-0-443-31486-5.00002-1

Olawade, D. B., Wada, O. Z., Ige, A. O., Egbewole, B. I., Olojo, A., & Oladapo, B. I. (2024). Artificial intelligence in environmental monitoring: Advancements, challenges, and future directions. *Hygiene and Environmental Health Advances*, 100114. https://doi.org/10.1016/j.heha.2024.100114

Piyathanavong, V., Olapiriyakul, S., Garza-Reyes, J. A., Kumar, V., Huynh, V., & Karnjana, J. (2024). Implementing industry 4.0 and circular economy through the developmental culture perspective—driving a competitive advantage in the manufacturing industry. *Business Strategy and the Environment*. https://doi.org/10.1002/bse.3967

Raghav, A., Singh, B., Jermsittiparsert, K., Raghav, R., & Yadav, U. (2024). Artificial intelligence in environmental and climate changes. In *Practice, progress, and proficiency in sustainability* (pp. 485–506). https://doi.org/10.4018/979-8-3693-6336-2.ch019

Rolnick, D., Donti, P. L., Kaack, L. H., Kochanski, K., Lacoste, A., Sankaran, K., Ross, A. S., Milojevic-Dupont, N., Jaques, N., Waldman-Brown, A., Luccioni, A. S., Maharaj, T., Sherwin, E. D., Mukkavilli, S. K., Kording, K. P., Gomes, C. P., Ng, A. Y., Hassabis, D., Platt, J. C., & Bengio, Y. (2022). Tackling climate change with machine learning. *ACM Computing Surveys, 55*(2), 1–96. https://doi.org/10.1145/3485128

Singh, R., Crowther, D., & Ravesangar, K. (2024). Harnessing human capital for a sustainable future: HR's pivotal role toward net zero policies. In R. Singh & D. Crowther (Eds.), *Transition towards a sustainable future. Approaches to global sustainability, markets, and governance.* Singapore: Springer. https://doi.org/10.1007/978-981-97-5756-5_2

Singh, P., Sharma, I., & Guleria, A. (2022a). Impact of socioeconomic parameters on adoption of climate resilient technology under varying vulnerability conditions: Evidences from Himalayan region. In *Elsevier eBooks* (pp. 229–246). https://doi.org/10.1016/b978-0-323-99714-0.00011-x

Singh, R., Khan, S., & Dsilva, J. (2022b). A framework for assessment of critical factor for circular economy practice implementation. *Journal of Modelling in Management, 18*(5), 1476–1497. https://doi.org/10.1108/jm2-06-2021-0145

Sovacool, B. K., Hook, A., Martiskainen, M., Brock, A., & Turnheim, B. (2019). The decarbonisation divide: Contextualizing landscapes of low-carbon exploitation and toxicity in Africa. *Global Environmental Change, 60*, 102028. https://doi.org/10.1016/j.gloenvcha.2019.102028

Tariq, M. U. (2024). Leveraging artificial intelligence for a sustainable and climate-neutral economy in Asia. In *Advances in finance, accounting, and economics book series* (pp. 1–21). https://doi.org/10.4018/979-8-3693-1942-0.ch001

Vinuesa, R., Azizpour, H., Leite, I., Balaam, M., Dignum, V., Domisch, S., Felländer, A., Langhans, S. D., Tegmark, M., & Nerini, F. F. (2020). The role of artificial intelligence in achieving the Sustainable Development Goals. *Nature Communications, 11*(1). https://doi.org/10.1038/s41467-019-14108-y

Yadava, P. K., Kumar, H., Singh, A., Kumar, V., & Verma, S. (2022). Impact of climate change on water quality and its assessment. In *Elsevier eBooks* (pp. 39–54). https://doi.org/10.1016/b978-0-323-99714-0.00002-9

Zeshan, M., Morelli, M., Rasool, S., Centobelli, P., & Cerchione, R. (2024). Empowering sustainable workplaces: A perspective on employee well-being in the light of the job demand resource model. *Sustainable Development*. https://doi.org/10.1002/sd.3179

Zhao, J., & Fariñas, B. G. (2022). Artificial intelligence and sustainable decisions. *European Business Organization Law Review, 24*(1), 1–39. https://doi.org/10.1007/s40804-022-00262-2

Chapter 8
The Role of Eco-Innovation in Achieving Climate Neutral Food Systems: The Middle East and North Africa Perspective

Jacinta Dsilva

Abstract Eco-innovation is an emerging transformative technology, aiming to reduce the negative environmental impacts industries have. The aim of this chapter is to explore the critical role of eco-innovation in addressing global challenges such as climate change and food insecurity. Interconnected challenges such as climate change, water scarcity, and growing populations highlight the need for systemic reforms to mitigate climate impacts and improve food system resilience. The four case studies from the MENA region demonstrate the importance of integrating eco-innovation across industries can drive socio-economic growth, competitive advantage, and ultimately help environmental preservation. The chapter highlights key strategies such as the adoption of green technologies, fostering public–private partnerships, implementing sustainable leadership, and promoting customer engagement. Furthermore, emphasising the importance of regulatory frameworks, financial incentives, and robust stakeholder collaboration in fostering sustainable practices. It also identifies challenges such as the inequitable access to technologies and the necessity of stakeholder engagement in emerging economies. Practices such as transitioning to plant-based diets, reducing food waste, and creating resilient food systems are also examined, underscoring the necessity of aligning organisational values with consumer expectations and beliefs for environmental responsibility. For emerging economies, improving technology accessibility is a step towards the effective integration of sustainable solutions; by investing in eco-technology infrastructure and offering training towards the same, these measures can create an environment that serves as a catalyst for innovation, thus addressing region-specific environmental challenges. The chapter concludes by advocating for a holistic approach balancing environmental, social, and economic priorities to work towards the SDGs and achieve global sustainability.

J. Dsilva (✉)
SEE Research Center, SEE Institute, The Sustainable City, Dubai, UAE
e-mail: jacinta@seeinstitute.ae

R. Singh and W. L. Filho (eds.), *Climate Neutrality Through Smart Eco-Innovation and Environmental Sustainability*, Climate Change Management,
https://doi.org/10.1007/978-3-031-83250-5_8

Keywords Eco-innovation · Circular economy · Socio-economic growth · Food systems

Introduction

The World Wildlife Fund (WWF) published the Living Planet Report (2024) stating that the average size of wildlife populations has fallen by 73% since 1970. The World Economic Forum (2018) expands upon this with statistics about our food system being the largest user of land, with over 34% of the Earth being used to produce food. The production of crops has cleared 40% of land that was once forested and uses 69% of all freshwaters (Campari, 2018). This has drained natural commodities like aquifers and deprives communities of traditional water sources. Furthermore, the food system also leads to a substantial amount of greenhouse gas emissions, around 25% (Campari, 2018). Despite the scale of production of food, in 2023, over 282 million people worldwide experienced high food insecurity. Out of these, extreme weather conditions caused 71.9 million people to face acute food insecurity.

Although having a naturally occurring arid climate, the Middle East and North Africa (MENA) region's growing demand for freshwater and land resources is causing elevated levels of stress (Mahmoud, 2024). From the 1950s to 2000, the MENA region's population grew around four times, from 100 to 380 million. Previously abundant resources and large-scale irrigation projects led to developments in the arid region. However, years of poor water management, rising temperatures, and a population explosion have caused land degradation and limited the Middle East's water resources. By 2050, every single country in the MENA region will face high water stress. If average temperatures rise by 4°C, the region will experience a 75% drop in freshwater availability, states the Center for Strategic and International Studies (Hall, 2024).

Countries most vulnerable to the ongoing climate change and least likely to be able to adapt are also those undergoing conflict, like Syria, Yemen, Iraq, Lebanon, and Jordan. Countries with limited freshwater sources try to mitigate this scarcity by expanding utilisation on non-conventional water sources such as desalination and recycling, however this is dependent on a country's financial and development capacity, and hence rare. The water scarcity crisis is also projected to reduce the GDP of many Middle Eastern countries by approximately 14% by 2050. Inspite of the urgency, many governments and stakeholders continue to pursue unsustainable water management practices due to decades of unchecked behaviors and policies that have been ingrained for generations. This has dried up massive waterways such as the Euphrates, the Tigris, and the Jordan Rivers. Over pumping and wastewater contamination has caused freshwater to become brackish and toxic. Regional effects of global warming on the land and sea have direct impacts on water resource management. This in turn, will affect agricultural water used to support food production (Hall, 2024; Mahmoud, 2024).

In 2020, 90.7% of water withdrawal in Yemen, and 87.5% of it in Syria, was for agricultural purposes (Hall, 2024). The increase in warming causes an increase in the rate of transpiration from irrigated crops, upon which irrigation requirements per crop type need to be increased to ensure sufficient water supplies (Mahmoud, 2024). Without regulation and governmental support, many citizens struggle to maintain a livelihood in agriculture, moving to the cities to find better opportunities.

Looking beyond insufficient water availability, this scarcity also has long-term consequences; concerns over water infrastructure and transboundary water cooperation may lead to an influx in the region's socioeconomic challenges (Mahmoud, 2024); Governments are also struggling with urban growth; in Jordan, 91.8% of the population lives in urban areas, and in Lebanon it is 89.3% (Hall, 2024). Thus, extreme climate change, conflicts, and ongoing pandemics have caused a high risk of global food insecurity soon (Kumar et al., 2024). Our current food systems are unsustainable and generate large amounts of food waste (Hassoun et al., 2022). A shift towards sustainable food sourcing and production is urgently needed to meet the Sustainable Development Goals (SDGs).

This chapter explains what current buzzwords in the field of sustainability, such as eco-innovation and climate neutrality, mean before delving into the exploration of eco-innovation technologies being used in the agri-food industry. Through the highlighted case studies, it examines the role of eco-innovation and how sustainable leadership contributes to addressing global climate challenges. The case studies also demonstrate the applicability of innovative strategies across various MENA regions, discussing the need for robust policy frameworks and financial regulation to aid in the development of green strategies (Nwokolo et al., 2023a, 2023b, 2023c). It also identifies challenges such as the inequitable access to technologies and the necessity of stakeholder engagement in emerging economies, before concluding with recommendation for actions towards achieving the SDGs and advocating for the inclusion of circular economy frameworks to balance environmental sustainability as well as the socio-economic growth of firms (Khan et al., 2024).

Literature Review

With attention drawn towards the environment in recent decades by both practitioners and academics alike, concerns towards environmental responsibility are driven by stakeholders. Through this, environmentally friendly products or manufacturing processes called eco-innovations arise. If companies are to make a substantial change in solving environmental problems within the current business model, there is a crucial need to understand how to effectively innovate towards sustainability (Luqmani et al., 2017). One of the earliest definitions of smart eco-innovation refers to it as a new product that offers value to both the company and its consumers while also reducing environmental impacts (Fussler & James, 1996). Such a definition does not account for the technology to reduce pollutants produced but mitigates environmental impacts.

'Eco-Innovation', (or green innovation, environmental innovation, sustainable innovation), includes detailing critical aspects of the process such as innovation product, market orientation, environmental aspect, material flow reduction, economical or ecological impulse, and finally, setting a green standard for the firm (Schiederig et al., 2012). Eco-Innovation can have either a technological or non-technological purpose; where product and process innovations would be considered technological, while marketing and organisational innovations are usually non-technological (Pichlak & Szromek, 2021). Definitions in more recent years have adapted to address the interrelationship between the competitive market and the dynamic interactions companies have during the innovation process. A broader view of eco-innovations extends to changes in socio-cultural norms and institutional structures (Carillo-Hermosilla et al., 2010).

Not all companies would have the resources to innovate independently; Hagedoorn and Schakenraad believe that cooperation is the most time and money efficient way to innovate. Through diversified resources, firms can improve their efficiency and increase profits. In this case, moving away from a more traditional 'closed' innovation model which focuses on internal research and development, an 'open' approach allows firms to utilise outward transfer of knowledge and technologies.

Since the Paris Agreement in 2015 came the steadily rising awareness and acknowledgement of climate change (Kachi et al., 2020). The European Green Deal is the European Union's pledge to achieve net zero greenhouse gas emissions by 2050. Thus, climate neutrality aims to balance emissions, so they are less than or equal to those removed, while accounting for regional or local effects of human activities. 'Climate Neutrality' as a buzzword has become increasingly rampant between consumers and investors alike. Although both governments and private companies target neutrality with varying claims and targets, they are not always transparent about aspects like current emission levels and reduction strategies (Kachi et al., 2020). This makes it difficult to comprehend climate neutrality targets and their impacts on a global level. Furthermore, multiple industries have differing ambitions about their environmental impacts and may misrepresent climate action (Joshi et al., 2024).

Rising temperatures, extreme weather events, and shifting precipitation patterns have reduced crop yields globally. Climate change also adversely affects livestock through heat stress, reduced forage availability, and spread of diseases, further reducing productivity (Rojas-Downing et al., 2017). Climate change poses significant constraints on food supply chains, impacting not only production but also transportation, processing, packaging, and retail. Climate-related shocks disrupt these activities, exacerbating issues of food access and nutrition. Events such as COVID-19 can lead to blockages and breakdowns within supply chains, resulting in adverse dietary and nutritional outcomes (Fanzo et al., 2018). Furthermore, emissions created by these industries exacerbate climate change, accounting for close to 30% of global greenhouse gas emissions (Fanzo & Miachon, 2023), creating a feedback loop that threatens food security. Fossil fuel dependency across food system activities further exacerbates carbon dioxide, methane, and nitrous oxide emissions, which drive global warming. Additionally, agricultural expansion leads to significant biodiversity loss, with tropical deforestation for soybeans, cattle, and palm oil displacing native species and habitats. These interconnected challenges highlight the need for systemic reforms to mitigate climate impacts and improve food system resilience. Without systemic reforms, including mitigation efforts like reducing food waste and transitioning to sustainable practices, achieving global climate and food security goals will be increasingly difficult (Smith et al., 2019).

Climate change poses two critical challenges for global food systems: adapting to climate variables and extreme events and significantly reducing greenhouse gas (GHG) emissions through supply- and demand-side measures. Farm production policies play a crucial role in promoting sustainable food systems. Subsidies for staple crops, such as wheat, rice, and maize, have historically made these calorie-dense foods more affordable and widely available (Zurek et al., 2021). However, these policies often neglect perishable and nutrient-rich foods, creating significant barriers for small-scale producers and farmers. Challenges such as inadequate infrastructure, lack of timely market information, and power asymmetries hinder their ability to deliver perishable foods that meet safety, quality, and pricing standards, particularly in the context of rising global temperatures (Nwokolo et al., 2023a, 2023b, 2023c; Zurek et al., 2021).

Reports emphasise that sustainable use of land and water resources is essential but challenging. Freshwater scarcity and pollution are growing concerns, compounded by the need for increased agricultural productivity and land expansion. Food accessibility is another major concern. Scenarios highlight disparities in food affordability and distribution. While sufficient food is produced globally, inequitable distribution remains a core issue, exacerbated by dependency dynamics between the Global South and Global North. Solutions require enhanced global coordination to share knowledge and technologies, fostering regional self-sufficiency and reducing poverty.

Additionally, the high cost of healthy diets compounds these challenges. Diets that meet nutrient needs and promote health are five times more expensive than those based on starchy staples, limiting access to nutrition for vulnerable populations. Addressing these disparities requires innovative approaches, such as nudging, which subtly influence consumer behavior without restricting choices. Nudges include offering healthier default options, like salads instead of fries at restaurants, or introducing alternative meats that replicate traditional meat products for environmentally conscious consumers (Thaler & Sunstein, 2021).

Technological innovation is central to sustainable food system transformation. Existing solutions, such as regenerative agriculture and precision breeding, alongside emerging technologies, must be coordinated to avoid unintended consequences across the system. The role of research funding and market dynamics in shaping technological access is critical. Finally, equity underpins all strategies for sustainable food systems. Reports stress global collaboration to prevent regional dominance and emphasise the development of accessible technologies to benefit underserved populations, ensuring progress aligns with sustainability objectives.

Research Methodology

No research project is exhaustive, nor does it exist independently; each piece of research links in some way to the environment within the scope of the research, the theories or concepts that predate the research. There are two broad data collection approaches–namely, quantitative and qualitative research; this study uses the latter. Through this intensive literature review, the researcher delves into various case studies regarding eco-innovation in Europe, focusing then on emerging economies in the MENA region, aiming to be aware of existing research on the topic. Gathering data indirectly via desk-based research exposes the researcher to a plethora of perspectives; hence qualitative results are relevant for this study as they provide a rich and detailed description of complex phenomena (Sofaer, 1999). However, the credibility and reliability of the literature needs to be verified before the process can be replicated.

For the purposes of this chapter, literature was sourced from credible academic journals such as Taylor and Francis Online, Science Direct, and Carnegie Endowment. Access to reliable information is essential to be able to form an opinion and make a solid claim about the opportunities that await the Middle East region. Harnessing existing information allows organisations to build a wealth of knowledge without the need for time-consuming and resource-utilising fieldwork (Gell, 2023). Secondary research also includes the essential aspect of being aware of interpersonal relations that may be brought up. Although desk-research is accessible due to the wide range of information readily available through online databases, there is a limited control researchers have over the quality of data and its accuracy and relevance to the subject matter at hand (Gell, 2023). Data could be outdated or may be researched with a biased perspective, leading to potentially skewed information. To mitigate this, data collected is evaluated for a certain criterion such as the purpose for collection, the date of publication (all case studies were published in 2024), and methodology of data collection. All literature was sourced online through Google Scholar, using keywords such as "Eco-innovation", "Sustainable innovation", "Climate Neutral strategies", "Adapting food industry", "food industry in the MENA region", amongst others. Alongside the specific case studies care was taken to source information not older than 5–10 years by using filters to find the most up to date, yet relevant information.

Recommendations and Implications

A study focusing on eco-innovation as a conceptual keyword in the context of the European Union disclosed three main findings: firstly, that eco-innovation programs in the EU have been constructed around the idea of eco-efficiency, secondly, eco-innovation is framed as a point of contention between the government and private actors while stakeholders in the third sector (such as non-governmental organisations, social enterprises, cooperatives, and community-based initiatives) are largely ignored, and thirdly, there is a new rising discourse taking prevalence over eco-innovation, namely 'circular economy' (Colombo et al., 2019; Singh et al., 2024).

Integrating Eco-Innovation in Omani Business Strategies

In Oman, a large-scale survey-based research design was conducted to investigate the relationships between eco-innovation, research and development investments, government involvement, and the firms' competitive advantage. The survey included 214 high level management entities such as founders, presidents, and vice-presidents from a range of industries and firms (Abdelfattah et al., 2024). This online survey collected data in the form of structured questionnaires where all variables were measured through a 5-point scale: green product innovation performance included measures for customer satisfaction, environmental performance, and adoption of smart technologies; research and development investment was measured through the financial resources dedicated to it, and the number currently engaged in ongoing R&D projects; AI adoption was assessed through the extent of its implementation on its level of smart integration and the use of AI in decision-making; intellectual capital had measures relating to human capital, structural capital, and relational capital; government involvement was assessed through incentives for the regulation of sustainable practices, public–private partnership initiatives, and government support for innovation; lastly, competitive advantage was measured relating to the firm's profitability, market growth, and growth (Abdelfattah et al., 2024).

The results revealed a positive correlation between all variables. Building on research by Chang (2011), this positive relationship between eco-innovation products and a firm's competitive advantage can be explained through factors supporting the idea that eco-innovation can increase customer loyalty (Le, 2022). As consumer awareness regarding the environment grows, companies that showcase their dedication to sustainability are more likely to attract and retain customers with the same values. This can be explained by the congruity theory, which states that customers are more likely to be loyal to brands that align with their personal values. By marketing green products, companies can enhance their brand image of being conscious towards sustainability and showcase corporate social responsibility which can contribute to competitive advantages. Furthermore, such engagement enables firms to predict and anticipate regulatory changes and adapt in accordance, thus outperforming less proactive competitors. Firms can also benefit from incentives by aligning themselves with government policies promoting sustainable practices (Abdelfattah et al., 2024).

These findings contribute to the literature on the strategic importance of sustainability and its impact on firm performance and can aid companies in the Middle East to develop and market eco-innovative products for their improved profit. Adding to research on AI by Krakowski et al. (2022), the study finds that adopting smart technologies can help firms to streamline their operations, optimise resource allocation, and improve efficiency. By leveraging such AI-driven insights, firms can make informed choices, adapt to market changes, and outperform competitors (Abdelfattah et al., 2024). Important to note is the importance of investing in human capital development. Through training, nurturing employees' competencies and creativity, organisations can utilise their intellectual capital to drive innovation even further (Bhushan & Singh, 2024).

Fostering an environment encouraging the sharing of knowledge and overall collaboration can help firms maximise the benefits of the R&D in eco-innovation, facilitating continuous improvement and adaption to sustain a competitive edge in the market (Abdelfattah et al., 2024). A study by Alqassabi (2020) noted how the Omani government could contribute to encouraging eco-innovation through supportive policies and funding incentives. Local firms can be aided to prioritise environmentally friendly innovation by providing a centralised regulatory framework they can adapt and replicate. The need for collaboration between government entities and private firms is of utmost importance in promoting innovation and competition (Al-Kharusi, 2022). An expanded sample size to make it more generalisable to the overall Omani population, as well as a different method of data collection such as longitudinal could offer a more in-depth understanding of how various factors interact over time. Although this study was based around Oman's business environment, its culture, and national policies, through exercised caution it can be replicated to other countries in the Middle East region.

Sustainability Management in Nigerian F&B Firms

The second study was conducted in the Enugu state of Nigeria amongst 297 staff members from Nigerian Breweries Plc, Pepsi Bottling Company, and Coca-Cola Bottling Company, examining the impact of environmental sustainability management on food, beverage, and tobacco manufacturing firms in the region. Primary data collected via a questionnaire was then analysed to test the hypothesis, including descriptive statistics like percentages and frequencies, and inferential statistics like regression. This analysis revealed a statistically significant positive correlation between various ecosystem management practices and the production quality of Nigerian F&B products. Regression analysis depicted that ecosystem management accounts for 44% of the difference in production quality, thus highlighting the important role of strategies focused to enhance production processes and the output quality of products in the three firms studied (Ugwuanyi & Oluka, 2024).

Moreover, eco-innovative engineering significantly impacted the acquisition of adaptive skills among employees. This showcases that sustainable engineering approaches contribute to fostering adaptability and innovation in employees' as they respond effectively to organisational and environmental demands. This makes it crucial for managerial teams to adhere to environmental policies to sustain and enhance production quality in the food industry. Embedding eco-innovations and considerations into operational frameworks will ensure compliance with global sustainability standards (Ugwuanyi & Oluka, 2024). By centrally institutionalising policies prioritising eco-innovation approaches, as well as investing in renewable energy technologies, employee training programs, and a sustainable production line, will improve organisational performance and equip employees with the skills necessary to answer to the ever-changing market. By making such sustainability-driven strategies formal and mandatory, companies can achieve improved outcomes.

Implementing Eco-Innovation in the Tanzania Agri-Food Sector

The following study aims to explore the determinants of eco-innovation developments in Tanzania's food and beverage industries across 102 industries. Cross-sectional survey data is used across six major regions of Tanzania, wherein self-administered questionnaires were filled out by managerial staff involved in eco-innovation decision-making such as production managers, sustainability officers, and senior executives. Upon completion of the survey, the data analysed revealed the relationships between internal and external factors that affected eco-innovation implementation. To elaborate, 45.1% of the eco-certified businesses were dedicated to food production (oilseeds, dairy, tea, coffee, and flour) (Buzohera, 2024). There was a significant positive influence on financing on eco-innovation, supporting evidence by Hasan and Rahman (2023) on the role of financial support driving the adoption of eco-innovative technologies.

Financing is crucial as it aids in R&D activities and supports the exploration of alternative materials and processes. It also allows for co-production, minimising waste and resource depletion while leading to innovations that have the potential to unlock revenue streams (Buzohera, 2024). Investing in training programs raises environmental awareness between employees, enabling them to make conscious decisions and contributing to the firm's sustainability efforts. The study added to the research by Edeh and Vinces (2023) which demonstrated the significance of external knowledge acquisition to improve eco-innovation in a firm by suggesting that internal knowledge within industries needs further investigation to measure their influence on eco-innovation implementation. However, there was a statistically significant and positive correlation for internal stakeholders, showcasing the importance of engaging and involving internal stakeholders such as employees and managers in the eco-innovation process instead of outsourcing data (Buzohera, 2024). The study also finds that investing in research for eco-innovations is crucial to promote the implementation of eco-innovation within the industry; findings remain consistent with previous literature emphasising the importance of dedicating resources to research endeavors (Ceptureanu et al., 2017).

Furthermore, the assertion of environmental policies and regulations is a crucial determinant for the implementation of eco-innovation within industries. Therefore, robust and supportive policies are an essential need to drive the development of eco-innovation. Compliance with regulatory frameworks, adherence to environmental standards for products and technology, and concerns over pollution taxes and penalties are some of the prospective policy areas to tackle. Overall, a multifaceted approach is essential to accelerate eco-innovation adoption in Tanzania's food sector. Financial support can alleviate barriers by offering subsidies, low-interest loans, and grants, reducing the costs of eco-friendly technologies and practices. Public–private partnerships can further enhance resource availability and innovation. R&D through collaborations between industries, universities, and innovation hubs is crucial for developing eco-friendly materials, energy-efficient processes, and waste management technologies tailored to local needs. Enforcing robust environmental standards, incentivising green certifications, and providing tax benefits to companies maintaining pollution control and renewable energy use should be areas targeted by policymakers to ensure compliance through regular audits. This also fosters collaboration between industry and academia to create a knowledge and expertise transfer (Buzohera, 2024).

Stakeholder engagement is of utmost importance when it comes to advancing eco-innovation in the food industry; creating networks among sub-industries, NGOs, as well as international actors, fosters collaborative participation, whereas internal training programs align with organisational practices as well as with global sustainability goals. The study's use of entrepreneurship and ecological studies offers a distinct perspective on leveraging both internal and external dynamics to shape the implementation of eco-innovation (Buzohera, 2024). For emerging economies, improving technology accessibility is a step towards the effective integration of sustainable solutions; by investing in eco-technology infrastructure and offering training towards the same, these measures can create an environment that serves as a catalyst for innovation, thus addressing region-specific environmental challenges.

Green Leadership's Role in Customer Satisfaction

The next study investigates the crucial role of green leadership and eco-innovation in enhancing sustainable tourism practices and customer satisfaction within Dubai's hospitality industry. While not directly related to the food industry, it emphasises the growing importance of sustainable practices for customer loyalty and organisational success in a market influenced by environmental awareness. Dubai has transformed from a small Gulf port into a global commerce hub, diversifying its economy through tourism, construction, and trade (Bagaeen, 2007). Increasing awareness towards sustainable trends makes this timely research important in understanding how eco-innovation strategies and sustainable sourcing contribute significantly to customer satisfaction. Customers with strong environmental values rated higher satisfaction with eco-friendly ventures, highlighting the alignment of organisational goals with customer values. The research shows a shift towards green policies, emphasising the importance of eco-innovation in the hospitality industry (Asadi et al., 2020).

A notable result of this research is that leadership plays a key role in successfully implementing eco-innovation strategies. Green leadership fosters a culture that prioritises sustainability, encouraging employee involvement and the integration of green policies into business operations. This includes adopting energy-efficient systems, promoting recycling initiatives, and offering employee training on sustainability practices. Furthermore, engagement is vital for strengthening the connection between hotels and their environmentally conscious customers. Actively promoting sustainability policies and offering incentives for participation in eco-friendly programs can enhance customer loyalty and trust. Dubai's future challenges include retaining its attractiveness while promoting sustainability, with the Dubai 2040 Urban Master Plan focusing on sustainable development. This study showcases the importance of integrating eco-innovation and green leadership to improve customer satisfaction and align business objectives with sustainable development goals.

Conclusions, Limitations, and Future Scope

Financial support plays a foundational role in enabling the adoption of eco-innovation technologies and practices. To overcome cost barriers, governments and financial institutions should offer subsidies, low-interest loans, and tax incentives to organisations investing in sustainable practices (Abdelfattah et al., 2024; Buzohera, 2024). Funding initiatives can specifically target research and development (R&D) efforts, enabling firms to explore alternative materials, optimise production processes, and reduce waste. For example, investments in co-production and resource-sharing technologies can create revenue streams while minimising environmental impact (Hasan & Rahman, 2023). Additionally, creating green innovation grants and competitive funding programs for start-ups in developing economies, such as those in Oman and Tanzania, can catalyse sustainable industrial growth (Alqassabi, 2020).

Robust environmental policies and regulatory frameworks are essential drivers for eco-innovation. Governments should implement long-term policies that prioritise renewable energy adoption, pollution tax relief, and financial incentives for firms with green certifications. Such measures can encourage widespread adoption of eco-innovation strategies (Buzohera, 2024). Regulatory frameworks must also integrate mechanisms for monitoring and evaluating compliance, including periodic audits and public reporting of environmental metrics. Policymakers in regions like the Middle East should look toward replicating successful international models, aligning local policies with global sustainability standards like the United Nations Sustainable Development Goals (UN SDGs). Programs like Dubai's 2040 Urban Master Plan provide examples of how urban and industrial policies can integrate eco-innovation goals into development plans.

Collaboration across government, industry, academia, and non-governmental organisations (NGOs) is pivotal to driving eco-innovation. Public–private partnerships (PPPs) can enable industries to access shared knowledge, resources, and financial assistance to develop eco-friendly solutions (Abdelfattah et al., 2024).

Universities and research institutions can play an instrumental role in advancing eco-innovation by contributing technical expertise and developing cost-effective technologies. Innovation hubs that promote the exchange of ideas and solutions can amplify these efforts (Buzohera, 2024). Moreover, governments can encourage collaboration by establishing centralised platforms for industries to engage with stakeholders, enabling a smoother transition toward sustainability (Nwokolo et al., 2023a, 2023b, 2023c; Nwokolo et al., 2024).

Green leadership within organisations is critical for implementing and sustaining eco-innovation strategies. Leaders must advocate for sustainable practices, drive cultural changes within the organisation, and empower employees to participate in green initiatives. A supportive leadership style fosters innovation and helps align organisational objectives with environmental goals. Investment in employee training and capacity-building programs is vital for nurturing skills necessary for managing and implementing eco-innovation technologies. Training programs should emphasise resource management, energy efficiency, and green engineering, equipping employees with the adaptability needed to navigate changing market dynamics (Abukhait et al., 2020). Leaders can also play a vital role in securing stakeholder buy-in and ensuring that sustainability is embedded into organisational policies and practices.

Engaging with customers is essential for firms aiming to integrate eco-innovation into their business models. Transparent communication about sustainability policies, such as through marketing campaigns and annual sustainability reports, builds trust and loyalty among environmentally conscious consumers. Companies can strengthen customer relationships by involving them in sustainability initiatives. For instance, incentivising customers to participate in recycling programs or offering eco-friendly packaging alternatives enhances consumer perception of a brand's commitment to sustainability (Le, 2022). Feedback loops, wherein customers provide input on eco-innovation initiatives, also allow firms to refine and align their strategies with customer expectations.

Access to advanced technologies is criticalfor effective eco-innovation. Governments and private organisations must invest in eco-technology infrastructure, such as renewable energy systems, AI-driven decision-making tools, and sustainable production lines (Krakowski et al., 2022). Offering training programs tailored to the use of these technologies can help firms optimise their integration and achieve operational efficiency. In developing economies, where technology gaps often hinder sustainability efforts, international collaborations and knowledge-sharing networks can bridge these divides (Buzohera, 2024). By implementing AI-driven systems, firms can streamline operations, adapt to regulatory changes, and improve resource allocation, thus enhancing overall competitiveness (Abdelfattah et al., 2024; Nwokolo et al., 2023d).

While these recommendations are informed by studies conducted in Oman, Tanzania, and Dubai, their broader applicability to other regions in the Middle East and Sub-Saharan Africa cannot be overlooked. Careful adaptation of eco-innovation frameworks to local cultural, economic, and regulatory contexts is essential for replicating success across diverse markets (Al-Kharusi, 2022; Nwokolo et al., 2023a,

2023b; Ugwuanyi & Oluka, 2024). Future research should include longitudinal studies to better understand the dynamic interactions between these variables over time, providing more nuanced insights into regional adaptations.

The implementation of a new discourse, circular economy (Colombo et al., 2019), could lead to more opportunities for eco-innovation and inclusive methods towards more developed sustainability initiatives as well as the inclusion of not-for-profit organisations and community programs. This approach should be promoted along with making efforts to find the right balance between eco-innovation and socio-economic sustainability, critical to achieving the United Nations' Sustainable Development Goals (SDGs) by 2030. This framework seeks to align natural resource management and food systems with equitable and sustainable practices. However, addressing key challenges such as food waste and its impact on climate and biodiversity remains paramount. Currently, about a third of global food production is wasted or lost, contributing to significant greenhouse gas (GHG) emissions and threatening the planet's ability to support a population projected to exceed 8.5 billion by 2030 (Hall, 2024).

Developed nations must lead by promoting a circular economy that ensures responsible resource use and equitable food distribution. Failing to do so risks worsening food inequality and undermining global sustainability efforts. One approach involves nationwide food-based dietary guidelines that encourage shifts towards plant-based diets by reducing red meat and dairy consumption. These measures aim to address GHG emissions—21–37% of which stem from the food supply chain, including crop and livestock production and processing—and the 8% attributed to food waste. Strategies to discourage the excessive consumption of terrestrial animal-sourced foods include public education campaigns on sustainable diets and policies to increase the cost of these products, thereby reducing demand. Complementing these efforts, applied research in food packaging is essential to transitioning from synthetic plastics to biopolymer-based materials, incentivising sustainability throughout the supply chain.

Green technologies play a transformative role in reducing food waste and valorising by-products. These technologies, which have minimal impact on nutritional and sensory properties, provide efficient alternatives for extracting valuable compounds from food waste, overcoming the limitations of traditional methods. Such innovations can significantly decrease the environmental footprint of food production while unlocking economic opportunities through biowaste valorisation. By integrating smart technologies, sustainable dietary practices, and circular bioeconomy principles, global food systems can reduce their environmental impact, address food inequality, and build resilience against future challenges. These strategies collectively ensure progress towards a more sustainable and equitable future. Across diverse case studies in Oman, Nigeria, Tanzania, and Dubai, the evidence strongly supports the adoption of environmentally responsible strategies as a driver for socio-economic growth, environmental preservation, and competitive advantage.

Eco-innovation, whether through green technologies, sustainable leadership, or collaboration, has demonstrated significant potential in enhancing resource efficiency, improving customer satisfaction, and driving organisational success. The findings from Oman underscore the importance of centralised regulation, public–private partnerships, and R&D investments in fostering eco-friendly practices. Similarly, the Nigerian study highlights the role of ecosystem management and employee adaptability in improving production quality. Tanzania's exploration of internal and external factors emphasises the necessity of financial support and stakeholder engagement in advancing eco-innovation.

Sustainable practices in the food supply chain, as seen in Dubai, reveal the growing consumer demand for environmentally conscious businesses and the critical role of green leadership in achieving these goals. The implementation of dietary guidelines, reduction in animal-based product consumption, and application of smart technologies further illustrate how systemic changes can mitigate environmental impacts while addressing food inequality and economic challenges. Despite these advancements, challenges remain. The persistence of food waste, inefficient resource utilisation, and lack of equitable access to eco-technologies underscore the need for a multifaceted approach. Governments, industries, and communities should work collaboratively to align policies, investments, and innovations with the United Nations' Sustainable Development Goals (SDGs). Initiatives such as biowaste valorisation, transitioning to plant-based diets, and leveraging AI and blockchain technologies offer pathways to more resilient and sustainable systems.

In conclusion, achieving global sustainability demands a holistic framework that balances environmental, economic, and social priorities. By integrating eco-innovation with circular economy strategies, fostering leadership, and enhancing global cooperation, nations can address critical challenges and secure a sustainable future for the generations to come. These efforts not only align with climate neutrality goals but also promise to create a more equitable and prosperous world.

Acknowledgements I would like to thank Ms. Nandita Mulay for her valuable contributions to this chapter.

References

Abdelfattah, F., Salah, M., Dahleez, K., Darwazeh, R., & Al Halbusi, H. (2024). *International Journal of Innovation Studies, 8*(2), 154–171. https://doi.org/10.1016/j.ijis.2024.02.001

Abukhait, R., Bani-Melhem, S., & Mohd Shamsudin, F. (2020). *International Journal of Innovation Management, 24*(07), 2050070. https://doi.org/10.1142/S136391962050070X

Al-Kharusi, S. (2022). *The Role of Inward FDI in Enhancing Sultanate of Oman's National Competitiveness*. King's College London.

Alqassabi, M. A. (2020). *International Journal of Economics and Financial Issues, 10*(1), 209–218. https://doi.org/10.32479/ijefi.9063

Asadi, S., Pourhashemi, S. O., Nilashi, M., Abdullah, R., Samad, S., Yadegaridehkordi, E., & Razali, N. S. (2020). *Journal of Cleaner Production, 258*, 120860. https://doi.org/10.1016/j.jclepro.2020.120860

Bagaeen, S. (2007). *International Journal of Urban Policy and Planning, 14*(4), 317–332. https://doi.org/10.1080/13563470701486372

Bhushan, M., & Singh, R. (2024). Talent management facilitates net-zero transition through employee green behaviour. In *CSR, Sustainability, Ethics & Governance* (pp. 117–129). https://doi.org/10.1007/978-3-031-55779-8_7

Buzohera, M. I. (2024). *Cogent Business & Management*. https://doi.org/10.1080/23311975.2024.2350801

Campari, J. (2018, November 16). *World Economic Forum*. https://www.weforum.org/stories/2018/11/we-must-rethink-our-food-system-from-planet-to-plate

Carrillo-Hermosilla, J., del Río, P., & Könnölä, T. (2010). *Journal of Cleaner Production, 18*(10), 1073–1083. https://doi.org/10.1016/j.jclepro.2010.02.014

Ceptureanu, E. G., Ion, S., Orzan, M. C., Bordean, O. N., & Radulescu, V. (2017). *Sustainability, 9*(10), 1779. https://doi.org/10.3390/su9101779

Chang, C.-H. (2011). *Journal of Business Ethics, 104*(3), 361–370. https://doi.org/10.1007/s10551-011-0914-x

Colombo, L. A., Pansera, M., & Owen, R. (2019). *Journal of Cleaner Production, 214*, 653–665. https://doi.org/10.1016/j.jclepro.2018.12.150

Edeh, J., & Vinces, J. P. (2023). *Journal of Business & Industrial Marketing, 39*(2), 318–335. https://doi.org/10.1108/JBIM-02-2023-0115

Fanzo, J., & Miachon, L. (2023). *Anthropocene, 42*, 100381. https://doi.org/10.1016/j.ancene.2023.100381

Fanzo, J., Davis, C., McLaren, R., & Choufani, J. (2018). *Global Food Security, 18*, 12–19. https://doi.org/10.1016/j.gfs.2018.06.001

Fussler, C., & James, P. (1996). Pitman Publishing. ISBN: 0273622072.

Gell, T. (2023). *Drive Research*. https://www.driveresearch.com/market-research-company-blog/desk-research-what-it-is-and-how-you-can-use-it/

Hall, N. (2024). *CSIS Features*. https://features.csis.org/surviving-scarcity-water-and-the-future-of-the-middle-east/

Hasan, M. J., & Rahman, M. S. (2023). *Heliyon, 9*(7), e18102. https://doi.org/10.1016/j.heliyon.2023.e18102

Hassoun, A., Prieto, M., & Carpena, M. (2022). *Food Chemistry, 364*, 130328. https://doi.org/10.1016/j.foodchem.2021.130328

Joshi, A., Singh, R., Kumar, V., & Rani, S. (2024). Net zero strategy: Exploring opportunities, confronting challenges for developing countries. In *Approaches to Global Sustainability, Markets, and Governance* (pp. 233–252). https://doi.org/10.1007/978-981-97-5756-5_11

Kachi, A., Mooldijk, S., & Warnecke, C. (2020). *Global Environmental Change, 63*, 102078. https://doi.org/10.1016/j.gloenvcha.2020.102078

Khan, S, Singh, R., Alnahas, J., Abbate, S., & Centobelli, P. (2024). Navigating the smart circular economy: A framework for manufacturing firms. *Journal of Cleaner Production*, 144007–144007. https://doi.org/10.1016/j.jclepro.2024.144007

Krakowski, S., Luger, J., & Raisch, S. (2022). *Strategic Management Journal, 43*(4), 711–738. https://doi.org/10.1002/smj.3287

Kumar, K, Singh, R., Sharma, A., & Lehri, B. P. S. (2024). Climate change adaptation strategies for achieving net-zero economy. In *Approaches to Global Sustainability, Markets, and Governance* (pp. 163–183). https://doi.org/10.1007/978-981-97-5756-5_8

Le, T. T. (2022). *International Journal of Emerging Markets, 17*(2), 198–215. https://doi.org/10.1108/IJOEM-06-2021-0814

Living Planet Report. (2024). World wildlife fund. https://livingplanet.panda.org/

Luqmani, A., Leach, M., & Jesson, D. (2017). *Environmental Innovation and Societal Transitions, 24*, 94–105. https://doi.org/10.1016/j.eist.2016.10.007

Mahmoud, M. (2024). *Carnegie endowment for international peace*. https://carnegieendowment.org/research/2024/04/the-looming-climate-and-water-crisis-in-the-middle-east-and-north-africa?lang=en

Nwokolo, S. C., Singh, R., Khan, S., Kumar, A., & Luthra, S. (2023a). Remedies to the challenges of renewable energy deployment in Africa. In *CSR, Sustainability, Ethics & Governance* (pp. 59–74). https://doi.org/10.1007/978-3-031-44514-9_3

Nwokolo, S. C., Singh, R., Khan, S., Kumar, A., & Luthra, S. (2023b). Impacts of climate change in Africa. In *CSR, Sustainability, Ethics & Governance* (pp. 237–262). https://doi.org/10.1007/978-3-031-44514-9_7

Nwokolo, S. C., Singh, R., Khan, S., Kumar, A., & Luthra, S (2023c). Global investment and development in Africa. In *CSR, Sustainability, Ethics & Governance* (pp. 15–58). https://doi.org/10.1007/978-3-031-44514-9_2

Nwokolo, S. C., Singh, R., Khan, S., Kumar, A., & Luthra, S. (2023d). Technological pathways to net-zero goals in Africa. In *Africa's path to net-zero. CSR, sustainability, ethics & governance.* Springer, Cham. https://doi.org/10.1007/978-3-031-44514-9_5

Nwokolo, S. C., Singh, R., Khan, S., & Kumar, A. (2024). Threats to the rapidity of sustainability transitions posed by technological changes. In *Elsevier EBooks* (pp. 39–75). https://doi.org/10.1016/b978-0-443-31486-5.00002-1

Pichlak, M., & Szromek, A. R. (2021). Journal of open innovation: Technology. *Market, and Complexity, 7*(2), 149. https://doi.org/10.3390/joitmc7020149

Rojas-Downing, M. M., Nejadhashemi, A. P., Harrigan, T., & Woznicki, S. A. (2017). *Environmental International, 106*, 185–194. https://doi.org/10.1016/j.envint.2017.06.017

Schiederig, T., Tietze, F., & Herstatt, C. (2012). *R&D Management, 42*(2), 180–192. https://doi.org/10.1111/j.1467-9310.2011.00672.x

Singh, R., Gupta, A., & Kandpal, V. (2024). *ESG Excellence in the Digital World* (pp. 193–204). https://doi.org/10.1002/9781394234035.ch10

Smith, P. J., & Pye, A. (2019). *Climate Change and Land: An IPCC Special Report on Climate Change, Desertification, Land Degradation, Sustainable Land Management, Food Security, and Greenhouse Gas Fluxes in Terrestrial Ecosystems.* https://www.ipcc.ch/site/assets/uploads/2019/08/Fullreport.pdf

Sofaer, S. (1999). Qualitative methods: What are they and why use them? https://pubmed.ncbi.nlm.nih.gov/10591275/

Thaler, R. H., & Sunstein, C. R. (2021). Yale University Press. ISBN: 9780300262285.

Ugwuanyi, C., & Oluka, K. U. (2024). Environmental sustainability management and performance of food, beverages and tobacco manufacturing firms in Enugu state, Nigeria. *Journal of Current Research in Business and Management Sciences, 12*(4), 1–12. https://doi.org/10.5281/zenodo.13891008

Zurek, M., Hebinck, A., & Selomane, O. (2021). *Q Open, 1*(1), qoaa001. https://doi.org/10.1093/qopen/qoaa001

Chapter 9
Voluntary Higher Education and Awareness as Catalysts for Eco-Innovation Adoption

Raed Awashreh

Abstract Higher education therefore plays a very important role in promoting eco-innovation for sustainable development and care for the environment. This review looks at how universities can foster the adoption and diffusion of eco-innovation through integration into the curriculum, research and development, skill development, and awareness. Universities are increasingly embedding sustainability principles into interdisciplinary curricula that would equip students with knowledge and skills to tackle complex environmental challenges. Additionally, institutions of higher learning serve as centers of R&D, where sustainable technologies and business models are developed with industrial partners. Notwithstanding all the efforts mentioned above, difficulties persist due to resistance, commercialization pressures, and general lack of funds for research. Universities contribute to the development in students of technical and leadership abilities that would be required in the application of eco-innovative solutions in various industries. Besides, voluntary education initiatives and grassroots programs are very important in bringing about awareness, behavioral change, and eco-friendly practices within the community, although scaling up remains an issue. The review emphasizes public–private partnerships and monitoring frameworks for overcoming these barriers to make education in eco-innovation effective and impactful. In turn, higher education can contribute to accelerating eco-innovation diffusion and the transition to sustainability by promoting collaborative efforts across sectors and levels of education. It thereby contributes to long-term environmental and social welfare. The findings underlined that a step toward scaling up eco-innovation in order to meet global environmental challenges requires an integrated approach to education through both formal institutions and voluntary initiatives.

Keywords Eco-innovation · Sustainability · Climate neutrality · Higher education · Interdisciplinary curriculum · Public–private partnerships

---

R. Awashreh (✉)
A'Sharqiyah University, Ibra, Oman
e-mail: Raed.raya2020@gmail.com

R. Singh and W. L. Filho (eds.), *Climate Neutrality Through Smart Eco-Innovation and Environmental Sustainability*, Climate Change Management,
https://doi.org/10.1007/978-3-031-83250-5_9

Introduction

Eco-innovation is a recent idea that has quickly become popular with education authorities and in society. There is a need to encourage sustainable activities as the society is plagued with climate change issues, resource shortages and extinction of species. With regard to this transition, Higher Education Institutions serve the means of integrating eco-innovation into the professional education of students with the aim of creating competencies that are crucial in addressing environmental issues (Hyytinen et al., 2023). Besides, the integration of voluntary education and awareness and public–private collaborations enhances the universal acceptance of eco-innovate solutions in different spheres of life. These aspects not only enhance sustainability but equally enable communities and industries to adopt environmentally friendly practices (Berchin et al., 2021). However, in spite of the existing success, eco-innovation in education is still faced with many challenges such as lack of financial support and lack of acceptance from the organizations themselves. It is though, addressing such challenges using different approaches to funding and new approaches to building the curriculum may expand the effectiveness of education on eco-innovation. In addition, peer-to-peer learning and bottom-up initiatives are also important in the process of implementation of environmentally sound technologies as it sparks the change witnessed (Serdyukov, 2017).

Research Questions

The research focus for this paper is: In what ways and how can education, be it through higher education institutions, be it through voluntary initiatives or public- private partnerships, encourage the communities and the industries to break the existing barriers and make the least carbon footprint in the process of utilizing and producing all devices and services. This central question encompasses the various elements outlined in the sub-questions, aiming to explore the integration of eco-innovation into curricula, the role of awareness campaigns, the contribution of partnerships, and the challenges faced by educational initiatives in advancing the goals of eco-innovation.

- How can higher educational institution(s) including universities, broaden the perhaps narrow interpretations of eco-innovation and mainstream it so as to develop a genuinely ecologically conscious society?
- What role does voluntary education and awareness campaigns contribute to the communities and industrial users' uptake of eco-implementable solutions?
- And what is the role so far of public- private partnerships in enhancing the success of eco-educations?
- Are the educational initiatives the weakest link in promotion of eco-innovation to communities and industrial enterprises, and how can this linkage be hermetically sealed?

- How does a peer and bottom-up approach help, contribute to the desired goal of adoption of ecologically sustainable mechanisms of operation that are economically viable?

The aim of this chapter is to explore the role of voluntary higher education and awareness initiatives in driving the adoption of eco-innovation. It seeks to emphasize the importance of integrating eco-innovation into educational systems, raise awareness about sustainable practices, and identify effective strategies for encouraging eco-innovation across various sectors.

Significance

This chapter is significant because it highlights the powerful potential of education in advancing sustainability. By examining the role of both formal and informal education in fostering eco-innovation, it contributes to understanding how knowledge dissemination, awareness campaigns, and community engagement can accelerate the adoption of environmentally sustainable solutions. Furthermore, it provides practical recommendations for policymakers, educators, and organizations looking to promote eco-innovation through education and awareness initiatives.

The purpose of this chapter is to analyze how voluntary higher educational institutions and centers acting for diffuse awareness support the embedding of eco-innovations. It stresses the need to embed eco-innovation in education, promote the most relevant sustainable concepts, and develop the ways of encouraging eco-innovation in the education sector and other sectors as well.

This chapter is significant in the way that it brings forward the great ability of education to enable progression toward sustainability. It also helps in the advancement of the theory that there is a need for the combined effort of official authorities, outreach and education, and community participation, followers of which can reduce the time to the implementation of eco-friendly practices. In addition to that, it offers viable approaches to policymakers, educators, and other institutions who want to harness eco-innovation via education and outreach strategies.

The main aim of this chapter is to examine the response of higher education institutions in promoting eco-innovation evolution. One of them is examining the role of integrating curriculum, developing research programs, and enhancing skills to improve eco-innovation efforts. A further aim is such objectives as the evaluation of the results of voluntarily undertaken educational and awareness activities on the level of eco-innovation adoption in various sectors.

The chapter also intends to analyses the role of public–private partnerships and policy tools in increasing the efficiency of education on eco-innovation. Furthermore, it aims at identifying the determinants and hindering factors of educational activities directed at the popularization of eco-innovation and suggest measures for addressing these barriers. Finally, the chapter will offer concrete suggestions on how to incorporate eco-innovation into the designs of higher education course content and communications strategies so as to help transition to a more innovative and sustainable economy.

Literature Review

The Role of Higher Education in Eco-Innovation Adoption

Higher education will continue to be of essence in promoting eco-innovation for the advance towards sustainability and environmental stewardship. Indeed, universities act as influential catalysts towards the adoption of eco-innovative practices at industry, community, and government levels in view of their responsibility for developing knowledge. This review attempts to investigate how higher education promotes eco-innovation via curriculum integration, research and development, the development of skills, and challenges and opportunities that come along with this effort (Giesenbauer & Müller-Christ, 2020). For the curriculum integration and interdisciplinary programs, higher education is positioned to be a driver of eco-innovation through integrating the principles of sustainability within their curriculum. Interdisciplinary programs in environmental science, engineering, and management equip students with a wide-ranging knowledge of environmental challenges and their solutions (Abo-Khalil, 2024). Certain universities, such as Stanford and MIT, have programs that integrate sustainability with technological innovation to equip students with skills for solving complex environmental issues. collaboration speeds up the dissemination of eco-innovation because the market needs will be adequately met (Graham, 2014). Some are concerned, however, about commercialization of academic research as, sometimes, the profit motive contradicts sustainability objectives (Kemper et al., 2019). This would raise difficult tensions in missions focused on truly sustainable innovation. According to Didham et al. (2024), interdisciplinary learning nurtures innovation and cross-sector collaboration, both very important features of eco-innovation. However, there are a number of challenges to integrating eco-innovation into curricula. Most traditional academic structures are strongly geared towards discipline-specific knowledge and can thus be resistant to developing interdisciplinary programs. Resistance from traditional discipline-based departments. Nevertheless, highlight that curricula should be holistic, covering business, economics, and policy in addition to environmental science (Didham et al., 2024).

A. Research and Development (R&D) as a Hub for Innovation

R&D in universities is one of the cornerstones for the development of eco-innovation. Universities represent a site of research that connects theory and practice in the development of sustainable technologies and business models (Arroyave et al, 2020). For example, UC Berkeley and the Technical University of Denmark have developed pioneering technologies in renewable energy and sustainable agriculture, respectively. University-industry collaborations are very important in scaling up academic research into practical, scalable eco-innovations (Hojeij, 2024). According to collaboration, the dissemination of eco-innovation speeds up since the needs in the market will be adequately met. Some are apprehensive, though, as commercialization in academic research sometimes directly goes against the motive for sustainability besides generating profits (Sjoer, Nørgaard & Goossens, 2013). This would bring up uneasy tension in truly sustainable focused missions.

B. Skills Development for Eco-Innovation Implementation

Higher education also provides students with skills needed to pursue eco-innovative projects. Education in sustainable design, environmental management, and eco-entrepreneurship streams prepares the students to overcome all the complex challenges in the sphere of eco-innovation (Awashreh, 2023). The curriculum of higher education should be developed continuously and keep up with the improved technology. Besides, training should combine technical and leadership aspects, as eco-innovation requires both expertise and managerial ability. Nevertheless, there are still challenges regarding the alignment of curricula with the newly emerging needs of the eco-innovation jobs market. Thus, institutions will need to update their program offerings in line with new innovative practices for sustainability and prepare students effectively for leadership and management roles in eco-innovation (Manning et al., 2020).

C. Voluntary Education and Its Impact on Awareness

Apart from formal education, voluntary education through workshops, online courses, and community programs plays an important role in increasing awareness about eco-innovation. These programs sometimes provided by NGOs or other environmental agencies reach a wide range of audiences, from adults to professionals (Gonçalves & Tilbury, 2024). Community-based education has been used very successfully in programs such as the Green Belt Movement in Kenya (Gregory, 2017). While the awareness they raise is successful, the challenges of voluntary education programs lie in scalability and sustainability. if not followed up appropriately, such programs may not sustain behavior change. Besides, they are mostly based on unstable external funding, which keeps changing with fluctuating political and economic priorities (Cogut et al., 2019).

Role of NGOs and Peer-To-Peer Learning

NGOs can play a very important role in the grassroots dissemination of eco-innovation knowledge. In community-based initiatives, peer-to-peer learning tends to be one of the strong ways to bring about behavior change (Mattijssen et al., 2024). According to Bandura's Social Learning Theory, the chances are that people will adopt new behavior if they see their peers doing so. Therefore, grassroots programs that encourage local knowledge sharing are bound to influence eco-innovation adoption (Kelder et al., 2017).

Awareness as a Driver for Eco-Innovation

Awareness campaigns are crucial to the adoption of eco-innovation through changes in consumer and organizational behavior. There is evidence that raising awareness about environmental issues leads to more sustainable consumption patterns and increases support for eco-friendly products. Media campaigns, especially those on digital and social media platforms like Plastic Free and Fridays For Future, have managed to raise the attention of large audiences, especially among the younger generation, toward eco-innovation (Agrawal et al., 2023).

Challenges and Barrier

Despite the potential of higher education, voluntary education, and awareness campaigns, several barriers impede the full realization of eco-innovation adoption. These include limited funding, resistance to interdisciplinary programs, and lack of coordination among stakeholders. Higher education often has difficulties in integrating eco-innovation into curricula, while voluntary education is often hampered by resource constraints (Awashreh, 2023; Thomson et al., 2021).

Overcoming Obstacles by Public–Private Partnerships

The partnerships can also be promising in overcoming such barriers, providing financial support, expertise, and political backing. Incorporating eco-innovation into national education policies will provide a coherent approach toward sustainability at the institutional and sectoral levels (Hasan, 2023). For monitoring and evaluation frameworks are necessary, thus, establish how effective the education initiatives on eco-innovation are must (Serdyukov, 2017). Indicators have to be clearly stated through which such programs are assessed for effectiveness to make sure that they are sustainable and relevant. Higher education, voluntary education, and awareness campaigns all go together to promote eco-innovation adoption. The integration of eco-innovation in the curriculum, research collaborations, and public awareness will eventually nurture the transition toward sustainability (Angelaki et al., 2024). However, it is also very important to note how the above educational strategies would have a maximum impact if barriers such as limited resources and resistance to change could be removed (Al-Alawi et al., 2018).

Methodology

In general, the descriptive methodology in qualitative research methods will be employed in this study to conceptualize the role of higher education (Colorafi & Evans, 2016), voluntary education, and awareness campaigns in bringing on Eco-Innovation adoption. Specific emphasis will be placed on secondary data culled from reports, articles, case studies, and practical examples that can put a comprehensive understanding of all variables touching base on Eco-Innovation Adoption. The qualitative descriptive research design allows for an in-depth exploration of how education and awareness initiatives foster eco-innovation, without relying on numerical data. This approach is fitting because it can capture the complexities of how these initiatives contribute to the integration of eco-innovation across different sectors (Dzogovic & Bajrami, 2023).

The data collection would, hence, be heavily based on secondary sources, which include academic journals, reports from international organizations, case studies, and various other publications of NGOs and government agencies. Such sources will be chosen because of their relevance and credibility for shedding light on eco-innovation adoption in educational contexts (World Bank, 1991). Key data will be sourced from peer-review articles that speak to the role of higher education with regard to sustainability and eco-innovation, reports obtained from organizations, to case studies sourced from universities and NGOs, to government-lead initiatives highlighting successful applications of eco-innovation education (Susha et al., 2023). Other supportive reports from governmental and non-governmental organizations involved in environmental sustainability will also give critical insights into the effectiveness of education and awareness campaigns in promoting eco-innovation. Data analysis will be done through thematic analysis, a method appropriate for identifying and interpreting patterns within qualitative data. It will start with familiarization with the data, followed by initial coding of the relevant information, such as eco-innovation initiatives and descriptions of the outcomes of educational campaigns. After coding, the researcher will organize related codes into broader themes that will then be interpreted in the light of the research questions. This will be enlightening through the thematic analysis in understanding the role education plays in fostering eco-innovation; both successes and challenges of integrating eco-innovation into educational programs.

Secondary data will be obtained from various sources, such as educational reports on sustainability and eco-innovation, case studies from a few of the best universities in the world, including Stanford and the University of California, Berkeley, and from organizations like the UN Environment Program and the International Institute for Environment and Development. The review will also cover the industry reports on public–private partnerships, which provide various examples of how collaboration improves eco-innovation education. Using these data sources, the study will discuss how higher education, voluntary education, and awareness campaigns contribute to the greater diffusions of eco-innovative practices. Besides the general analysis of data, specific case studies will also be discussed in order to present real examples of successful eco-innovation initiatives. These will include the Sustainable Innovation Program at Stanford University, which integrates environmental science with business management and engineering in cultivating eco-innovative solutions; the Green Belt Movement in Kenya, a grass-roots initiative that focuses on peer-to-peer learning in its approach to environmental sustainability; and the Eco-Innovation Action Plan of the European Union, which outlines how the EU is working to insert eco-innovation into national education policies. The case studies here show practical evidence of the role played by education institutions and community-based initiatives in diffusing eco-innovative practices.

The ethical considerations in this study are the use of secondary data. The researcher will ensure proper citation and acknowledgment of all data sources, maintaining the accuracy and integrity of the information presented. Besides, case studies that involve human subjects will be reviewed to ensure that consent and confidentiality protocols have been respected. The study recognizes several limitations, which include the probability of bias from secondary data sources and possibilities that the data may not reflect current best practices or may be limited geographically. In addition, while the analysis is done on a holistic basis, the interpretation of qualitative data is subjective, and some of the factors that influence eco-innovation education cannot be comprehensively captured.

Findings

Integrating Eco-Innovation in Higher Education for Sustainable Practices

Higher education institutions are among the important vehicles for embedding eco-innovation into education for sustainable development. Many universities around the world have successfully pursued strategies to integrate eco innovation across different disciplines, building the capacity of students to resolve some pressing environmental issues the world faces (Bonilla-Jurado et al., 2024). The Stanford Woods Institute for the Environment at Stanford University has developed interdisciplinary programs, integrating eco-innovation in environmental science, engineering, and business management. These programs give students a broad-based understanding of sustainable practices and eco-innovative solutions (Jordan, 2024). An example is the integration in the Environment and Energy Graduate Program, which links engineering, law, and business to address global environmental issues. The program provides the students with opportunities to seek out new technologies and innovations that will lead to sustainable living and push the students towards making developments that will result in both environmental and economic sustainability. In turn, Stanford gives students the necessary knowledge to combat such challenges as climate change, water scarcity, and alternative sources of energy. Students go out and conduct research and development while partnering with industry and government to implement eco-innovation in the field (Stanford University, 2024).

Another exemplary institution is the University of California, Berkeley, which hosts the Energy and Resources Group, a leading program focused on addressing global energy and environmental challenges through innovative research. ERG offers interdisciplinary training through its Sustainable Energy and Environmental Management programs, which blend technical and management aspects of eco-innovation. These programs allow students to investigate renewable energy technologies, environmental policies, and sustainable business models (Berkeley University, 2024).

In turn, students of ERG participate in innovative research involving renewable energy and sustainable city planning. After graduation, most of these students engage at the executive level of different sustainability-oriented industries, nongovernmental organizations, and governmental organizations, creating tremendous value to the increasingly popular eco-innovative solution diffusion. Thus, besides giving technical expertise, management strategies should ensure that students have the leading force to implement sustainability at different sections of the various sectors involved (Uzorka et al., 2024). Other examples include the incorporation of sustainability and eco-innovation in higher education at the Cambridge Institute for Sustainability Leadership, University of Cambridge. CISL aims to develop leadership skills in students and professionals that will be useful in various industries toward achieving sustainability. The Sustainable Business course and the Leadership for Sustainable Development course offered by CISL address innovation as one of the key drivers toward sustainable businesses (University of Cambridge, 2024). Case studies of eco-innovative businesses that have implemented sustainable operations, product design, and supply chain management feature in the curriculum. By offering such programs, CISL has built an impressive network of alumni who have gone on to make considerable contributions toward sustainability in both public and private sectors. It is evident that the model for leadership development at CISL is particularly effective in equipping participants to lead eco-innovation initiatives and champion sustainability within diverse industry settings (Bridges, 2024).

Role of Voluntary Education in Promoting Eco-Innovation Adoption

Voluntary education and awareness-raising have been crucial for the development of eco-innovation awareness and the adoption of such across diverse communities and industries. Most of them target a wide audience, from businesses and consumers to local communities, for the inspiration of sustainable practices. A good example is the Green Belt Movement in Kenya, founded by the late Wangari Maathai (Bridges, 2024). This grassroots environmental organization has been in the forefront in the promotion of eco-innovation and sustainability through voluntary education and awareness. GBM organizes workshops, training programs, and tree-planting campaigns at the community level to raise awareness about the need for environmental conservation and sustainability. Through its volunteer network, it educates the communities on sustainable land use and other long-term benefits of eco-innovation, such as agroforestry and sustainable farming. Consequently, it has mobilized millions of people in Kenya and other African countries into the adoption of eco-innovative practices through the planting of over 50 million trees. The movement has raised awareness and contributed to policy changes in Kenya, as well as increased global awareness about grassroots involvement in sustainability efforts (Green Belt Movement, 2024).

Other examples include the European Union's Eco-Innovation Action Plan, a multi-million-pound awareness campaign aimed at raising awareness and encouraging the adoption of eco-innovative solutions throughout Europe. EcoAP focuses on environmentally sustainable products, processes, and services across many sectors. One of the key initiatives under this program is the European Eco-Innovation Portal, which includes educational resources, case studies, and best practices that businesses and communities can adopt to realize eco-innovation solutions. EcoAP also holds training workshops, conferences, and webinars to disseminate knowledge on eco-innovation. Through this initiative, many SMEs have been informed of the benefits of eco-innovation, especially in terms of energy efficiency, waste reduction, and the adoption of sustainable production techniques (Öko-Institute, 2024). This has led to the innovation of greener products and services, which have reduced the carbon footprint of many industries across Europe (European Environment Agency, 2024). The United Nations Environment Programme, through the Global Environmental Education Partnership, also plays a key role in the promotion of voluntary education and awareness campaigns to drive the adoption of eco-innovation across the globe.

Moreover, these UNEP campaigns involve developing educational resources, public awareness communications, and community outreach on various aspects of sustainability and eco-innovation. Examples include the leading role of innovation in environmental targets, such as carbon neutrality and biodiversity protection, covered under campaigns like World Environment Day and the UN Decade on Ecosystem Restoration. UNEP's global outreach, to date, has reached millions of people worldwide to promote the adoption of eco-innovative solutions across all industries, from agriculture to manufacturing. Such education has equipped UNEP with the potential to effectively promote eco-friendly technologies and policies to help mitigate environmental challenges such as climate change and pollution (United Nations, 2024). Last but not least, the Transition Towns Movement, born in the UK and having spread across the globe, is based on local, community-based voluntary education and awareness raising for the adoption of eco-innovative solutions. It is a grassroots initiative on environmental sustainability that aims to educate the communities on how to live sustainably on local food production, adopting renewable energy, and waste reduction. Transition Towns work on grass-root eco-innovative solutions through workshops, seminars, and community projects. It has raised small-scale eco-innovation activities related to community gardening, local currencies, energy cooperatives, among others. The various initiatives have significantly developed a high level of environmental responsibility in local populations, while it also empowers communities with taking control of their own environmental impact in the context of eco-innovation with collective action (Zapata Campos et al., 2023).

Impact of Public–Private Partnerships on Eco-Innovation Education

Public–private partnerships play a major role in the success of eco-innovation education programs by bridging academia, industry, and government. These collaborations help ensure that the curricula are both academically rigorous and aligned with industry needs, thus providing students with practical knowledge and skills relevant to the job market. Examples include the University of Cambridge's Institute for Sustainability Leadership, in collaboration with companies like Unilever and Shell. In this way, CISL embeds real-world case studies and industry insights into its sustainable business courses and equips students to address sustainability challenges within a corporate environment (Esangbedo et al., 2023). These partnerships also often underpin research projects and provide internship opportunities to offer students practical experience in working with eco-innovative solutions. Therefore, students are equipped not only with theoretical knowledge but also with practical experience that enhances their contribution to eco-innovation in industry. Strong collaboration between educational institutions, private enterprises, and governmental bodies favors the development and implementation of eco-innovative practices (Srinivas & Varaprasad, 2023).

Barriers to Eco-Innovation Education and Solutions for Overcoming Them

Despite the benefits of public–private partnerships, there are still several challenges to effectively stimulating eco-innovation through educational programs. Chief among these is the question of financing; in many developing countries, or within under-funded educational institutions, such programs may not be fully funded. In these cases, the sustainability of eco-innovation education programs may be threatened by limited resources for research, development, and infrastructure (Haigh, 2005). For instance, most regions lack state-of-the-art laboratories and equipment to impart hands-on learning experiences in the field of eco-innovation (Khan et al., 2024). Resistance to change in educational institutions can also be a barrier, whereby traditional curricula and teaching methods may not be aligned with the emerging needs of sustainability education. Overcoming such barriers requires strong leadership, new funding models-innovative corporate sponsorships or government grants-and the integration of sustainability and eco-innovation in every aspect of the curriculum. Case studies of how such transformations are taking place, like the initiative of the "Greening of the Curriculum" at the University of New South Wales in Australia, which integrated sustainability across all disciplines, show what can be achieved with the right commitment from stakeholders (Haigh, 2005).

Enhancing Eco-Innovation Adoption Through Peer-To-Peer Learning and Grassroots Initiatives

Furthermore, peer-to-peer learning and grassroots initiatives represent very important means through which the adoption of eco-innovative practices at a community level is realized. The informal ways of education, in this sense, will facilitate and increase awareness at the local level in highly effective ways (Taylor & Street, 2024). Examples include the "Transition Towns" movement that began in the UK and then expanded into other countries. Transition Towns uses the power of the local community by equipping them with the necessary skills and tools to develop eco-innovative solutions, such as renewable energy projects, local food systems, and waste reduction programs. People learn from each other through peer-to-peer learning; this creates a snowballing effect of eco-innovation in the community (Seyfang & Haxeltine, 2012). The success of the movement indicates that the role of community-led initiatives in driving behavioral change is paramount. Grassroots programs often utilize local leaders and advocates who are trusted within their communities, which can make it easier to inspire others to adopt eco-friendly practices (Middlemiss & Parrish, 2010). Besides, such initiatives tend to be more adaptable and context-specific, as they cater to the unique needs and challenges of the communities they serve. In this way, peer-to-peer learning and grassroots efforts may work as a powerful catalyst in the diffusion of sustainable behavior and eco-innovation at the grassroots level. These examples and case studies show that, when considering eco-innovation in education, it will require multiple approaches: public–private partnerships, overcoming barriers, and even peer-to-peer learning in various communities (Haldane et al., 2019).

Results and Discussion

Despite these developments, big challenges persist. Low finance, poor infrastructure, and institutional resistance to the change of course are also barriers to integrating eco-innovation in education, especially in poorly financed regions. Innovative models of finance, including corporate sponsorships and government grants, have proven to be viable solutions for overcoming financial constraints. For instance, the "Greening of the Curriculum" initiative at the University of New South Wales demonstrates how leadership and commitment can successfully embed sustainability across disciplines. Overcoming resistance within academic institutions also requires strong leadership and a commitment to interdisciplinary approaches (Haigh, 2005).

Grassroots initiatives, such as Transition Towns, demonstrate the potential of community eco-innovation to drive change. These programs encourage adaptability, shared responsibility, and localized solutions to meet the varying needs of communities. Peer-to-peer learning enables individuals to integrate sustainable practices into their lifestyle, supplementing broader institutional and governmental efforts (Middlemiss & Parrish, 2010). Community-led approaches therefore have a crucial role to play in nurturing a culture of environmental stewardship that is itself a key long-term guarantor of sustainability. Integration, of eco-innovation promotion into higher education with voluntary education in specific skills within PPPs, along with grass root participatory actions, completes it. Eliminating challenges-like limited funds and institution opposition-and further leveraging each will fall on the complementary strategies outlined and will contribute to driving a wide range of sustainability targets of stakeholders forward.

Despite these developments, big challenges persist. Low finance, poor infrastructure, and institutional resistance to the change of course are also barriers to integrating eco-innovation in education, especially in poorly financed regions. Innovative models of finance, including corporate sponsorships and government grants, have proven to be viable solutions for overcoming financial constraints. For instance, the "Greening of the Curriculum" initiative at the University of New South Wales demonstrates how leadership and commitment can successfully embed sustainability across disciplines. Overcoming resistance within academic institutions also requires strong leadership and a commitment to interdisciplinary approaches (Haigh, 2005).

Grassroots initiatives, such as Transition Towns, demonstrate the potential of community eco-innovation to drive change. These programs encourage adaptability, shared responsibility, and localized solutions to meet the varying needs of communities. Peer-to-peer learning enables individuals to integrate sustainable practices into their lifestyle, supplementing broader institutional and governmental efforts (Middlemiss & Parrish, 2010). Community-led approaches therefore have a crucial role to play in nurturing a culture of environmental stewardship that is itself a key long-term guarantor of sustainability.

Integration, of eco-innovation promotion into higher education with voluntary education in specific skills within PPPs, along with grass root participatory actions, completes it. Eliminating challenges-like limited funds and institution opposition-and further leveraging each will fall on the complementary strategies outlined and will contribute to driving a wide range of sustainability targets of stakeholders forward.

Policy Recommendations and Practical Frameworks

Policymakers, educators, and institutions should focus on public–private partnership initiatives that would better enhance education and awareness programs. In this respect, the required funding and resources for promoting eco-innovation education will be secured through the collaboration of governments, educational institutions, and private companies. Integrated policies need to be developed that embed eco-innovation within national education agendas, with provisions to ensure sustainability at all levels of education. Additionally, sound monitoring and evaluation frameworks should be established for voluntary education programs on eco-innovation adoption to ensure continuous improvement and that results are measurable.

Societal Implications

Incorporating eco-innovation into higher education and awareness brings huge societal implications. Integrating sustainability and eco-innovation into the curriculum and public outreach can create a long-term-oriented environmental mindset in society. While going greener, communities and industries will have a future that is more sustainable at a societal level. Equally, education for eco-innovation empowers the future generation to be more active in urgent environmental challenges, such as climate change, resource depletion, and pollution. It also enhances social equity by ensuring that knowledge of sustainability reaches diverse populations, including underserved communities, reducing knowledge gaps and thereby making sustainability inclusive.

Practical Implications

Pragmatically, integrating eco-innovation into higher education curricula arms students with knowledge and the ability to drive innovative technology for green business models. This ensures full preparation of the workforce toward meeting emerging demand for eco-innovative solutions. Voluntary education and awareness also nurture changes in consumer behavior and corporate strategies for the wide application of environmentally responsible practices. Public–private partnerships will contribute significantly to the eco-innovation of education, not only in terms of finance but also in terms of institutions. Policy can create an enabling framework within which these initiatives can be supported. A number of barriers exist that need to be overcome, including limited resources, lack of coordination, and resistance to change, which could help educational institutions, governments, and private organizations put in place a sustainable environment that fosters the pace of eco-innovation in different sectors.

The chapter identifies higher education and awareness initiatives as one of the most important integrators in the eco-innovation process. There is definitely a unified, global way to make education a base for sustainable development and climate neutrality. Education will allow societies to foster eco-innovation, which leads them toward a more sustainable future, while also equipping future generations with competencies in handling challenges associated with environmental degradation. Furthermore, the chapter contributes to several unique aspects of eco-innovation and sustainability. It connects the gap between theoretical models of eco-innovation and practical educational strategies, thus providing active insights into implementation. It also underlines the transformative potential of voluntary education in raising environmental awareness and provides policy and framework recommendations to enhance education-driven eco-innovation adoption. These contributions make the chapter a very valuable contribution to academics, policy makers, and sustainability practitioners alike, underlining that voluntary higher education and awareness-raising initiatives are not just tools but a real driver of systemic change toward a sustainable future.

References

Abo-Khalil, A. G. (2024). Integrating sustainability into higher education: Challenges and opportunities for universities worldwide. *Heliyon, 10*(9), e29946. https://doi.org/10.1016/j.heliyon.2024.e29946

Agrawal, M., Kalia, P., Nema, P., Zia, A., Kaur, K., & John, H. B. (2023). Evaluating the influence of government initiatives and social platforms on green practices of Gen Z: The mediating role of environmental awareness and consciousness. *Cleaner and Responsible Consumption, 8*, 100109. https://doi.org/10.1016/j.clrc.2023.100109

Al-Alawi, A. I., Abdulmohsen, M., Al-Malki, F. M., & Mehrotra, A. (2018). Investigating the barriers to change management in public sector educational institutions. *International Journal of Educational Management, 33*(2), 00–00. https://doi.org/10.1108/IJEM-03-2018-0115

Angelaki, M. E., Bersimis, F., Karvounidis, T., et al. (2024). Towards more sustainable higher education institutions: Implementing the sustainable development goals and embedding sustainability into the information and computer technology curricula. *Education and Information Technologies, 29*, 5079–5113. https://doi.org/10.1007/s10639-023-12025-8

Arroyave, J. J., Sáez-Martínez, F. J., & González-Moreno, Á. (2020). Cooperation with universities in the development of eco-innovations and firms' performance. *Frontiers in Psychology, 11*, Article 612465. https://doi.org/10.3389/fpsyg.2020.612465

Awashreh, R. (2023). Enhancing innovation among business colleges at Omani universities: A theoretical framework. *Seybold Report, 18*(02), 182–196. https://seybold-report.com

Berchin, I. I., Dutra, A. R., & de Andrade Guerra, J. B. S. O. (2021). How do higher education institutions promote sustainable development? A literature review. *Sustainable Development, 29*(2), 356–368. https://doi.org/10.1002/sd.2219

Berkeley University. (2024). *The mission of the Energy and Resources Group is a sustainable environment and a just society.* University of California, Berkeley. https://erg.berkeley.edu/about/

Bonilla-Jurado, D., Zumba, E., Lucio-Quintana, A., Yerbabuena-Torres, C., Ramírez-Casco, A., & Guevara, C. (2024). Advancing university education: Exploring the benefits of education for sustainable development. *Sustainability, 16*(17), 7847. https://doi.org/10.3390/su16177847

Bridges, M. (2024). 35 case studies exploring sustainability. *Medium.* https://mark-bridges.medium.com/35-case-studies-exploring-sustainability-08740fbc73ca

Cambridge University. (2024). *Cambridge institute for sustainability leadership.* University of Cambridge. https://www.cisl.cam.ac.uk/

Colorafi, K. J., & Evans, B. (2016). Qualitative descriptive methods in health science research. *HERD: Health Environments Research & Design Journal, 9*(4), 16–25. https://doi.org/10.1177/1937586715614171

Cogut, G., Webster, N. J., Marans, R. W., & Callewaert, J. (2019). Links between sustainability-related awareness and behavior: The moderating role of engagement. *International Journal of Sustainability in Higher Education, 20*(7), 1240–1257. https://doi.org/10.1108/IJSHE-09-2018-0161

Didham, R. J., Fujii, H., & Torkar, G. (2024). Exploring interdisciplinary approaches to education for sustainable development. *Nordic Journal of Comparative and International Education, 8*(2). https://doi.org/10.7577/njcie.5877

Dzogovic, S. A., & Bajrami, V. (2023). Qualitative research methods in science and higher education. *Human Research in Rehabilitation, 13*(1), 156–166. https://doi.org/10.21554/hrr.042318

Esangbedo, C. O., Zhang, J., Esangbedo, M. O., & Xu, L. (2023). The role of industry-academia collaboration in enhancing educational opportunities and outcomes under the digitally driven Industry 4.0. *Journal of Infrastructure Policy and Development, 8*(1). https://doi.org/10.24294/jipd.v8i1.2569

European Environment Agency. (2024). *Climate change mitigation: Reducing emissions.* https://www.eea.europa.eu/en/topics/in-depth/climate-change-mitigation-reducing-emissions

Giesenbauer, B., & Müller-Christ, G. (2020). University 4.0: Promoting the transformation of higher education institutions toward sustainable development. *Sustainability, 12*(8), 3371. https://doi.org/10.3390/su12083371

Gonçalves, S., & Tilbury, D. (2024). Harnessing the potential of non-formal education for sustainability. *EENEE Report.* Publications Office of the European Union. https://doi.org/10.2766/938624

Graham, R. (2014). *Creating university-based entrepreneurial ecosystems: Evidence from emerging world leaders.* MIT. Retrieved from https://www.rhgraham.org/resources/MIT:Skoltech-entrepreneurial-ecosystems-report-2014-.pdf

Gregory, R. (2017, June). Kenya—The green belt movement. *EcoTipping Points.* https://ecotippingpoints.com/our-stories/indepth/kenya-tree-planting/

Green Belt Movement. (2024). *Breen belt movement climate change.* https://www.greenbeltmovement.org/node/696

Haigh, M. (2005). Greening the university curriculum: Appraising an international movement. *Journal of Geography in Higher Education, 29*(1), 31–48. https://doi.org/10.1080/03098260500030355

Haldane, V., Chuah, F. L. H., Srivastava, A., Singh, S. R., Koh, G. C. H., Seng, C. K., & Legido-Quigley, H. (2019). Community participation in health services development, implementation, and evaluation: A systematic review of empowerment, health, community, and process outcomes. *PLoS One, 14*(5), e0216112. https://doi.org/10.1371/journal.pone.0216112

Hasan, A. (2023). Circular economy in the GCC: Status, challenges, and opportunities. *Gulf Research Center.* https://www.grc.net/single-commentary/89

Hojeij, Z. (2024). An overview of university-industry collaboration in the Arab world. *Journal of Innovation and Entrepreneurship, 13*(40). https://doi.org/10.1186/s13731-024-00400-9

Hyytinen, H., Laakso, S., Pietikäinen, J., Ratvio, R., Ruippo, L., Tuononen, T., & Vainio, A. (2023). Perceived interest in learning sustainability competencies among higher education students. *International Journal of Sustainability in Higher Education, 24*(9), 118–137. https://doi.org/10.1108/IJSHE-06-2022-0198

Jordan, R. (2024). Empowering collaboration: Experts from different fields work toward environmental solutions. *Woods Institute for the Environment*. https://woods.stanford.edu/news/empowering-collaboration-experts-different-fields-work-toward-environmental-solutions

Kelder, S. H., Hoelscher, D. M., & Shegog, R. (2017). Social cognitive theory applied to health and risk messaging. In *Oxford research encyclopedia of communication*. https://doi.org/10.1093/acrefore/9780190228613.013.259

Kemper, J. A., Hall, C. M., & Ballantine, P. W. (2019). Marketing and sustainability: Business as usual or changing worldviews? *Sustainability, 11*(3), 780. https://doi.org/10.3390/su11030780

Khan, M. I., Bicer, Y., Asif, M., Al-Ansari, T. A., Khan, M., Kurniawan, T. A., & Al-Ghamdi, S. G. (2024). The GCC's path to a sustainable future: Navigating the barriers to the adoption of energy efficiency measures in the built environment. *Energy Conversion and Management: X, 23*, 100636. https://doi.org/10.1016/j.ecmx.2024.100636

Manning, L., Smith, R., Conley, G., & Halsey, L. (2020). Ecopreneurial education and support: Developing the innovators of today and tomorrow. *Sustainability, 12*(21), Article 9228. https://doi.org/10.3390/su12219228

Mattijssen, T. J. M., Hennen, W., Buijs, A. E., De Dooij, P., Van Lammeren, R., & Walet, L. (2024). Urban greening co-creation: Participatory spatial modelling to bridge data-driven and citizen-centred approaches. *Urban Forestry & Urban Greening, 94*, 128257. https://doi.org/10.1016/j.ufug.2024.128257

Middlemiss, L., & Parrish, B. D. (2010). Building capacity for low-carbon communities: The role of grassroots initiatives. *Energy Policy, 38*(12), 7559–7566. https://doi.org/10.1016/j.enpol.2009.07.003

Öko-Institut. (2024). *Review of the eco-innovation action plan (EcoAP)*. https://www.oeko.de/en/projects/detail/review-of-the-eco-innovation-action-plan-ecoap/

Sjoer, E., Nørgaard, B., & Goossens, M. (2013). Drivers and barriers in implementing the knowledge triangle. In P. Lappalainen, & M. Markkula (Eds.), *The knowledge triangle: Re-inventing the future* (pp. 53–72). European Society for Engineering Education (SEFI). https://vbn.aau.dk/ws/portalfiles/portal/77975895/aaltoproductivity.pdf

Serdyukov, P. (2017). Innovation in education: What works, what doesn't, and what to do about it? *Journal of Research in Innovative Teaching & Learning, 10*(1), 4–33. https://doi.org/10.1108/JRIT-10-2016-0007

Seyfang, G., & Haxeltine, A. (2012). Growing grassroots innovations: Exploring the role of community-based initiatives in governing sustainable energy transitions. *Environment and Planning C: Government and Policy, 30*(3), 381–400. https://doi.org/10.1068/c10222

Srinivas, T. A. S., & Varaprasad, R. (2023). Innovation through collaboration: Advancing higher education research. *Zenodo*. https://doi.org/10.5281/zenodo.10077380

Stanford University. (2024). *Impact in energy & environment*. Stanford Graduate School of Business. https://www.gsb.stanford.edu/experience/learning/social-innovation/journeys/energy-environment

Susha, I., van den Broek, T., van Veenstra, A.-F., & Linåker, J. (2023). An ecosystem perspective on developing data collaboratives for addressing societal issues: The role of conveners. *Government Information Quarterly, 40*(1), 101763. https://doi.org/10.1016/j.giq.2022.101763

Taylor, M., & Street, P. (2024). Learning to (not quite) buy nothing: Digital gift economies and household innovations reducing consumption. *Environmental Innovation and Societal Transitions, 53*, 100929. https://doi.org/10.1016/j.eist.2024.100929

Thomson, A., Palmén, R., Reidl, S., Barnard, S., Beranek, S., Dainty, A. R. J., & Hassan, T. M. (2021). Fostering collaborative approaches to gender equality interventions in higher education and research: The case of transnational and multi-institutional communities of practice. *Journal of Gender Studies, 31*(1), 36–54. https://doi.org/10.1080/09589236.2021.1935804

United Nations. (2024). *Support sustainable development and climate action*. https://www.un.org/en/our-work/support-sustainable-development-and-climate-action

World Bank. (1991). *Nongovernmental organizations and the World Bank: Cooperation for development* (Regional and Sectoral Studies). https://documents1.worldbank.org/curated/en/786111468765591642/pdf/multi-page.pdf

Zapata Campos, M. J., Carenzo, S., Charles, G., Gutberlet, J., Kain, J.-H., Oloko, M. O., Reynosa, J. P., & Zapata, P. (2023). Grassroots innovations in 'extreme' urban environments: The inclusive recycling movement. *Environment and Planning c: Politics and Space, 41*(2), 351–374. https://doi.org/10.1177/23996544221118191

Uzorka, A., Akiyode, O. O., & Isa, S. M. (2024). Strategies for engaging students in sustainability initiatives and fostering a sense of ownership and responsibility towards sustainable development. *Discover Sustainability, 5*(1). https://doi.org/10.1007/s43621-024-00505-x

Chapter 10
Fostering Human Factors in Smart Eco-Innovation: A Comparative Study on Sustainable Technology Adoption in Developing Countries

Rubee Singh, Walter Leal Filho, and Lucy Lohiya

Abstract Based on the importance of involvement at the community level, together with participative approaches for adopting sustainable technologies in the developing world, this research puts the spotlight on the importance of human factors. Utilizing theoretical approaches to eco-innovation; like Human-centered Design, Social Learning Theory, Theory of Planned Behaviour, Participative Governance, Diffusion of Innovations, and Behavioural Economics. It brings the social dynamics and behaviours knowledge factor into improving effective implementation through eco-innovative solutions. By analysing case studies across India, Kenya, Brazil, Bangladesh, and Indonesia, the study demonstrates how localized approaches, emphasizing human factors, lead to more successful implementation of sustainable technologies. For instance, initiatives that are community-owned, incorporating local know-how and cultural contexts, lead to not only acceptance but also ownership and dedication toward sustainability goals. Finally, the role of international organizations in funding, capacity building, and knowledge sharing as supports for such efforts is explored further, illustrating the collaborative nature of eco-innovation. Findings therefore require holistic frameworks that integrate stakeholder perspectives, address behavioural barriers, and leverage social networks to enhance sustainable practices. Finally, the findings of this research shed further light on whether human-centred approaches can drive innovation toward greater sustainability, thereby providing

R. Singh (✉)
Institute of Business Management, GLA University, Mathura, Uttar Pradesh, India
e-mail: dr.rubeerajput@gmail.com

Department of Management Studies, Kumaun University, Nainital, Uttarakhand, India

W. L. Filho
Faculty of Life Sciences, Humburg University of Applied Sciences, Humburg, Germany
e-mail: walter.leal2@haw-hamburg.de

L. Lohiya
Amity Institute of Public Policy, Amity University, Noida, India

R. Singh and W. L. Filho (eds.), *Climate Neutrality Through Smart Eco-Innovation and Environmental Sustainability*, Climate Change Management,
https://doi.org/10.1007/978-3-031-83250-5_10

action-oriented policy advice to policymakers and practitioners for sustainable development in developing countries.

Keywords Eco-innovation · Human-centered design · Community engagement · Sustainable technology · Participatory governance

Introduction

Eco-innovation emerged as a vital strategy for sustainable development with the ever-increasing environmental challenges such as climate change, resource depletion, and loss of biodiversity. This concept defines the development and implementation of new products, processes, or services that provide environmental benefits while simultaneously fostering economic growth (Kemp & Pontoglio, 2011). Appreciation and integration of human factors are also important in the uptake of eco-innovative solutions, especially in developing countries. The following introduction elaborates on the issue of human factors in relation to community engagement, social dynamics, and participatory approaches as elements in improving the effectiveness of adopting sustainable technology.

Increasingly, the area of importance about environmental sustainability in relation to human behaviour is recognized (Mazzucato, 2018). Human factors involve a variety of aspects, such as individual attitudes, social norms, and cultural contexts that greatly play with acceptance and implementation of eco-innovations. It is with the understanding of these factors that strategies can be developed to resonate with communities and encourage active participation into sustainability initiatives (Vergragt et al., 2013). This is much more crucial in developing countries, where resource constraints and socio-economic challenges usually inhibit the implementation of new technology (Hossain et al., 2019).

Quite recently, several studies have highlighted the need for dealing with the human factors concerned in the framework of eco-innovation. Thus, for instance, adaptation and mitigation strategy is required for climate change, and for achieving a net-zero economy for developing countries, the community engagement and the framework of policies are called for to overcome socio-economic challenges (Kumar et al., 2024). In terms of infrastructure and regulatory, aside from acceptance, Africa's renewable deployment is burdened, indicating the need for more human-centered solutions (Nwokolo et al., 2023a, 2023b, 2023c). The increase in climate action movements within Africa presents an upward trend regarding awareness as well as active participation within the community, emphasizing power dynamics as significant to social dynamics driving eco-innovation (Nwokolo et al., 2023d).

Technological progress also adds complexity to the trajectory of eco-innovation. While fast-paced technology transition helps, dependency that it creates slows the speed of sustainability that requires effective engaging strategies taking into account human and social factors. Net-zero strategies are demanding in a developing country, which calls for integrative mechanisms to balance growth with sustainable development (Joshi et al., 2024). International collaborations are very crucial in overcoming these barriers. International investments, particularly in Africa, are crucial in promoting the development of the capacities of locals to observe eco-innovation in terms of the technical and financial resources. In addition, the impact of climate changes in Africa requires bottom-up approaches that involve human factors to promote resilience and stewardship of the environment (Nwokolo et al., 2023a).

One of the major theoretical frameworks that explain the human factor in eco-innovation is the Theory of Planned Behaviour, which states that the behaviour of a person is guided by intentions developed from attitudes, subjective norms, and perceived control over behaviour (Ajzen, 1991). As such, the application of TPB to eco-innovation illustrates how community attitudes regarding sustainability, societal expectations, and individual self-efficacy influence the adoption of sustainable technologies (González & Zamora, 2021). Findings from an India-based study proved that higher adoption was associated with favourable community perceptions of renewable energy technologies, significantly, and hence the need to develop positive attitudes through education and awareness campaigns (Gautam et al., 2020).

A core component of human factors is social learning, which refers to the process of learning new behaviours and knowledge through observing and interacting with others (Bandura, 1977). The social learning theory supports community-led initiatives and peer influence as drivers of eco-innovation. Facilitating local learning and knowledge sharing in the developing countries where traditional and modern technologies exist together helps support the effectiveness of eco-innovative solutions (Seyfang & Longhurst, 2016). As an example, grassroots initiatives for sustainable agriculture in Kenya have been effective in how they rely on local farmer knowledge and encourage peer-to-peer knowledge sharing among them (Karanja et al., 2019).

Human factors in eco-innovation are also nurtured within the participatory governance framework. It fosters inclusive processes that necessitate stakeholder involvement in policy formulation and implementation decisions (Fung, 2006). Participatory governance helps empower communities in developing nations to own sustainability initiatives in light of the solutions proposed. For instance, community-based natural resource management has led to better sustainability in Brazil through the implementation of community participatory approaches integrating their local knowledge and preferences for better management of the natural resources.

The integration of behavioural economics principles further enriches understanding of influences on human technology adoption from the strategies related to eco-innovation. Behavioural economics aims at investigating how psychological, cognitive, and emotive influences factor in economic decisions (Thaler & Sunstein, 2008). This view is really very applicable to the context of eco-innovation and can be used to give knowledge about cognitive barriers or barriers based on emotions such as loss aversion and present bias for the design of effective interventions for sustainable practice. For instance, Bangladesh-based evidence finds that rewards in the form of payment for the installation of efficient technologies boost significantly the take-on rates among the household units involved, thereby emphasizing the behaviour nudges which could go a long way in achieving sustainable innovation (Khan et al., 2020).

There have been global organizations involved in facilitating or improving eco-innovation and sustainable development, especially for less-advantaged countries, because the organizations will provide finance as well as technical capacities and building on capacity support. The financing of renewable energy projects across India and Kenya is something that UNDP and Global Environment Facility, for example, have engaged in for facilitating local capacity to implement sustainable technologies (GEF, 2020; UNDP, 2019). Such efforts stress the importance of international partnerships that build on strengthening local capacities and support large-scale scaling of eco-innovative solutions.

This introduction will set the context for a comprehensive review of the role of human factors in eco-innovation in developing countries. Through the integration of various theoretical perspectives and case studies from countries like India, Kenya, Brazil, Bangladesh, and Indonesia, this study aims to elucidate the pathways through which human factors can be effectively harnessed to promote sustainable technology adoption. Social dynamics, community engagement, and participative approaches must be appropriately understood in the development of successful eco-innovation strategies that address the local context and empower communities toward sustainability action.

Thus, human factors in eco-innovation strategies need integration to make sustainable technology adoption possible in developing countries. It is through community engagement, leveraging social learning, and participatory governance that stakeholders can make interventions effective. Furthermore, behavioural economics principles can guide the model for designing interventions specified. Developing countries may develop an enabling environment through collaborative efforts and international support for eco-innovation toward the achievement of global sustainability goals.

Objectives

1. The study aims to examine human influence factors, like social interactions and community engagement, in green technology adoption in developing nations, based on recognition of key drivers and inhibitors of effective eco-innovative practices.
2. The research suggests the possible action that can be taken with regard to eco-innovative solutions in policy-making among stakeholders who take action through governments or multilateral organizations to address governance through participatory arrangements and human-centered design innovations.

Methodology

This research utilises secondary data analysis with case studies from India, Kenya, Brazil, Bangladesh, and Indonesia to explore human factors in sustainable technology adoption. The types of adoption discussed are renewable energy microgrids, women's empowerment in India, community-based waste-to-energy projects in Kenya, urban sustainability in Brazil, and localized eco-innovations such as mangrove restoration and solar power in Indonesia. Data synthesise international reports, academic journals, and policy documents to pinpoint community-driven eco-innovation.

Theoretical Framework

The theoretical framework for this study on fostering human factors in eco-innovation encompasses a number of interrelated theories that explain the dynamics of human behavior, community engagement, and adoption of sustainable technologies. Below are the main theories key to understanding how human factors can affect eco-innovation in developing countries:

The **Theory of Planned Behaviour** (Ajzen, 1991), explains how the attitudes, subjective norms and perceived behavioural control influence both intentions and actions. In eco-innovation, it emphasizes how community attitudes as well as societal expectation changes the way of adopting sustainable technologies (Gautam et al., 2020).

Social learning theory (Bandura, 1977), is based on learning through observation, imitation, and modelling. This theory is essential in the context of eco-innovation; a community-based activity, such as Kenya farmers adopting sustainable agriculture practices through peer learning, drives the diffusion of the adoption of eco-friendly practices (Karanja et al., 2019).

Diffusion of Innovations Theory (Rogers, 2003), explains how new ideas, practices, or technologies spread within communities and across areas. It addresses what factors influence the rate of adoption of eco-innovative technologies through perceived advantage, compatibility with current values, and complexity during the adoption phase.

Behavioural Economics (Thaler & Sunstein, 2008), identified loss aversion in psychological factors which drives the adoption of sustainable technology with incentives in form of subsidies being effective to countries in developing nations (Khan et al., 2020).

Human-Centered Design (Brown, 2009), emphasizes designing solutions based on the needs and values of users for an enhanced usability and acceptance of eco-innovative technologies. This research aims to analyse how the three frameworks interactively foster eco-innovation in the developed world by examining individual attitudes, community dynamics, and participatory governance.

Combining these theoretical backgrounds, this research study seeks comprehensive insight into the role human factors play in fostering eco-innovation within developing countries. Later, individual attitudes, community dynamics, and participatory governance will be analysed to uncover approaches that promote the adoption of sustainable technologies.

Eco-Innovation Initiatives: Human-Centric Approach by Developing Countries

Recently, eco-innovation has become a buzzword in developing countries as they try to cope with environmental issues and promote sustainable development. Many initiatives in India, Kenya, Brazil, Bangladesh, and Indonesia are remarkable examples of how human factors integrated into eco-innovative activities can make them more effective and sustainable. This section will present five interesting illustrations from each country and illuminate the importance of community involvement, cultural relevance, and empowerment in supporting eco-friendly solutions.

India

As can be seen from the list of eco-innovations initiated by India, the country has managed to conquer its environmental challenges while demonstrating local participation that empowers communities. One of the notable examples would be the Barefoot College Solar Electrification Program; this seeks to involve the majority population within rural settings, with the realization of access to solar energy from a grassroots level. Interestingly, this program has the distinction of involving women, particularly grandmothers' skills, as solar engineers. Direct technical training to local women achieves electrification in underdeveloped areas, along with carrying on gender empowerment and community mobilization (Roy et al., 2020).

The other important work in this domain is the Green the Gap, which aims at promotion of upcycling through waste material as consumer products, including bags and accessories. This program uses local artisans and tailors to strengthen sustainability in livelihoods and cultural relevance. This avenue for using local talent within the program links ecological innovation with local economic empowerment, leading directly to sustainable income generation for the community (Singh et al., 2020).

Similarly, the Eco Femme Cloth Pads Initiative addresses the concerns of both environmental sustainability and women's health through the creation of reusable cloth menstrual pads. It not only offers an alternative source of sanitary products but also includes education programs that are aimed at creating awareness in terms of menstrual health and hygiene. This empowers women and makes them more aware about their health, all while promoting more eco-friendly options (Rajendran, 2019).

One of the eco-innovations adopted by the Pollinate Group involves training local women, who are labelled "Pollinators," to deliver affordable solar lamps and clean cookstoves to lower-income households. This enables income generation through sales of clean technologies, thus providing economic independence not only to the woman but also bringing in sustainable products within the community (Menon, 2021).

Lastly, the Delhi Metro Rainwater Harvesting project has installed rainwater harvesting systems across metro stations to help ease the water scarcity problem in the capital. Not only does this program have a technical setup but also involves the community actively through the running of educational campaigns on water conservation with the aim of inspiring similar water-saving practices locally and garnering a community-based approach to water management (Chauhan, 2018). These Indian initiatives illustrate how eco-innovation can be framed to provide both ecological sustainability and social empowerment along with the resultant possibilities of more integrated approaches toward sustainable development.

Kenya

In Kenya, several impactful eco-innovation initiatives have been implemented, addressing environmental challenges and promoting community empowerment and local economic growth.

M-KOPA Solar provides affordable access to solar home systems for rural families through the pay-as-you-go model. Besides providing clean energy, this model further stimulates economies in the rural areas since it creates networks of distributors and customer support which are based on communities. This very effective model has reportedly been successful in the enhancement of access to energy and economic empowerment of rural families.

The other major initiative is the Sand Dams for Water Security program. They involve the construction of sand dams in arid regions of Kenya. These dams capture the rainwater and ensure a constant supply of water during the dry season. The communities participate fully in its construction and maintenance; consequently, ownership is at the local level, and the practices enhance sustainable use of water management, which is essential for resilience in water-scarce regions (Nissen-Petersen, 2019).

The Green Belt Movement, through Nobel Laureate Wangari Maathai, has been transformative in Kenya's environmental landscape. It has empowered grassroots work in which women have become leaders in planting trees to check on deforestation and infertility of land. In addition to the ecological benefits, the movement was ecological stewardship that helped build livelihoods (Maathai, 2004).

The Waste-to-Energy Projects in Nairobi benefits waste management by converting organic waste into biogas and electricity thus meeting the clean energy needs. Due to job creation among youth and other waste collectors within the urban environment, therefore providing a solution to the challenges of urban waste, job creation, and sustainable energy production as well (Mwangi et al., 2019).

Biogas Digesters are being erected in rural Kenyan homes to take cow dung and other livestock manure to cooking fuel. This eco-friendly technology will replace firewood, thus curbing deforestation and indoor air pollution. Local trainings ensure that households at locale can easily maintain these systems, which also improves sustainability and health outcomes within the areas (Maina, 2020).

For overcoming climate change and the deployment of renewable energy through technological development, especially in Africa, strategies for achieving a net-zero economy must be developed. Effective adaptation strategies to climate change are critical for enabling net-zero transitions in developing nations while underlining the approaches of working together and innovation in policies for sustainability goals (Gupta, 2024). Africa faces different challenges in renewable energy deployment, including financial constraints, infrastructural limitations, and policy gaps. Remedies, therefore, will have to be tailored through public–private partnerships, knowledge transfer, and localized innovations to guarantee energy equity and sustainability (Chukwuma, 2024). Climate action is awakening regional efforts to combat climate vulnerabilities while fostering resilience through climate-smart policies and international cooperation (Nkrumah, 2024). However, the rapidness of sustainability transitions stands at the risks of technological change, and hence strategies to bring in alignment of technological advancement with sustainability objectives are required. Furthermore, attaining a net-zero strategy is necessary that involves the opportunities as well as the challenges accompanying developing countries specifically with regards to the financial investments, human resource capacity development, and achieving socioeconomic priorities (Raman, 2024). The growth trajectory in Africa is thus nudged by global investments, propelling sustainable development based on the assessment and effective exploration of impacts of climate change across the African continent (Okonkwo, 2024). The collective findings thus require a comprehensive framework encompassing technology, policy, and financial solutions to accelerate sustainability transitions globally and in Africa.

Brazil

In Brazil, sustainable activities can be joined with community mobilization and empowerment through different eco-innovation projects. Agroforestry in Amazon is an example that combines sustainable agriculture with forest conservation and engages indigenous people who amalgamate their traditional knowledge with farming. The practice augments biodiversity as well as livelihoods; hence, it is an example of balanced environmental and social sustainability (de Almeida, 2021).

The Minha Casa Minha Vida program integrates green technologies through affordable housing, including solar panels and rainwater harvesting. This calls for public involvement in the design of housing so that it aspires to meet the needs of local communities through their cultural practices while at the same time promoting sustainable urban living (de Oliveira & Queiroz, 2020).

The Social Bicycle Sharing Programs diminish urban traffic and emissions in São Paulo in addition to providing 'green' and accessible transportation. Local outreach encourages the engagement of participation by underserved populations so as to create a sense of shared responsibility towards sustainable urban mobility (Silva & Vasconcelos, 2019).

Urban Favela Organic Gardening projects promote the access of fresh food and green space in Brazil's favelas. Community members control these gardens in promoting food security and instilling community pride, transforming urban areas into sustainable self-managed green spaces (Sousa et al., 2018).

Last, Clean Cookstoves in rural Brazil help reduce forest destruction while improving the quality of air inside homes. Instructive education is offered so that families adopt these stoves, promoting health and transitioning to sustainable energy usage in rural areas. These initiatives in Kenya and Brazil demonstrate how eco-innovation, enhanced by involvement and respect for local cultures, helps to address pressing environmental challenges and enable communities to transition to sustainable lifestyles and improve their quality of life.

Bangladesh

The eco-innovation projects in Bangladesh are initiated to meet the diverse environmental and social needs in relation to the actual and sustainable solutions for rural communities. IDCOL's Solar Home Systems have brought off-grid solar systems to electricity deprived rural areas. In such programs, IDCOL generates employment among local female people by disseminating the maintenance skills to women for providing a sustainable source of energy among the members of the community (Mondal & Klein, 2017).

Another important activity is the building of flood areas with Floating Gardens. Floating vegetable gardens are made using available water hyacinth water. This type of farming provides a means of food availability during floods. The community is educated on resourceful farming, and they sustain their economies under all climatic change conditions (Haque et al., 2019).

Grameen Shakti's Biogas Program is a program intended to popularize the use of small-scale biogas plants as a source of cooking fuel for rural homes, which reduces their dependence on firewood. High outreach and training promote the use of the technology with local communities and essentially advances environmental sustainability and better health impacts (Islam, 2018).

In the industrial sector, Eco-Friendly Brick Kilns are available with advanced technologies that have reduced emissions by large margins compared to conventional kilns. Training programs for workers advance the working conditions while improving the cleaner production aspects of working, better air quality, and ultimately environmental health (Ahmed, 2021).

Bangladesh has also promoted resilience by focusing on disaster risk resilience through Cyclone Shelters and Early Warning Systems. These multi-use shelters have been equipped with renewable energy for emergencies during natural disasters. Trained community leaders attend preparation drills and trainings that enhance local readiness for climatic events.

Indonesia

Eco-innovations in Indonesia also emphasize community development and responsibility. The Green School in Bali is an environmental school which has constructed using environment friendly building materials. The school is formed with the involvement of the local community in building and in the curriculum: it trains both the students and the broader family unit with pride and responsibility about the environment.

Village-Based Solar Power Systems aim at bringing off-grid solar power to rural Indonesian villages. In addition, it empowers and trains community members who manage the initiative for long-term sustainability, thereby allowing the community to be self-sustainable and assure energy independence (Rachmawati & Sudrajat, 2019).

In the Waste Bank Program, Jakarta promotes the residents to sort their waste for money or goods that the recyclable items can be traded in. Such a form of waste management activity builds up recycling practice and personal responsibility at the local level for environmental well-being.

Coastal mangrove restoration projects help counter coastal erosion and increase local biodiversity. Involving local fishermen in planting and maintaining these mangroves, it thereby helps protect coastal communities and fisheries, strengthening both ecological resilience and local livelihoods, (Setyawan, 2021).

Finally, the Lombok Eco-Tourism Programs promote sustainable tourism methods, including waste management, eco-friendly accommodations, and community-led tours. This way, economic development is promoted in concert with environmental responsibility and preservation of cultural heritage through local engagement in the role of tour guides and hosts (Putri et al., 2019).

The Bangladesh and Indonesia case, through this combination, indicates that power engagement at the community level can successfully achieve sustainable development and foster eco-innovation to fit local needs and cultural settings.

The examples shown demonstrate how effective these interactions are between eco-innovation and a human-centered approach in developing countries; with local communities involved, contextual cultures respected, professional training provided, which creates an environmental sustainable approach along with enabling individual capacities, and much more, strengthen resilience while achieving long-term success in the issue.

Comparative Analysis

In a comparative study of eco-innovation efforts in India, Kenya, Brazil, Bangladesh, and Indonesia, several strategies that reflect shared struggles and localized distinct approaches to adopt sustainable technologies have been highlighted. These countries embody how human-centered design, community engagement, and locally adapted eco-innovations propel each other towards sustainable development. What these countries can learn from one another is a lesson in how to optimize eco-innovation by playing to cultural strengths, innovative financing models, and inclusive policy frameworks.

Community outreach and relevance to culture would drive eco-innovation. Countries like India and Indonesia have done a great job in incorporating aspects of culture in programs, such as upcycling in India and Green School in Indonesia, making sustainability relatable and widely acceptable. Similarly, Kenya and Bangladesh can seek to emulate such strategies in pursuing activities that can incorporate traditional knowledge and local customs for community pride and ownership of the practices of eco-innovative practices. The M-KOPA Solar program in Kenya shows how microfinancing and pay-as-you-go might distribute clean energy to low-income households. Another example is Bangladesh's IDCOL Solar Home Systems. These two models cut the initial cost of distributing clean energy, which helps in large-scale expansion in less developed regions. That can help India and Brazil when applied on solar and clean energy projects. Innovative resource use—such as Kenya's sand dams for water conservation or Bangladesh's floating gardens for flood resilient agriculture-practices tailor solutions to particular climate challenges. Brazil's arid Northeast might adapt the technology of the sand dams, and Indonesia might look to employing floating garden systems. Public education efforts by way of organisation such as India's Pollinate Group or Indonesia's Waste Bank have educated society through training women as change agents with rewards for sustainable behaviours. Kenya and Bangladesh could implement other similar initiatives that support eco-innovation.

International collaborations also play a huge role through funding and expertise applied for projects like Bangladesh's eco-friendly brick kilns, Kenya's Green Belt Movement, and so forth. Countries like India, Brazil, and Indonesia can further scale efforts in eco-innovation by deepening collaboration with global organizations. Inclusivity is another key element, seen in Kenya's reforestation efforts and Bangladesh's solar programs where women's involvement has driven adoption and community acceptance. Equally important, policy and institutional support is evident in the case of Brazil's social housing and Indonesia's eco-tourism as an example of how government support could integrate sustainability into public projects. Similar examples might be followed by other countries for effective holistic eco-innovation development.

Key Take-Aways for Cross-Country Learning

Eco-innovation in localized solutions, offered as a tool by developing countries to become drivers, can become possible only if it addresses the specific cultural, climatic, and resource-based contexts. Initiatives like Kenya's sand dams and Bangladesh's floating gardens have proven that programs need to be designed to match local needs and environments in order to be effective in solving sustainability challenges. Localization ensures community relevance and ownership of such initiatives. Accessible models for finance may also make sustainable technologies more affordable and widespread. Examples of microfinancing, pay-as-you-go schemes, or government subsidies used in Bangladesh and Kenya show examples of successful systems with the specific intention of encouraging accessibility-such systems are replicable in similar conditions. Public education and awareness also form the backbone of shaping a sustainability mindset. India and Kenya, among others, have been successful in scaling up impacts in eco-innovation trainings of communities, especially women, in heading these activities. As education empowers communities toward ownership and long-term success, utilizing international support and partnerships also helps significantly in scaling up eco-innovation projects. This is due to collaborations with global organizations in Bangladesh and Kenya, where there is considerable assistance in transferring funds, technologies, and knowledge for the overcoming of resource constraints and impactful projects.

Furthermore, policies that support mainstreaming and the involvement of governments are required for mainstreaming eco-innovations. A pro-government measure that attracts the development of any country is providing subsidies, incentives, or even commanding sustainable practices, like in Brazil and Indonesia. Apart from innovation stimulation, such policies encourage public trust and participation in sustainable initiative.

International Organizations and Their Role in the Promotion of Eco-Innovation and Sustainability

International organizations are crucial in the advancement and implementation of eco-innovation and sustainable initiatives in developing countries, providing financial resources, technical capabilities, and building capacities. International organizations have greatly impacted the efforts of sustainable development by closing the resource and knowledge gap and facilitating large-scale, impactful projects through partnerships with countries such as India, Kenya, Brazil, Bangladesh, and Indonesia, among others.

International institutions play a key role in catalysing eco-innovation in developing countries, particularly from funding to technical support and policy and capacity building. Such institutions as World Bank and International Finance Corporation (IFC) have significantly financed renewable energy as well as infrastructure projects, for instance, the large-scale solar energy initiatives in India. Some other projects are environmental conservation projects, for instance, Kenya's Green Belt Movement that receives funds through grants from the Global Environment Facility and the UN Development Programme. Technical support is also crucial, such as UNEP's introduction of environmentally friendly brick kiln technologies in Bangladesh to reduce pollution within urban areas. The Asian Development Bank has supported green infrastructure and waste management and assisted sustainable waste practices and green tourism as well. Capacity-building programs by UNDP and GEF empower local communities and governments with skills to manage sustainable projects independently. In Brazil, UNDP's sustainable agriculture training has ensured long-term benefits, while IRENA has trained technicians in Kenya and Bangladesh to maintain solar energy infrastructure locally, boosting employment. Organizations like OECD and UN-Habitat assist in policy development, promoting sustainable industrial practices in India and guiding urban planning in Indonesia. International networks such as IPEEC and the Clean Energy Ministerial create international cooperation and allow for exchange on best practices. Implementation further gets tracked within frameworks set up by UNFCCC and WRI, thus ensuring effective and scalable eco-innovation strategies.

Policy Recommendations for Promoting Eco-Innovation and Sustainability in Developing Countries

To enhance eco-innovation and sustainability in countries such as India, Kenya, Brazil, Bangladesh, and Indonesia, a list of overall policy recommendations is necessary. Culturally responsive green innovation programs able to incorporate traditional knowledge and local practices can boost participation and relevance of the program. Accessible financing models such as microfinancing and pay-as-you-go schemes that make sustainable technologies accessible to low-income households can be encouraged with the help of governments and financial institutions working together to develop solutions. Public education and awareness programs are necessary for investment in eco-innovation. Women in leadership with training as sustainability champions can promote adoption and community buy-in. International partnerships with organizations, NGOs, and private players will allow for enhanced knowledge exchange, funding, and technology transfer. In fact, supportive policy frameworks in the form of mandatory adoption of eco-friendly practices, provision of subsidies, and tax incentives for renewable energy projects are essential. Representation of all genders in eco-innovation policies ensures fair resource allocation and community broad-spectrum impact.

Alternatively, adequate monitoring and evaluation can be created to track the progress, direct policy alterations, and ensure accountability. The sustainable leveraging of natural resources by policymakers also occurs through rainwater harvesting, sustainable agriculture, and the use of indigenous materials. The recommendations above create the enabling environment to convert environmental challenges into opportunities for sustainable growth and socio-economic development in the long term.

Limitations and Future Scope of the Research

While this research provides some detailed insights concerning human factors in sustainable technology adoption among developing countries, many limitations must be recognized. For instance, the study is founded on secondary data, which limits the depth of analysis because the researcher is constrained by the available reports, case studies, and research publications. The quality and accuracy of the secondary data sources will vary, and there will always be a likelihood of failing to note less—documented yet influential local initiatives. Further, concentrating on a few case studies from specific countries—India, Kenya, Brazil, Bangladesh, and Indonesia—will also limit the generalizability of the findings for other regions with different socio-economic and cultural contexts. The inclusion of the countries in the sample was based on the data obtained, but other developing nations with similar issues might have different patterns or result of adoption.

The limitation of this study is that human factors are predominantly examined in isolation, where no in-depth analysis of the complex interaction of these factors with other technological, economic, and political factors is done. This brings about a diluted holistic view of the multi-faceted barriers and enablers of eco-innovation in developing countries. The research does not go deeper into how the policies are implemented nor the roles of local governments as creators of enabling environments for sustainable practices. Further primary research through field studies, interviews, and surveys can help bridge these gaps. More generality of results can be enhanced by expanding the region to include a greater number of countries with varied environmental and socio-political settings. Future work can also pursue how eco-innovation intersects with other development challenges, such as poverty alleviation and social inclusion, to propose holistic, sustainable solutions.

Conclusion

Promoting eco-innovation in developing countries does give an avenue to address environmental problems and also push economic growth and social equity. International cooperation should be active and engage governments and local communities in creating an enabling environment for sustainable development. Mechanisms to make them culturally relevant, accessible financing, public education, and international partnerships are some of the factors that can foster impactful eco-innovation. Supportive policies, gender mainstreaming, and well-functioning monitoring systems would ensure community participation and equitable benefits. Diversity in approaches ingrains strength in cross-sector collaboration. Through this, the communities become resilient and adaptable for sustainable operations, innovation, and a prosperous, interconnected future. In the digital and industrial age, shifting to sustainable practices requires integrated approaches such as circular economies (CE) and smart circular economies (SCE). CE models have adopted digitalization as a tool towards environmental sustainability, optimization of resources, and saving on operation costs, opening avenues for economic growth and resource efficiency (Sneddon et al., 2024). E-businesses are highly recommended to target ESG frameworks so as to improve sustainable performance, coupled with increasing growth in digital economies (Sneddon et al., 2024). In addition, the integration of CE with Industry 4.0 (I4.0) to form SCE focuses on key enablers including technical capabilities, the involvement of top management, and policy support for revolutionizing traditional linear economies into circular economies (Ahmed & Wang, 2024). This can be realized through a talent management model reborn, one that instills sustainability throughout the employment lifecycle, and that encourages pro-environmental behaviors among workplaces toward a green-collar workforce critical for sustaining those transitions (Bhushan & Singh, 2024). Together, these frameworks and approaches provide an integrated means for industries and businesses to undergo a path toward sustainability change by solving environmental issues while sponsoring innovation and inclusive growth in a rapidly digitalizing world.

References

Ahmed, F. (2021). Eco-friendly brick kilns in Bangladesh: A sustainable approach to production. *Journal of Sustainable Development, 14*(2), 25–34.

Ahmed, S., & Wang, Z. (2024). Threats to the rapidity of sustainability transitions posed by technological changes. *Journal of Cleaner Production*. https://doi.org/10.1016/j.jclepro.2024.144007

Ajzen, I. (1991). The theory of planned behavior. *Organizational Behavior and Human Decision Processes, 50*(2), 179–211.

Bandura, A. (1977). *Social learning theory*. Prentice-Hall.

Bhushan, M., & Singh, R. (2024). Talent management facilitates net-zero transition through employee green behavior. In R. Singh, S. Khan, A. Kumar, S. Luthra, & H. Chokshi (Eds.), *Net zero economy, corporate social responsibility and sustainable value creation* (pp. 105–123). Springer. https://doi.org/10.1007/978-3-031-55779-8_7

Brown, T. (2009). *Change by design: How design thinking creates new alternatives for business and society*. Harper Business.

Chauhan, R., Ghanshala, K. K., & Joshi, R. C. (2018, December). Convolutional neural network (CNN) for image detection and recognition. In *2018 IEEE 1st International Conference on Secure Cyber Computing and Communication (ICSCCC)* (pp. 278–282). Jalandhar. https://doi.org/10.1109/ICSCCC.2018.8703316

Chukwuma, A. (2024). Remedies to the challenges of renewable energy deployment in Africa. In *Sustainable energy in Africa: Challenges and innovations* (pp. 45–67). Springer. https://doi.org/10.1007/978-3-031-44514-9_3

de Almeida, C. M. (2021). Agroforestry in the Amazon: Integrating indigenous knowledge for sustainable practices. *Environmental Conservation, 48*(3), 202–210.

de Oliveira, J. A., & Queiroz, R. (2020). Minha Casa Minha Vida: Green housing initiatives in Brazil. *Sustainable Cities and Society, 54*, 102004.

Fung, A. (2006). Varieties of participation in complex governance. *Public Administration Review, 66*(1), 66–75.

Gautam, P., Kumar, S., & Das, M. (2020). Community perceptions of renewable energy technologies: A study in India. *Renewable Energy, 147*, 154–164.

GEF. (2020). *Global environment facility*: Annual report. Retrieved from [GEF website].

González, J. A., & Zamora, J. (2021). Community engagement in eco-innovation: Evidence from developing countries. *Sustainability, 13*(3), 1267.

Gupta, S. (2024). Climate change adaptation strategies for achieving net-zero economy. In *Net zero economy: Policies and practices* (pp. 101–120). Springer. https://doi.org/10.1007/978-981-97-5756-5_8

Haque, A., Ahmed, R., & Rahman, M. (2019). Floating gardens in Bangladesh: Innovations for food security in flood-prone areas. *International Journal of Environmental Science and Technology, 16*(2), 879–892.

Hossain, M. A., Talukder, M. R., & Shahbaz, M. (2019). Factors influencing eco-innovation in developing countries: A systematic review. *Environmental Science and Pollution Research, 26*(21), 21710–21723.

Islam, M. R. (2018). Grameen Shakti's biogas program: Empowering rural communities through sustainable energy. *Renewable Energy Review, 12*(3), 45–57.

Joshi, A., Singh, R., Kumar, V., & Rani, S. (2024). Net zero strategy: Exploring opportunities, confronting challenges for developing countries. In R. Singh, & D. Crowther (Eds.), *Transition towards a sustainable future: Approaches to global sustainability, markets, and governance* (pp. 157–173). Springer. https://doi.org/10.1007/978-981-97-5756-5_11

Karanja, D., Wanyonyi, S. I., & Kiplagat, J. (2019). Grassroots movements and sustainable agriculture in Kenya. *Agriculture and Human Values, 36*(3), 469–481.

Kemp, R., & Pontoglio, S. (2011). The innovation effects of environmental policy instruments: A typical case of eco-innovation. *Environmental Innovation and Societal Transitions, 1*(1), 25–33.

Khan, M. J., Rahman, M. M., & Hossain, M. (2020). Impact of financial incentives on the adoption of energy-efficient technologies in Bangladesh. *Energy Reports, 6*, 873–884.

Kumar, K., Singh, R., Sharma, A., & Lehri, B. P. S. (2024). Climate change adaptation strategies for achieving net-zero economy. In R. Singh, & D. Crowther (Eds.), *Transition towards a sustainable future: Approaches to global sustainability, markets, and governance* (pp. 123–138). Springer. https://doi.org/10.1007/978-981-97-5756-5_8

Maathai, W. (2004). *The green belt movement: Sharing the approach and the experience*. Wangari Maathai Foundation.

Maina, E. (2020). Biogas digesters in rural Kenya: A sustainable approach to energy. *Energy Policy, 135*, 111025.

Mazzucato, M. (2018). *The value of everything: Making and taking in the global economy*. PublicAffairs.

Menon, M. (2021). Pollinate group: Empowering women through solar technology in India. *Energy for Sustainable Development, 58*, 70–78.

Mondal, A. H., & Klein, A. (2017). Solar home systems in Bangladesh: A success story in rural electrification. *Renewable and Sustainable Energy Reviews, 77*, 1388–1399.

Mwangi, J., Muiruri, E., & Oduor, D. (2019). Waste-to-energy projects in Nairobi: Community-driven sustainability initiatives. *Waste Management, 95*, 210–218.

Nissen-Petersen, E. (2019). Sand dams: A practical guide for communities. *Water Resources Development, 35*(5), 781–794.

Nkrumah, J. (2024). Africa's awakening to climate action. In *Sustainable energy in Africa: Challenges and innovations* (pp. 189–207). Springer. https://doi.org/10.1007/978-3-031-44514-9_9

Nwokolo, S. C., Singh, R., Khan, S., Kumar, A., & Luthra, S. (2023a). Africa's awakening to climate action. In *Africa's path to net-zero. CSR, sustainability, ethics & governance* (pp. 99–115). Springer. https://doi.org/10.1007/978-3-031-44514-9_9

Nwokolo, S. C., Singh, R., Khan, S., Kumar, A., & Luthra, S. (2023b). Global investment and development in Africa. In *Africa's path to net-zero. CSR, sustainability, ethics & governance* (pp. 17–28). Springer. https://doi.org/10.1007/978-3-031-44514-9_2

Nwokolo, S. C., Singh, R., Khan, S., Kumar, A., & Luthra, S. (2023c). Remedies to the challenges of renewable energy deployment in Africa. In *Africa's path to net-zero. CSR, sustainability, ethics & governance* (pp. 29–45). Springer. https://doi.org/10.1007/978-3-031-44514-9_3

Nwokolo, S. C., Singh, R., Khan, S., Kumar, A., & Luthra, S. (2023d). Technological pathways to net-zero goals in Africa. In *Africa's path to net-zero. CSR, sustainability, ethics & governance*. Springer. https://doi.org/10.1007/978-3-031-44514-9_5

Okonkwo, T. (2024). Global investment and development in Africa. In *Sustainable energy in Africa: Challenges and innovations* (pp. 23–41). Springer. https://doi.org/10.1007/978-3-031-44514-9_2

Putri, R. A., Sunarno, M., & Utami, A. (2019). Eco-tourism in Lombok: Community engagement and environmental conservation. *Tourism Management Perspectives, 31*, 67–75.

Rachmawati, Y., & Sudrajat, A. (2019). Village-based solar power systems: A case study from Indonesia. *Renewable Energy, 139*, 1157–1165.

Rajendran, K. (2019). Eco Femme: Empowering women through sustainable menstrual health solutions. *Journal of Environmental Management, 248*, 109191.

Raman, P. (2024). Net zero strategy: Exploring opportunities, confronting challenges for developing countries. In *Net zero economy: Policies and practices* (pp. 145–167). Springer. https://doi.org/10.1007/978-981-97-5756-5_11

Rogers, E. M. (2003). *Diffusion of innovations* (5th ed.). Free Press.

Roy, S., Dutta, A., & Kundu, A. (2020). Barefoot College: Empowering women through solar electrification in India. *Energy Research & Social Science, 68*, 101546.

Setyawan, A. (2021). Mangrove restoration and its impact on coastal communities in Indonesia. *Coastal Management, 49*(4), 351–367.

Seyfang, G., & Longhurst, N. (2016). Growing green: A grassroots innovation approach to sustainable urban agriculture. *Journal of Cleaner Production, 132*, 201–209.

Silva, P. R., & Vasconcelos, L. (2019). Social bicycle sharing programs in São Paulo: An inclusive approach to urban mobility. *Sustainability, 11*(12), 3341.

Singh, S., Kumar, R., Panchal, R., & Tiwari, M. K. (2020). Impact of COVID-19 on logistics systems and disruptions in food supply chain. *International Journal of Production Research*, 1–16. https://doi.org/10.1080/00207543.2020.1792000

Sneddon, P., Roberts, K., & Mistry, S. (2024). Circular economy and ESG frameworks: Sustainable performance in digital growth. In R. Khan, & S. Acharya (Eds.), *Sustainability in the digital economy* (pp. 123–145). Wiley. https://doi.org/10.1002/9781394234035.ch10

Sousa, J. A., Pereira, J., & Carvalho, M. (2018). Favela organic gardening projects: Urban sustainability and community empowerment in Brazil. *Sustainable Cities and Society, 41*, 419–427.

Thaler, R. H., & Sunstein, C. R. (2008). *Nudge: Improving decisions about health, wealth, and happiness*. Yale University Press.

UNDP. (2019). *UNDP India: Annual report*. Retrieved from [UNDP website]. https://annualreport.undp.org/2019/

Vergragt, P., Akenji, L., & Dewick, P. (2013). Sustainable production, consumption, and livelihoods: Global and regional research perspectives. *Journal of Cleaner Production, 63*, 1–12. https://doi.org/10.1016/j.jclepro.2013.09.028

Part III
Technology Driving Environmental Sustainability

Chapter 11
A Bibliometric Analysis of Alternative Energy: Evaluating Technologies, Policies, and Strategies for a Sustainable Future

Hiranya Dissanayake, Valentin Penchev Vasilev, and Shahbaz Khan

Abstract The global transition to alternative energy systems is essential to addressing climate change, energy insecurity, and sustainable development challenges. Over the past three decades, alternative energy research has expanded significantly, driven by the need for decarbonization, resilience, and inclusivity. This bibliometric analysis examines the evolution of alternative energy research from 1996 to 2024, analyzing 91 documents sourced from 78 academic journals. Using advanced bibliometric techniques, including keyword co-occurrence, thematic mapping, and co-citation analysis, the study identifies key trends, thematic clusters, and critical research gaps. The findings highlight a dynamic progression in alternative energy research, with established themes such as biomass and bioenergy serving as foundational pillars, while emerging areas like hydrogen technologies, energy storage systems, and life cycle assessments (LCA) illustrate the field's innovative trajectory. Despite significant advancements, several critical gaps remain. These include the limited scalability and affordability of energy storage technologies, insufficient integration of sustainability metrics into policymaking, geographic imbalances in research contributions favoring developed nations, and a lack of comprehensive techno-economic feasibility studies for emerging technologies. This study emphasizes the importance of interdisciplinary collaboration and equitable representation to address these gaps. By synthesizing insights from leading contributors and mapping thematic clusters, the analysis provides actionable recommendations for advancing

H. Dissanayake (✉)
Department of Accountancy, Wayamba University of Sri Lanka, Kuliyapitiya, Sri Lanka
e-mail: hiranya@wyb.ac.lk

V. P. Vasilev
Higher School of Security and Economics, Plovdiv, Bulgaria
e-mail: valentin.vasilev@vusi.bg

S. Khan
Department of Industrial Engineering, University of Tabuk, Tabuk, Kingdom of Saudi Arabia
e-mail: shahbaz.me12@gmail.com

R. Singh and W. L. Filho (eds.), *Climate Neutrality Through Smart Eco-Innovation and Environmental Sustainability*, Climate Change Management,
https://doi.org/10.1007/978-3-031-83250-5_11

alternative energy solutions. The findings serve as a roadmap for researchers, policymakers, and industry stakeholders to accelerate the adoption of sustainable energy systems, bridging the gap between technological innovation, policy frameworks, and global sustainability goals. This study underscores the critical role of alternative energy research in fostering a sustainable future.

Keywords Alternative energy · Bibliometric analysis · Policies · Life cycle assessments (LCA) · Sustainable future

Introduction

Transition to other energy systems is central to solving climate change, energy vulnerability and ultimate sustainable development (Apostu et al., 2022; Nwokolo et al., 2023a, 2023c; Rahman Mohamed & Lee, 2006). For years fossil has been used due to its effectiveness and cheapness but they play a major role in green house emission, hence the need to embrace renewable and other forms of energy (Anto et al., 2020). Renewable energy sources including solar, wind, biomass, hydrogen, and storage technologies are now recognized more as important solutions to transformation of energy systems toward low-carbon, resilient, and inclusive. These energy solutions are designed to lower harms to the environment and strive to enhance the access to energy for the developed and the developing countries to ensure more effective solution towards the future (Nwokolo et al., 2023d).

Interest in research on alternative energy sources has increased over the past three decades with global policies such as the Paris Agreement, the European Green Deal, and the United Nations' Sustainable Development Goals (SDGs) (Bhushan & Singh, 2024; Joshi et al., 2024). This proliferation of literature is a sign of a systems attempt to seek technical, economic, as well as policy solutions (Nwokolo et al., 2023b). Technologies like hydrogen production and storage (Fernández-González et al., 2022), Life Cycle Assessment (LCA) as an metric of the overall sustainability performance (Hampo et al., 2022), and adoption of circular economy practices show the constantly evolving nature of the subject area. Although there are well-defined topics such as bioenergy and combustion, modern topics like techno-economic analysis show that new ideas, processes, and concepts' deployment is important for optimized solutions.

Prior studies have mapped trends in renewable energy technologies, but the fragmented approach in much of this research underscores the need for more holistic evaluations. These gaps provide critical opportunities for future research to address pressing challenges and foster sustainable energy transitions. Existing literature frequently examines renewable energy technologies and policies independently, with minimal focus on their interdependence. A holistic approach that explores how policies can effectively drive technological innovation and adoption is crucial to understanding and facilitating energy transitions (Bortoluzzi et al., 2021; García-Lillo et al., 2023).

In addition, the rapid expansion of renewable energy sources has highlighted challenges related to energy storage and grid integration, yet these remain under-explored in bibliometric studies. Addressing these issues is critical for ensuring the reliability and sustainability of renewable energy systems (Kut & Pietrucha-Urbanik, 2023).Empirical literature tend to disproportionately focus on specific regions, such as Poland and Germany, limiting the generalizability of findings. Broadening the geographical scope to include diverse regions—particularly underrepresented areas in the Global South—can enhance the understanding of global trends, policy effectiveness, and challenges in renewable energy adoption (Kut & Pietrucha-Urbanik, 2023).

Despite significant advancements in renewable energy technologies, there is limited research on the role of business innovations in complementing these technologies. Exploring how innovative business models and practices can align with sustainability goals represents a critical gap in the literature (Mentel et al., 2023).

Addressing these research gaps is pivotal for advancing both the academic discourse and practical applications in alternative energy. By integrating technological, policy, and socio-economic perspectives, current bibliometric study can provide more comprehensive insights into the drivers and barriers of renewable energy adoption. This approach will be instrumental in achieving sustainable development goals and tackling global challenges such as climate change and energy security.

Accordingly, the primary objectives of this bibliometric analysis are:

1. To investigate trends in the research on alternate energy sources based on time, thus giving a picture of equal active time periods of interest.
2. To assess the contributions of influential journals, authors and geographic regions.
3. To use thematic mapping to categorize research themes into basic or foundational, motor, specific or niche, emerging or declining clusters.
4. To reveal key research gaps in energy storage technologies; geographical distribution; sustainability indicators; techno-economic analysis.
5. To provide practical suggestions for developing and improving scholarship in the area of alternate energy.

This bibliometric analysis that offers the best approach to evaluating the research on the performance of alternative energy research from 1996 to 2024. Based on 91 documents from 78 academic sources, this research investigates the intellectual map, thematic orientation, and development patterns of the field. Keyword co-occurrence analysis, thematic mapping and co-citation analysis are used with suitable categories derived to discover the trends and themes and gaps that need to be filled. This strategy is beneficial to get an idea of the development of the field or industry and to distinguish strengths and future avenues of research. The discoveries of this study should benefit researchers, policy makers and humanity, particularly in the improvement of policy framework that will encourage greater investment towards the achievement of sustainable energy future and energy justice worldwide.

Next section of this chapter explores the methodology.

Methodology

This bibliometric approach used both quantitative and qualitative analysis to assess systematically the research trend, thematic analysis, and intellectual map of the alternative energy studies. The following steps outline the methodological framework:

Data Collection

The Web of Science stands as the most extensive and comprehensive database for scientific literature. It encompasses over 11,000 peer-reviewed, high-impact academic journals spanning a wide range of disciplines, including the life and physical sciences, technology, medicine, and related fields (Dissanayake et al., 2022). This study utilized the Web of Science database for its research.

The search criteria for this study were "Alternative Energy Sources". Initially 158 articles obtained and only selected 91 articles taken from 78 sources. It covered the time period of 1996–2024.

Trend Analysis

To control for the temporal aspect in the published research, the longitudinal prevalence of research activity was explored.

Source Analysis

From the study, it was realized that the assessment of the impact of academic journals and other contributors was on analysis of research on sources of alternative energy. The impact in terms of total citations, h-index, g-index, and m-index were used to evaluate the Journals.

Author Analysis

The productivity and impact of the authors in the articles collected were assessed to identify stakeholders supporting the field. Author citation impact was measured using the number of publications, h-index, g-index, and m-index values.

Geographic Analysis

The distribution of performed research contribution by countries was then compared to determine leading players in the field of alternative energy. According to the metrics of total publication, total citation, and average citation per document the impact of specific countries identified.

Keyword Analysis

In order to determine the major themes and novel issues in research of the area of interest, 400 Keywords Plus and 377 author Keywords were used. Keyword cross tabulation approach was used in order to determine relationships between the specific terms and identify the field's thematic similarities and connections.

Thematic Mapping

Thematic mapping techniques were employed to categorize research topics into four clusters.

Basic Themes: Specialist areas with high gross density but low gross centrality, for example the biomass and bioenergy.

Motor Themes: In fact, the research field, the density degree and the centrality of issues that are more directly connected to micro-generation, renewable energy and transesterification are far the higher.

Niche Themes: Specialist content embracing low-visibility zones associated with the web, such as hydrogen and life cycle assessment.

Emerging or Declining Themes: High-scatter, low-hub activities, like biodiesel and wastewater, meaning topics that are not very significant or have moderate relevance.

The mapping has been carried out with the help of centrality and density analyses to determine the timeliness and significance of the presented research topics.

Co-citation Analysis

Using co-citation analysis, the nature of connections between documents that are frequently cited in the field was determined to provide an understanding of the intellectual structure of the field. When clustering, the studies were stratified and based on the results of this exercise, the foundational works, the emerging fields, and interdisciplinarity could be identified.

To enhance the interpretability of the findings, the study employed visualization tools in biblioshiny including Word Clouds, Trend Charts, Thematic Maps and Co-Citation Networks (Aria & Cuccurullo, 2017).

The chosen methodology offers an extensive and complex view at the subject of the research, which is the alternative energy. Such a systematic approach guarantees that the study not only presents a history of the research on the development of alternative energy sources, but also the findings indicate the key gaps and prospects for further research.

Results

Trend Analysis

The trend analysis of publications over the period from 1996 to 2024 in Fig. 11.1 shows the growth of research activity, and especially in recent years. The first contributions can be defined as scarce: one article appeared in 1996 and the rest of the publications were published in between 2006. The first three articles started to appear in 2011 and the frequency started increasing in subsequent years. The highest number of publications was contributed in the year 2020 with 11 articles which might be work published during this period because interest and activity in this area was high. Since 2018 the annual output of articles has been fairly constant falling between 6 and 11 suggesting a consistent scholarly interest in the topic. Such increase of attention to the topic is evident and could be attributed to emerging global challenges and priorities of alternative energy sources.

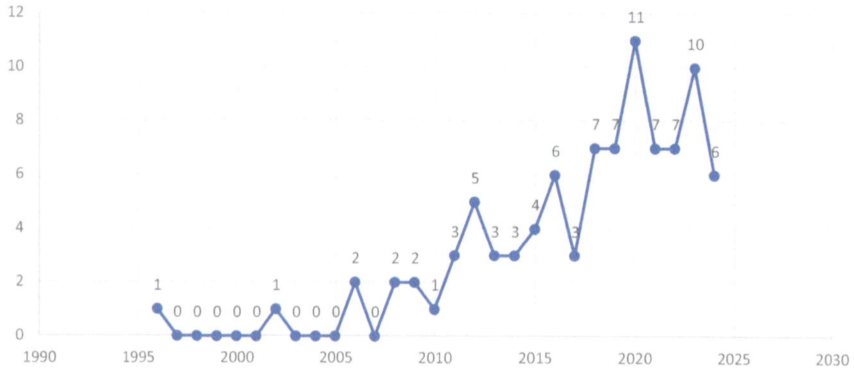

Fig. 11.1 Trend analysis

Source Analysis

The analysis of the top sources in Table 11.1 shows that these are quite a diverse array of journals that focus on alternative energy sources. CHEMOSPHERE gives the most total citations (159) to any single publication after 2020, which signals the journal's impact. Second and third are SCIENTIFIC REPORTS and ENERGY with 147 and 113 total citations respectively and ENERGY has contributed three papers in the journal since its inception in 2009. Among the sources, FUEL has one of the highest sources h-index figures of 3 and the highest g-index of 4, and has published four articles since 2018. Among the top 30 venues, the INTERNATIONAL JOURNAL OF HYDROGEN ENERGY and FUEL CELLS show particular interest in hydrogen technologies, presenting two articles each, and high citation scores. Relatively new and exciting journals include BIORESOURCE TECHNOLOGY, ACS SUSTAINABLE CHEMISTRY & ENGINEERING along with RENEWABLE & SUSTAINABLE ENERGY REVIEWS that surprisingly exhibit improved Institute of Scientific Information (ISI) outputs without compromising on their growth rates; it means fewer publication outlets with more concentrate rates. This distribution also shows that renewable research is inherently cross-disciplinary and growing in demand.

Table 11.1 Source analysis

Source	h_index	g_index	m_index	TC	NP	PY_start
Chemosphere	1	1	0.2	159	1	2020
Scientific Reports	1	1	0.091	147	1	2014
Energy	3	3	0.188	113	3	2009
Fuel	3	4	0.429	106	4	2018
International Journal of Hydrogen Energy	2	2	0.154	89	2	2012
Fuel Cells	1	1	0.125	74	1	2017
Bioresource Technology	1	1	0.091	69	1	2014
Renewable & Sustainable Energy Reviews	1	1	0.25	67	1	2021
ACS Sustainable Chemistry & Engineering	1	1	0.083	56	1	2013
Journal of Renewable and Sustainable Energy	1	1	0.167	49	1	2019

Author Analysis

The author analysis in Table 11.2 highlights significant contributions to the database, with notable performances in terms of citation metrics and research impact. Several authors, including *Anto S*, *Deviram G*, *Kumar SS*, *Mathimani T*, *Mukherjee SS*, *Muthappa R*, *Pugazhendhi A*, and *Verma TN*, have contributed single articles in 2020, each achieving an h-index, g-index, and m-index of 1 and receiving an impressive 159 citations per publication. These metrics underline the high impact of their work despite the limited number of publications. *Ahmad M* emerges as a standout contributor with four publications since 2019, earning a higher h-index and g-index of 4, along with a robust m-index of 0.667, and accumulating 148 citations overall. Similarly, *Abdel-Rahman M* contributed a publication in 2014 that achieved a noteworthy 147 citations, reflecting the sustained relevance of their research. This data showcases both the depth of influence and collaborative potential within this academic network.

The analysis of the top contributing countries indicates a disparity in the quantity and quality of the published work. The leading country in terms of both published articles is China with 16 numerous publications and the total citations, 238, Vietnam is in the second place with 4 publications, but it gets the highest average article citation 90.50. Another G20 country, Saudi Arabia, published 7 articles and on average, the articles attracted 83 citations each. The USA published 12 articles and has the highest overall number of citations, 154, which gives an average of nearly 31 articles per publication. Out of 23 publications from Pakistan, 23 publications came from India too but Pakistan has more total citation 152 than India with 86 and average citation for Pakistan is more (19.00) than for India (10.80) as well. Other countries are as follows: Brazil, Spain, Germany and Japan with mixed but significant presence of articles, Spain made an average of 35.00 citations, Germany 22.00 and Japan 25.50, meaning they do focus and directed research even though they have not published many of them (Table 11.3).

Table 11.2 Author analysis

Author	h_index	g_index	m_index	TC	NP	PY_start
Anto S	1	1	0.2	159	1	2020
Deviram G	1	1	0.2	159	1	2020
Kumar Ss	1	1	0.2	159	1	2020
Mathimani T	1	1	0.2	159	1	2020
Mukherjee SS	1	1	0.2	159	1	2020
Muthappa R	1	1	0.2	159	1	2020
Pugazhendhi A	1	1	0.2	159	1	2020
Verma TN	1	1	0.2	159	1	2020
Ahmad M	4	4	0.667	148	4	2019
Abdel-Rahman M	1	1	0.091	147	1	2014

Table 11.3 Country analysis

Country	Total publications	TC	Average article citations
China	16	238	47.60
Vietnam	4	181	90.50
Saudi Arabia	7	166	83.00
USA	12	154	30.80
Pakistan	23	152	19.00
Brazil	4	113	37.70
India	23	86	10.80
Spain	3	70	35.00
Germany	8	66	22.00
Japan	6	51	25.50

Keyword Frequency Analysis

Table 11.4 keyword frequency analysis shows that a large attention is paid to the application of the non-traditional energy sources: the terms 'biodiesel' and 'biomass' being mentioned 8 times each. According to the literature, Biodiesel is defined as a renewable and biodegradable fuel produced through the transesterification process, which involves the chemical conversion of organic materials such as vegetable oils and animal fats into fatty acid methyl esters (FAME) and glycerol using alcohol and a catalyst (Khairati, 2023; Malabadi et al., 2023). In the same way, Biomass is defined as solid materials of organic origin, encompassing a variety of sources such as agricultural residues and wood, which are utilized for energy production Biomass includes non-fossilized biological materials like wood, agricultural residues, and waste from various sources (Popa, 2021). Both terms importance kin emphasis on the bio-based solutions to energy issues.

The six occurrences of biofuels as well as renewable energy indicates the need for green sources as compared to the fossil energies. Biodiesel and bioethanol come from biological feeds and play a central role in emission mitigation Renewable energy including solar, wind and hydro power symbols the overarching move towards decentralization of power generation (Paraschiv & Paraschiv, 2023). In addition, bioethanol which appears five times is commonly understood as ethanol derived from fermentable sugar crops including corn and sugarcane and may be mixed with gasoline to reduce emissions (Anto et al., 2020).

Specific product-related terms include 'biobon' 'bioenergy' (appearing thrice) and 'biogas'. Bio-energy on the other hand is energy derived from biomass, whereas biogas is methane rich from the anaerobic digestion of organic substances for waste management and energy generation (Deviram et al., 2020). The forward-looking ideas include hydrogen and energy storage each of which appeared twice indicating the direction of development of various renewable energy sources. Besides, hydrogen, as a clean energy carrier, has recently been reported as attracting a large

Table 11.4 Key word frequency analysis

Words	Occurrences
Biodiesel	8
Biomass	8
Biofuels	6
Renewable energy	6
Bioethanol	5
Energy	5
Bioenergy	3
Biogas	3
Climate change	3
Macroalgae	3
Optimization	3
Transesterification	3
Activated carbon	2
Barriers	2
Biofuel	2
Combustion	2
Emission	2
Energy storage	2
Hydrogen	2
Life cycle assessment	2

interest in term of its application for minimizing greenhouse gases emissions (Rahim Malik et al., 2023). Battery systems combined help reduce the intermittent nature of renewable energy sources and energy storage contributes to the stability of the energy grid (Datta et al., 2021).

Last of all, the keyword frequently used twice is the 'life cycle assessment," specifying that new practices are being adopted intending to evaluate the environmental footprint of the energy systems over their life span contributing to the sustainable growth of the energy industries. This research demonstrates that the research community's interest in identifying a wide range of renewable energy sources and their effects on the environment, the economy, and technologies.

Word Cloud Analysis

The word cloud in Fig. 11.2 demonstrates the most popular topics and the trends based on renewable and alternative energy research. The most frequent terms are 'biodiesel', 'biomass', 'biofuels' and 'renewable energy', indicating that this research field largely involves the production of bio-based and green energy. Biodiesel, which

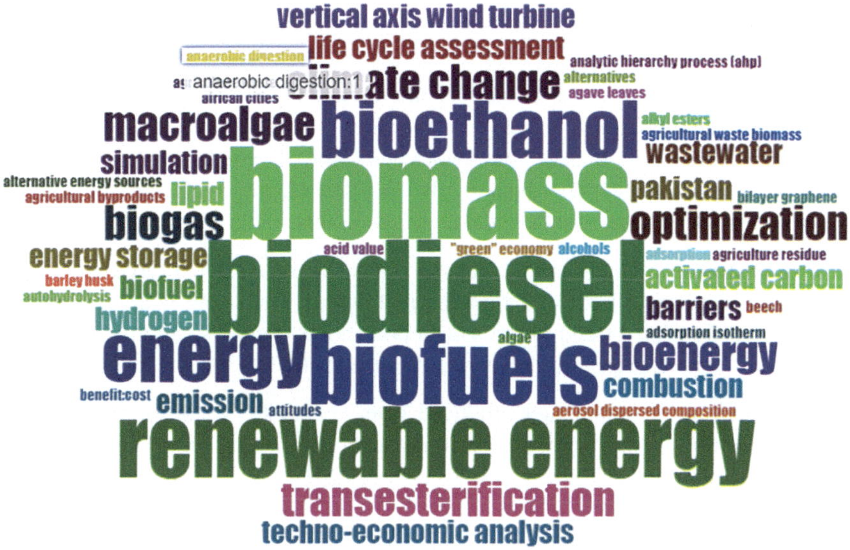

Fig. 11.2 Wordcloud

is produced from natural oils and fats and biomass, the organic material used in the generation of energy, epitomize the policy on low-carbon and renewable energy sources. The rising profile of bioethanol and biogas indicates research into other fermentable and fermentable fuels such as bioethanol and biogas. Sophisticated ideas such as "hydrogen," "energy storage," and "life cycle assessment" indicate the rising concern for clean energy carriers to meet the intermittency challenge, and preserving the ecosystem.

In addition, using the terms such as "optimization" of the process, "transesterification" of biodiesel, and the utilization of "macroalgae" as feeds give the paper an efficiency slant as well as expanding the search for biomass feedstock for energy generation. Terms like 'climate change' or 'emissions' relate to the wider process of reducing greenhouse effect through transition to green energy sources. Thus, the regulatory role is supported by such terms as 'activated carbon', 'barriers' and 'techno-economic analysis' pointing to complex measures needed to address the difficulties and evaluate the effectiveness of sustainable energy systems. Altogether, the word cloud provides a picture of a set of topics, some of which renders the variety of categories and their interconnectedness – the network of innovation and collaboration.

Trend Topic Analysis

The trend topic visualization offers information regarding the shifts in the focus area of renewable energy research over a period. Some of these ground terms include ‘renewable energy’, ‘biodiesel’, ‘biomasses and ‘biofuel’ which have been terms of prominence in recent years because of advocacy towards renewable and bio energy sources. These terms represent critical concepts in renewable energy.

Renewable Energy: This term covers any naturally renewable source of energy including; solar, wind, hydro, and geothermal power. The rising interest in recent years reflects the global trend towards decarbonization of energy supply systems for managing climate change (Ahmad et al., 2014) (Fig. 11.3).

Biodiesel: An international fuel produced from vegetable oils or animal fat by a process called transesterification used in replacement of petroleum Diesel and has lesser emissions of greenhouse gases (Patel et al., 2019).

Biomass: Including agricultural residues, woods employed as the energy source major concentrate in the bioenergy studies due to the depletion of the fossil fuel.

Biofuels: Biological derived biodiesel bioethanol liquid fuel classified as having a significant impact of cutting on greenhouse gas emissions in the transportation and energy industries (Mathimani & Pugazhendhi, 2019).

Novel ideas such as “hydrogen” and “energy storage” were mentioned as modern issues which have risen in the center stage in the recent past. Hydrogen a carbon-free energy carrier is being considered for addressing GHG emissions cutting especially in sectors including industries and transport (Kumar et al., 2024). Future renewable energies like the wind or solar power are only available partially during the day, lithium-ion batteries are a stable energy storage solution thus advancing energy systems (Datta et al., 2021).

Moreover, special emphasis is made on the “Life Cycle Assessment” as the area in development. LCA also assesses energy system’s cradle to grave environmental footprint hence promoting sustainable practices in renewable energy system deployment. Words as ‘transesterification’, ‘macroalgae’, and ‘optimization’ militate to the process controls and the search for other biomasses.

This study unveils an evolutionary shift of focus of bioenergy research with aging technologies including, hydrogen and storage indicating that the field is in concert with sustainable development goals.

Thematic Map

We then employ the thematic mapping analysis that categorizes them in terms of their centrality and density and forms clusters within the domain of research interest which

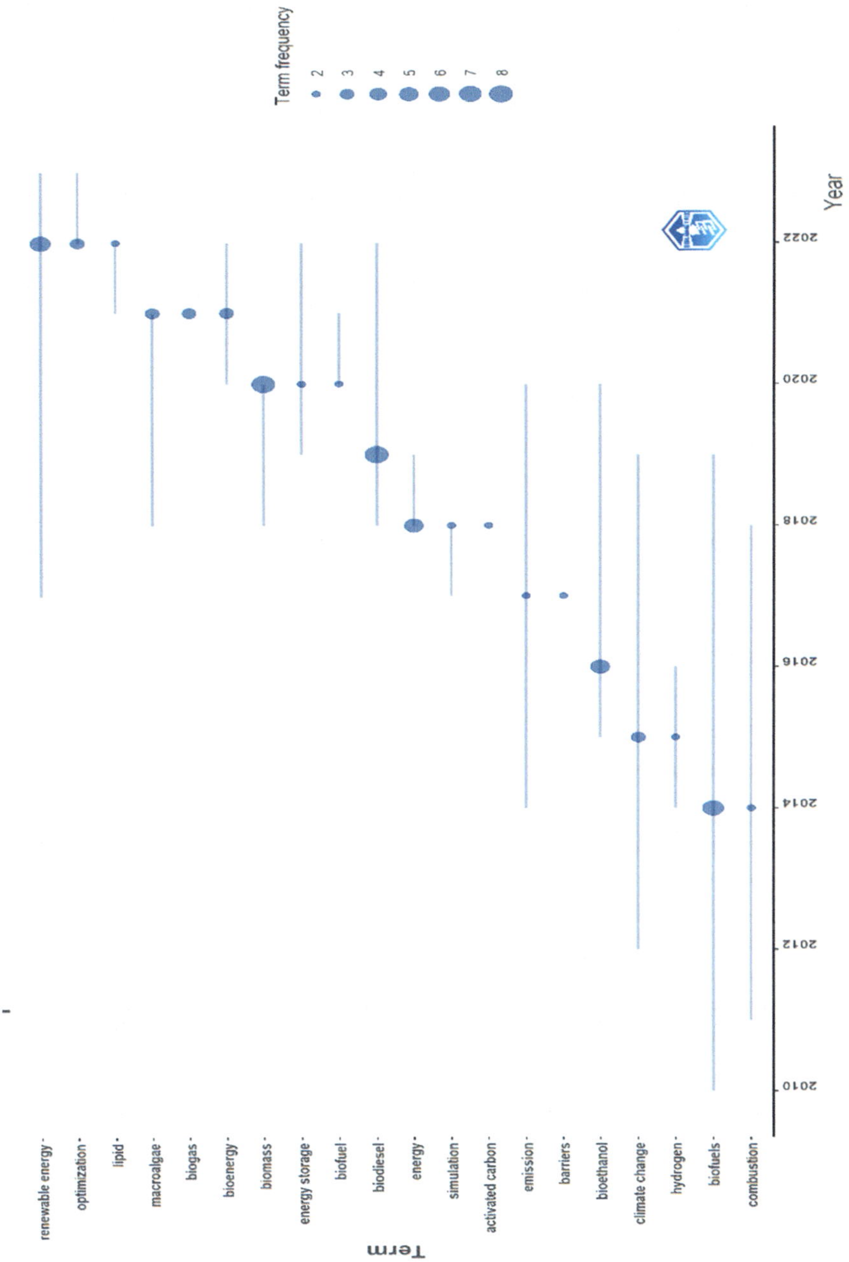

Fig. 11.3 Trend topics analysis

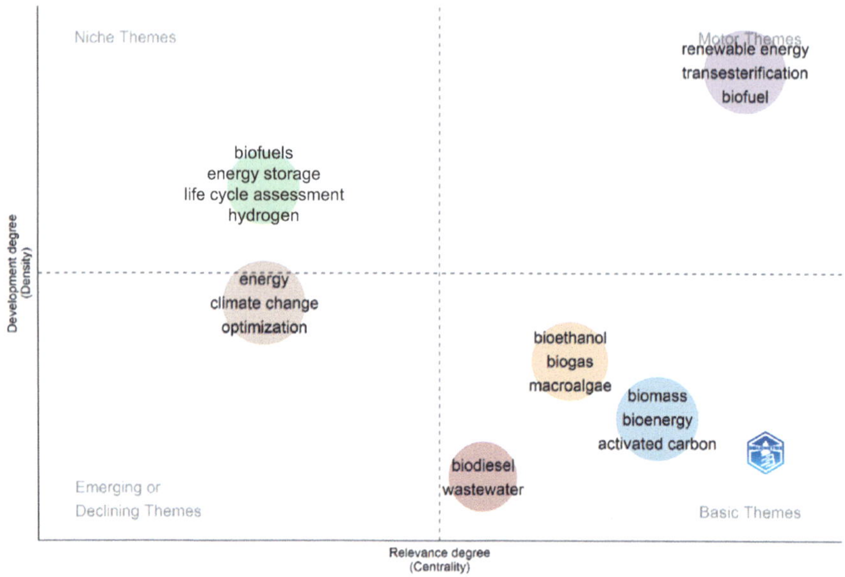

Fig. 11.4 Thematic map

is renewable energy. Such clusters raise the focus on the importance, sustainability, and interrelations of particular fields in the area (Fig. 11.4).

A. Biodiesel Cluster

The next set of terms includes the biodiesel-related terms such as "wastewater"; the betweenness centrality measure for this cluster is high, 792.00, and the closeness centrality, 0.0015. This shows its applicability as a bridging theme in renewable energy research appropriately high. Biodiesel production and its link with wastewater treatment demonstrates the necessity of biodiesel more as a waste management solution than as an energy source. The frequent occurrences of the cluster (10) testify to its constant concern in the scientific literature (Rozina et al., 2019).

B. Bioethanol and Emerging Technologies Cluster

The signature terms of the bioethanol cluster include 'biogas', 'macroalgae', and 'techno-economic analysis' pointing to the linkages between different types of fuels and sustainable biofuel production and evaluation processes. This cluster has been mentioned 13 times and has a betweenness centrality of 736.78 and highlights the increasing concern of diversifying feedstock beyond woody bioresources and assessing the utilitarian viability of renewable power systems (Mathimani & Pugazhendhi, 2019). The moderate value of closeness centrality equal 0.00125 add to notion that it might contain ability to correlate some specific topic to overarching renewable energy systems.

C. Biofuels and Hydrogen Cluster

In the case of the biofuels cluster, the energy storage and hydrogen concepts show their interest in advanced energy carriers and storage technologies. This ranks hydrogen in the forefront of this cluster justifiably as it is a versatile clean energy vector on the path towards achieving net-zero outcomes. With a betweenness centrality of 400.00 and occurrences equal to 10 this cluster concern the connection of biofuels and innovative storage systems to improve the efficiency of renewable energy systems (Kumar et al., 2024).

D. Biomass Cluster

For example, a biomass cluster comprises of simple words such as ‘bioenergy’, ‘activated carbon’, ‘combustion’ because the biomass cluster forms the root or fundamental focal area of research in renewable energy. This cluster has the highest frequencies of (15) and is also knowledge-intensive, with a high betweenness centrality of 784.72. Bioenergy is mainly derived from biomass, and activated carbon demonstrates its use as an energy storage mechanism and in environmental science (Ahmad et al., 2014).

E. Niche and Motor Themes

Thematic mapping also reveals fragmentary themes, alike ‘’life cycle assessment” and ‘’hydrogen”, which are set on sustainability measurements and green transition, and motor themes alike ‘’renewable energy” and ‘’transesterification”, which are set on innovation and practice. These themes are indeed innovative and are at the cutting edge of renewable energy exploration; they have great policy, innovative and sustainability implications for both the material assessment and the research on environmental impact assessments (Verma et al., 2020).

The cluster analysis indicates that while there is emphasis on well developed areas such as biomass and biodiesel, there is a meaningful emphasis on new niche like hydrogen, energy storage and techno economic analysis. These findings represent the past and present development of the renewable energy research to provide the fundamental issues and state of the art methods for sustainability.

Co-Citation Cluster Analysis

Co- citation analysis compares documents which are cited together and determines the thematic groups and the important intellectual schema existing in the field of renewable energy research. The clusters could reflect the fields of related research works which serve as the core for particular topics or methods in the field. Based on the provided data, the co-citation clusters are analyzed in Fig. 11.5.

Cluster 1: The Studies on Biomass and Bioenergy in Literature

The works included in Cluster 1 are: McKendry (2002) and Hanegraaf et al. (1998). These papers form the basis of biomass and bioenergy research and offer fundamental understanding of biomass conversion technologies, feedstock and environmental impacts of biomass and other bio-based energy systems. Also, their closeness centrality of 0.2 and PageRank of 0.053 indicate their importance in the networking of the distinct subareas of renewable energy. Such research has provided the basis for further improvements in bioenergy, with the main aim of increasing the commercially viable aspects of biomass-based systems.

Cluster 2: Technologies That Are Taking Shape in Renewable Energy

This cluster includes papers of authors like Anwar et al. (2010) and (Leung et al., 2010) and focuses on new promising renewable technologies with a focus on integration of solar and wind energy. These studies respond to important questions in sizing and configuring the power, transmission, storage, and generation systems, in improving the conversion efficiency of energy and reducing the cost of energy. The greater PageRank score (0.053) and centrality values evidence the relevance of these works to develop the technical character of renewable energy systems. This

Fig. 11.5 Co- citation analysis

cluster continued and further developed the relationship between theoretical developments and pragmatic solutions with focus on shifting from legacy energy systems to modern, complex systems.

Cluster 3: Waste to Energy

The largest cluster, the Cluster 3 consists of authors (Ariunbaatar et al., 2014; Bohutskyi et al., 2014; Ehimen et al., 2013) among others. In these studies, they concentrate on waste management using energy technologies, the circular economy model, and anaerobic digestion. The closeness centrality calculated as 0.02 similarly indicates that these works target focused but expanding topics in renewable energy. For instance, Ehimen's study on the conversion of waste-organic waste to energy both for research and practical purposes concerning policies on waste management have been enriched. This cluster addresses the escalating significance of resource minimalism and coordination of renewable power solutions with waste treatment measures.

Cluster 4: Renewable Energy in Developing Regions

Indeed, the works in Cluster 4 like Baloch et al. (2016) and Mirza et al. (2007) help to explore the usage of renewable energy technologies in the developing context. These studies relate to issues including access to energy, financial constraints, and policies especially in the emerging market. This high value of 0.2 for closeness centrality reveals their connection with other energy equity and global sustainability discussions as well. This cluster points to the importance of differentiated RE solutions touching on socio-economic, and geographic conditions of the developing countries.

Cluster 5: Techno-Economic Analysis

The fifth cluster includes two papers, Gul et al. (2015) and Huang et al. (2016), which are devoted to techno-economic analysis of renewable power technologies. These works assess the feasibility, applicability, and sustainability of energy systems, as well as giving useful information to investors, policy makers and scholars. The centrality measures of these works are rather high: the authors equal their closeness centrality to 0.2 and PageRank score to 0.053, which indicates that these works perform the function of a crucial link between technical studies and economic decisions. In particular, this cluster emphasizes mobile adaptable renewable energy innovations and their connection to the perceived market to promote global acceptance.

Five groups are identified in the co-citation analysis, and these are crucial significant themes in renewable energy research. Preliminary researches regarding biomass as well bioenergy are fundamental for future further progress and new technologies are being developed further consistently. Technologies for transforming waste to energy, and waste management using a circular economy viewpoint are emerging topics focusing on both environmental and resource efficiency concerns. On one end, the literature on RE adoption in developing regions focuses on socio-economic perspectives of transition, as on the other end the techno-economic aspects steer the adoption process through technological realities in relation to market and policy context.

Collectively, these clusters emphasize the interdisciplinarity and evolution of the field of renewable energy research, understanding its knowledge landscape and ongoing development. This analysis shows how research themes are connected and how the concerns of the renewable energy community have shifted over time by mapping out the citation connections between the most cited papers.

Discussion

Despite much development of alternative energy research over the past three decades, the following five issues stand as obstacles to the creation of sustainable, equitable, and scalable energy solutions. These include the technology gaps, the geographic gaps, the economic gaps and some gaps between the principles of sustainability and their implementation. A comprehensive discussion of these gaps is outlined below.

Energy Storage Integration

Energy storage has always been an essential part of renewable energy systems especially where the source of energy is unreliable as is the case with solar and wind power. The hydrogen energy and battery storage technologies have made certain progress; however, these technologies have some drawbacks such as high costs, low scalability and energy conservation drawbacks (Anto et al., 2020; Kumar et al., 2024). Today, the emphasis is being placed mainly on the technological development and testing of new solutions at laboratory level, while few work directly on the problem of how such an integration can be done at the national level. This lack underscores a research need in overall low-cost and high-capacity models as well as multisourced systems that include different technologies for enhanced dependability.

Spatial Inequalities in Research

Despite the global growth of interest in renewable energy, specific countries tend to stand out as leaders in research, as measured either by the volume of publications or by impact, including China, the USA, and Saudi Arabia. On the other hand, the majority of Asia and Africa continents which has ample potentials in the development of renewable energy systems is under represented (Nwokolo et al., 2023d). Such a gap restricts the universality of renewable energy and does not consider regional disadvantages including infrastructure constraints and energy poor populations. To close this gap, foreign affairs need increased capacity building, research funding, and partnership strengthening in areas that represent the less involved regions.

Metrics Pertaining to Sustainability

Life cycle assessment (LCA) together with other sustainability metrics have been increasingly adopted as instruments for measuring environmental effects of renewable energy systems. Nevertheless, they only have a partial role in policymaking and integration with the industrial processes (Verma et al., 2020). This gap prevents synchronization of renewable energy systems with sustainability goals and leaves important actors without adequate information for their actions. The widening of the integration of the LCA findings in policy domains and industrial norms is crucial for enhancing the knowledge of sustainable energy systems.

Techno-Economic Feasibility Studies

Due to its relevance to the cost, technical feasibility, and market potential of renewable energy technologies, techno-economic analysis is important. Although there has been research on these aspects especially with regard to hydrogen and biofuels, there seems to lack a more holistic research that assesses new developments in these technologies in life cycle system perspectives (Mathimani & Pugazhendhi, 2019; Nwokolo et al., 2025). This gap makes it impossible for the stakeholders to be in a position to make right investment and hence retards the commercialization of viable innovations. Subsequent investigations should target general systems where technological and economic analyses of producing technologies, distribution infrastructure, and end use applications should be made.

Circular Economy Systems

We found that circular economy-oriented technologies, namely Waste to Energy technologies, are not well-studied in the context of renewable energy sources, although it can be potentially beneficial to boost resource utilization efficiency and decrease environmental impacts (Anto et al., 2020; Khan et al., 2024; Singh et al., 2025). For example, the generation of biodiesel from waste oil or biological gas from organic waste at the same time solves the problem of waste disposal and provide energy. However, research in this area remains scattered with little emphasis placed on how such models can be scaled and how policies can be incorporated. Future research studies should be oriented to the systems level and take the combination of circular economy and renewable energy technologies.

Policy Alignment and Governance

As efforts have been directed towards availing new technologies for service delivery, the task of integrating these technologies with supportive policies has not been easy. This is because poor or skewed policies create a barrier on the implementation of these renewable energy systems. For example, the fossil fuel subsidies in some areas are still driving down the cost competitiveness of renewable technologies. This means cross-cutting research which combines technical, economic and governance approaches to develop policies that are compatible and effective.

Conclusion

This bibliometric investigation gives an appraisal of the growth, theme, and research gap of the AE literature from 1996 to 2024. The study provides an overview of progress accomplished in important fields like biomass, biofuels, hydrogen and LCA and signposts new areas of interest like energy storage, circular economy and technoeconomic evaluation. This analysis shows understanding of the direction and historical patterns of the field of research area by integrating findings from 91 articles sourcing from 78 research papers.

Key Findings includes established and Emerging Themes such as bioenergy and biodiesel are still introduced as major topics, however topics such as hydrogen, the role of storage in energy, and circular economy represent new areas which have evolved the field. Secondly, G8 countries dominate, and while Asia's oil-importing nations have more publishing power, China, the USA and Saudi Arabia are most influential. Special appreciation to authors including Ahmad M., Anto S., and Abdel-Rahman M. Pave way to rising acknowledgment of the camaraderie of the academic society. Thematic mapping classifies work according to basic, motor, niche, and emerging themes, which define the degree of development and interest in specified subject matters.

Incoming opportunities in enhancing critical gaps including the energy storage technologies, geographic origin, sustainability characteristics, assessment through techno economic analysis and conformity to policies were seen as hurdles towards the achievement of sustainable energy future. To address the identified gaps, future research should prioritize Affordable and Scalable Energy Storage, Equitable Geographic Representation, Policy-Driven Sustainability Integration, Comprehensive Techno-Economic Assessment, and Circular Economy Synergies.

The results of this study useful for the researchers, policymakers, and the members of the relevant industries. It can be applied by researchers to target the gaps and stimulate interdisciplinarity. The analysis can be used by policymakers to promote favorable conditions under which innovative solutions can be created and implemented while eliminating regional inequalities. The identified facts can be helpful for industry stakeholders to coordinate the technological development with the market requirements and sustainability goals.

Therefore, the present bibliometric examination underlines the significance of academics and studies within the energy context for global sustainability. Although new challenges require addressing current gaps and encouraging innovation and collaboration as well as equitable access to renewable energy sources, it is possible to present some ways in which the renewable energy sector can play a significant role in combating climate change, improving energy security, and nurturing a better future for humanity.

References

Ahmad, M., Rajapaksha, A. U., Lim, J. E., Zhang, M., Bolan, N., Mohan, D., Vithanage, M., Lee, S. S., & Ok, Y. S. (2014). Biochar as a sorbent for contaminant management in soil and water: A review. *Chemosphere, 99*, 19–33. https://doi.org/10.1016/J.CHEMOSPHERE.2013.10.071

Anto, S., Mukherjee, S. S., Muthappa, R., Mathimani, T., Deviram, G., Kumar, S. S., Verma, T. N., & Pugazhendhi, A. (2020). Algae as green energy reserve: Technological outlook on biofuel production. *Chemosphere, 242*, 125079. https://doi.org/10.1016/J.CHEMOSPHERE.2019.125079

Anwar, J., Shafique, U., uz-Zaman W, Salman, M., Dar, A., & Anwar, S. (2010). Removal of Pb(II) and Cd(II) from water by adsorption on peels of banana. *Bioresource Technology, 101*(6), 1752–1755. https://doi.org/10.1016/J.BIORTECH.2009.10.021

Apostu, S. A., Panait, M., & Vasile, V. (2022). The energy transition in Europe—A solution for net zero carbon? *Environmental Science and Pollution Research, 29*(47), 71358–71379. https://doi.org/10.1007/S11356-022-20730-Z/TABLES/14

Aria, M., & Cuccurullo, C. (2017). Bibliometrix: An R-tool for comprehensive science mapping analysis. *Journal of Informetrics, 11*(4), 959–975. https://doi.org/10.1016/J.JOI.2017.08.007

Ariunbaatar, J., Panico, A., Esposito, G., Pirozzi, F., & Lens, P. N. L. (2014). Pretreatment methods to enhance anaerobic digestion of organic solid waste. *Applied Energy, 123*, 143–156. https://doi.org/10.1016/j.apenergy.2014.02.035

Baloch, M. H., Kaloi, G. S., & Memon, Z. A. (2016). Current scenario of the wind energy in Pakistan challenges and future perspectives: A case study. *Energy Reports, 2*, 201–210. https://doi.org/10.1016/J.EGYR.2016.08.002

Bhushan, M., & Singh, R. (2024). Talent management facilitates net-zero transition through employee green behavior. *CSR, Sustainability, Ethics and Governance, Part F, 2880*, 117–129. https://doi.org/10.1007/978-3-031-55779-8_7

Bohutskyi, P., Betenbaugh, M. J., & Bouwer, E. J. (2014). The effects of alternative pretreatment strategies on anaerobic digestion and methane production from different algal strains. *Bioresource Technology, 155*, 366–372. https://doi.org/10.1016/J.BIORTECH.2013.12.095

Bortoluzzi, M., Correia de Souza, C., & Furlan, M. (2021). Bibliometric analysis of renewable energy types using key performance indicators and multicriteria decision models. *Renewable and Sustainable Energy Reviews, 143*, 110958. https://doi.org/10.1016/J.RSER.2021.110958

Datta, U., Kalam, A., & Shi, J. (2021). A review of key functionalities of battery energy storage system in renewable energy integrated power systems. *Energy Storage, 3*(5), e224. https://doi.org/10.1002/EST2.224

Deviram, G., Mathimani, T., Anto, S., Ahamed, T. S., Ananth, D. A., & Pugazhendhi, A. (2020). Applications of microalgal and cyanobacterial biomass on a way to safe, cleaner and a sustainable environment. *Journal of Cleaner Production, 253*, 119770. https://doi.org/10.1016/J.JCLEPRO.2019.119770

Dissanayake, H., Iddagoda, A., & Popescu, C. (2022). Entrepreneurial education at universities: A bibliometric analysis. *Administrative Sciences, 12*(4), 185. https://doi.org/10.3390/ADMSCI 12040185

Ehimen, E. A., Holm-Nielsen, J. B., Poulsen, M., & Boelsmand, J. E. (2013). Influence of different pre-treatment routes on the anaerobic digestion of a filamentous algae. *Renewable Energy, 50*, 476–480. https://doi.org/10.1016/j.renene.2012.06.064

Fernández-González, R., Puime-Guillén, F., & Panait, M. (2022). Multilevel governance, PV solar energy, and entrepreneurship: The generation of green hydrogen as a fuel of renewable origin. *Utilities Policy, 79*, 101438. https://doi.org/10.1016/J.JUP.2022.101438

García-Lillo, F., Sánchez-García, E., Marco-Lajara, B., & Seva-Larrosa, P. (2023). Renewable energies and sustainable development: A bibliometric overview. *Energies, 16*(3), 1211. https://doi.org/10.3390/EN16031211

Gul, M., Kazaz, S., Baydar, H., & Sirikci, B. S. (2015). A study about technical, economical situation, problems and improvement of oil rose (*Rosa damascena Mill.*) in Turkey. *Journal of Essential Oil Bearing Plants, 18*(3), 613–626. https://doi.org/10.1080/0972060X.2014.998723

Hampo, C. C., Ojo, D. O., Olatunde, D. E., Isioma, O. J., Oni, O. O., Echendu, A. J., Mathew, M., Adediji, M. A., Oladipo, S. F., & Aro, P. O. (2022). Life cycle assessment of renewable energy technologies in Northern Africa: A critical review. *Energy Sources, Part A: Recovery, Utilization, and Environmental Effects, 44*(4), 10248–10269. https://doi.org/10.1080/15567036. 2022.2143960

Hanegraaf, M. C., Biewinga, E. E., & Van Der Bijl, G. (1998). Assessing the ecological and economic sustainability of energy crops. *Biomass and Bioenergy, 15*(4–5), 345–355. https://doi.org/10. 1016/S0961-9534(98)00042-7

Huang, H., Long, S., & Singh, V. (2016). Techno-economic analysis of biodiesel and ethanol co-production from lipid-producing sugarcane. *Biofuels, Bioproducts and Biorefining, 10*(3), 299–315. https://doi.org/10.1002/BBB.1640

Joshi, A., Singh, R., Kumar, V., & Rani, S. (2024). *Net zero strategy: Exploring opportunities, confronting challenges for developing countries* (pp. 233–252). https://doi.org/10.1007/978-981-97-5756-5_11

Khairati, M. (2023). Biodiesel: An overview. *International Journal of Research and Review, 10*(11), 228–236. https://doi.org/10.52403/IJRR.20231127

Khan, S., Singh, R., Alnahas, J., Abbate, S., & Centobelli, P. (2024). Navigating the smart circular economy: A framework for manufacturing firms. *Journal of Cleaner Production, 480*, 144007. https://doi.org/10.1016/J.JCLEPRO.2024.144007

Kumar, K., Singh, R., Sharma, A., & Lehri, B. P. S. (2024). *Climate change adaptation strategies for achieving net-zero economy* (pp. 163–183). https://doi.org/10.1007/978-981-97-5756-5_8

Kut, P., & Pietrucha-Urbanik, K. (2023). Bibliometric analysis of renewable energy research on the example of the two European countries: Insights, challenges, and future prospects. *Energies, 17*(1), 176. https://doi.org/10.3390/EN17010176

Leung, D. Y. C., Wu, X., & Leung, M. K. H. (2010). A review on biodiesel production using catalyzed transesterification. *Applied Energy, 87*(4), 1083–1095. https://doi.org/10.1016/J.APE NERGY.2009.10.006

Mathimani, T., & Pugazhendhi, A. (2019). Utilization of algae for biofuel, bio-products and bio-remediation. *Biocatalysis and Agricultural Biotechnology, 17*, 326–330. https://doi.org/10.1016/ J.BCAB.2018.12.007

McKendry, P. (2002). Energy production from biomass (part 1): Overview of biomass. *Bioresource Technology, 83*(1), 37–46. https://doi.org/10.1016/S0960-8524(01)00118-3

Mentel, G., Lewandowska, A., Berniak-Woźny, J., & Tarczyński, W. (2023). Green and renewable energy innovations: A comprehensive bibliometric analysis. *Energies, 16*(3), 1428. https://doi. org/10.3390/EN16031428

Mirza, U. K., Ahmad, N., Majeed, T., & Harijan, K. (2007). Wind energy development in Pakistan. *Renewable and Sustainable Energy Reviews, 11*(9), 2179–2190. https://doi.org/10. 1016/J.RSER.2006.03.003

Nwokolo, S. C., Singh, R., Khan, S., & Kumar, A. (2025). Threats to the rapidity of sustainability transitions posed by technological changes. *Technological Pathways for Africa' s Net-Zero Economy*, 39–75. https://doi.org/10.1016/B978-0-443-31486-5.00002-1

Nwokolo, S. C., Singh, R., Khan, S., Kumar, A., & Luthra, S. (2023a). Africa's awakening to climate action. *CSR, Sustainability, Ethics and Governance, Part F, 1850*, 299–310. https://doi.org/10.1007/978-3-031-44514-9_9

Nwokolo, S. C., Singh, R., Khan, S., Kumar, A., & Luthra, S. (2023b). Global investment and development in Africa. *CSR, Sustainability, Ethics and Governance, Part F, 1850*, 15–58. https://doi.org/10.1007/978-3-031-44514-9_2

Nwokolo, S. C., Singh, R., Khan, S., Kumar, A., & Luthra, S. (2023c). Impacts of climate change in Africa. *CSR, Sustainability, Ethics and Governance, Part F, 1850*, 237–262. https://doi.org/10.1007/978-3-031-44514-9_7

Nwokolo, S. C., Singh, R., Khan, S., Kumar, A., & Luthra, S. (2023d). Remedies to the challenges of renewable energy deployment in Africa. *CSR, Sustainability, Ethics and Governance, Part F, 1850*, 59–74. https://doi.org/10.1007/978-3-031-44514-9_3

Paraschiv, L. S., & Paraschiv, S. (2023). Contribution of renewable energy (hydro, wind, solar and biomass) to decarbonization and transformation of the electricity generation sector for sustainable development. *Energy Reports, 9*, 535–544. https://doi.org/10.1016/J.EGYR.2023.07.024

Patel, A., Sartaj, K., Pruthi, P. A., Pruthi, V., & Matsakas, L. (2019). Utilization of clarified butter sediment waste as a feedstock for cost-effective production of biodiesel. *Foods, 8*(7), 234. https://doi.org/10.3390/FOODS8070234

Popa, V. I. (2021). Biomass and sustainability. In *Sustainability of biomass through bio-based chemistry* (pp. 1–33). https://doi.org/10.1201/9780429347993-1

Rahim Malik, F., Yuan, H. B., Moran, J. C., & Tippayawong, N. (2023). Overview of hydrogen production technologies for fuel cell utilization. *Engineering Science and Technology, an International Journal, 43*, 101452. https://doi.org/10.1016/J.JESTCH.2023.101452

Rahman Mohamed, A., & Lee, K. T. (2006). Energy for sustainable development in Malaysia: Energy policy and alternative energy. *Energy Policy, 34*(15), 2388–2397. https://doi.org/10.1016/J.ENPOL.2005.04.003

Malabadi, R. B., Sadiya, M. R., Kolkar, K. P., & Chalannavar, R. K. (2023). Biodiesel production via transesterification reaction. *Open Access Research Journal of Science and Technology*, 9(2), 010–020. https://doi.org/10.53022/OARJST.2023.9.2.0064

Rozina, Ahmad, M., Zafar, M., & Ali, N. (2019). Biodiesel synthesis and physiochemical analysis of Taraxacum officinale F.H.Wigg seed oil. *International Journal of Environmental Science and Technology, 16*(8), 4103–4112. https://doi.org/10.1007/S13762-019-02336-0/METRICS

Singh, R., Gupta, A., & Kandpal, V. (2025). ESG excellence in the digital world. In *Fintech for ESG and the circular economy* (pp. 193–204). https://doi.org/10.1002/9781394234035.CH10

Verma, M. K., Murmu, M., Ahmad, I., Shrivastava, P., Verma, M. K., Murmu, M., & Ahmad, I. (2020). Life cycle assessment modelling for a developing city's water system. *IOP Conference Series: Materials Science and Engineering, 814*(1), 012029. https://doi.org/10.1088/1757-899X/814/1/012029

Chapter 12
Environmental Policies: Institutional Response to Smart Eco-Systematic Realities

M. C. R. Jayalath, P. G. S. Amila Jayarathne, and Ali Najeeb

Abstract This chapter examines how institutions are pivoting to the "Smart eco-systemic Realities" through which the interdependent umbrella of natural environmental, social, and economic sustainability can be facilitated. The policies of Engergiewende in Germany and the European Union Green Deal (EUGD) are focused on the transition to renewable energy, emission reduction, and eco-innovations by developed nations such as Germany, Sweden, and Japan. On the other side of the environmental benefits, this chapter demonstrates how climate policies primarily focus on social objectives, building the green economy, and enhancing social benefit, as seen in Sweden's Smart Cities projects. Economic incentives, such as green bonds or tax credits in the US and Canada, reinforce eco-friendly industrial practices. Stockholm and Amsterdam are helping to ensure more effective environmental data interpretation and resource allocation through artificial intelligence (AI) and the Internet of Things (IoT) as their smart grids. Global corporations with developed countries continue to play a crucial role in donating and providing technology and financial assistance for developing countries to switch to low carbon emissions. By looking into how these policies are implemented in these countries, the inscription of integrated policies to attain ecologically sustainable and socially inclusive climate neutrality is relevant today. It also looks into the barriers to implementing sustainability policies and imaginative eco-systematic approaches towards achieving climate neutrality. It explains that current global policies are fragmented, and any action has to face financial constraints and social inequalities. It also proposes a

M. C. R. Jayalath
Department of Commerce and Financial Management, University of Kelaniya, Kelaniya, Sri Lanka

P. G. S. A. Jayarathne (✉)
De Montfort University, Leicester, UK
e-mail: amila.jayarathne@dmu.ac.uk

A. Najeeb
Qasim Ibrahim School of Business, Villa College, Maldives
e-mail: najeeb@villacollege.edu.mv

R. Singh and W. L. Filho (eds.), *Climate Neutrality Through Smart Eco-Innovation and Environmental Sustainability*, Climate Change Management,
https://doi.org/10.1007/978-3-031-83250-5_12

few strategies, including international cooperation, technological innovation, and inclusive policy as drivers of sustainable change.

Keywords Climate neutrality · Eco-systematic realities · Sustainability · Environmental policies · Institutional response

Introduction

Recently, there has been a louder plea for climate neutrality, shifting the focus to reducing greenhouse gas emissions and the need for sustainable development. Climate neutrality or net-zero emissions is the achievement of emissions from the goal group exceeded by the amount of greenhouse gases removed from or otherwise controlled in the atmosphere (Rogelj et al., 2018; Dewasiri et al., 2024). Achieving this target is of utmost importance to achieving the goals of the society and environment so that the temperature increase is not more than 1.5°. Even those economies that have had significant gas emissions in the past are now taking the lead in ensuring that some policies and innovations enable countries to be net zero by 2050 or earlier (Keys, 2024; Bhushan & Singh, 2024).

Policies that address what can be called "eco-systematic realities "must affect the institutions. These realities identify that the environment, economy, and society are interconnected and have a core relationship, requiring solutions that benefit all three areas (Isopahkala & Nori, 2021). The degradation, global warming, and their associated implications emerge as critical global priorities of emerging and advanced economics as environmental challenges (Shabir et al., 2023). These advantages stem heavily from rapid industrialisation and economic development, specifically in identifying economic reasons which have significantly escalated the consumption of fossil fuels and other energy sources that humans use. The greenhouse gas (GHS) emissions have escalated, and the basic to catastrophic climate phenomena like hurricanes, rising sea levels, and heatwaves are consequently (Obobisa et al., 2022).

Among GHGs, CO_2 contributes the most to global warming. Since the 1960s, its emissions have tripled and depend on solid, liquid, and gaseous fuels. Therefore, this remains a contentious issue on the global policy agendas (Ahmad & Zheng, 2021). Acknowledging this problem dire, nations increasingly emphasise promoting environmental-related technological innovations and institutional responses. These measures aim to mitigate the adverse effects of CO_2 on the environment and public health and accelerate the transition toward a sustainable, low-carbon economy (Halperin et al., 2022).

Technological innovation has become identified as the cornerstone of most environmental challenges. Innovating eco-friendly products, processes, and systems should guarantee reduced energy consumption and lesser pollution with improved environmental quality. Also, it has been noted that the growth of environmentally-based technologies serves to redefine the shape of industries, optimise resources, and develop the green economy. In this line, continued utilisation by countries would provide a sure path toward reducing fossil fuel dependency, improving production efficiency, and contributing to the fight against climate change (Zhang et al., 2016).

In this chapter, we examine how institutions in developed countries respond to these eco-systematic realities, exploring policies and innovations that integrate environmental, social, and economic sustainability, addressing the need to discuss them highlighted in the literature, for instance, by Dellmuth and Gustafsoon (2021). This chapter examines how policy frameworks are changing to motivate and how nations like Sweden, Japan and Germany are in the lead in investing in sustainable technologies and generating green jobs. This also discusses the stage of reaching a carbo-neutral future by examining these occasions. The intersection of environmental policies and institutional responses with the bright eco-systematic realities emphasised in this chapter is technological innovation and institutional frameworks that support climate neutrality, ensure green economic development, and reach larger environmental sustainability goals (Nwokolo et al., 2024). Based on the policy discussion, case studies, and technological interventions, the chapter highlights pathways to align financial and environmental priorities in an era of climate change challenges (Kumar et al., 2024).

The chapter is organised in multiple sections. The introduction is followed by the methodology used to explore the insights presented as significant themes in the literature discussion of this domain. The major themes explored through the content analysis of the existing knowledge in this domain are comprehensively discussed before making the concluding remarks.

Methodology

This section explains the method used in the study to explore the institutional enactment of the bright eco-systematic realities within environmental policies. Published sources are the primary data sources used in the chapter according to the novelty of the study domain. The primary data analysis technique employed in this present study is content analysis, which involves more practical analysis and review of the available literature through the systematic literature review approach as adopted by Vijerathne et al. (2024), Bandara et al. (2024), Chathuranga et al. (2024), Mishra et al. (2024), Karunarathna et al. (2024), and Rathnasiri et al. (2024). This approach allows the involvement of existing literature and practices in evaluation to find patterns, trends, and insights in the existing knowledge domain and paw future investigations. The study's research design was qualitative, and a content analysis was deemed appropriate for systematically evaluating the findings obtained from the different data sources. To this end, the study establishes how institutions respond to environmental challenges with innovative eco-systematic solutions. The key steps involved in the research design are as follows:

Selection of Sources

An extensive search of the published literature was conducted to find the relevant document. This comprised scholarly journal articles, policy documents, government documents, and case studies. Keywords, namely, 'environmental policies', 'smart eco systematic realities', 'institutional response', 'climate neutrality', 'economic sustainability ', 'social sustainability', 'eco-innovations', 'technological innovations', 'climate change', and 'renewable energy' in the internet search helped in sourcing the literature. As inclusion criteria, peer-reviewed articles containing information about environmental policies, institutional responses, and innovative eco-systematic solutions were considered. Both theoretical and empirical works were included. Scholars who did not relate their work to the research questions or had insufficient information about institutional reactions and innovative eco-systematic solutions were excluded. Non-English sources were also excluded.

Data Collection

Such published material was actively searched for and classified. The data collection included academic databases, governmental archives, and information on organisation websites. One hundred twenty-four sources were found initially and then refined based on the defined inclusion/exclusion criteria to focus on 57 sources.

Content Analysis Procedure

The content was systematically analysed in several steps to improve the reliability of the findings.

Coding: All the chosen sources were analysed, and important topics, concepts, and patterns were determined. A coding scheme was also created to group the data into meaningful units.

Categorisation: The collected coded data were clustered according to the research questions to facilitate the analysis. These included issues related to institutional feedback, the availability of policies and technology, and success stories.

Analysis: The data was then categorised so that overall trends, relationships, and gaps in the current research literature could be observed. Standard cross-sectional comparison was employed to identify similarities and differences between developed and developing nations.

Validation and Reliability

To minimise coder bias and drift, inter-researcher reliability checks were completed. The first author and a research assistant coded a subset of the data separately. Inter-coder reliability was computed to determine the level of cooperation between the coders. Triangulation was used to ensure the credibility of the results, where related findings were compared with others or with the opinion of experts on the issue in question.

Discussion of the Findings

The major themes derived from the content analysis and validated through inter-researcher and inter-coder reliability are discussed as follows.

Environmental Sustainability and Institutional Role

Policy Drives in Developed Countries

With organisations approving laws promoting pollution prevention, resource efficiency and renewable energy, environmental sustainability has become a top concern for developed countries (Singh et al., 2024). The European Union's Green Deal is a bold plan to achieve carbon neutrality by 2050 by reducing emissions by 55% by 2023. This universal plan offers incentives for developing sustainable industries, infrastructure based on the green concept and pure and clean energy. The EU has created an environment driven by eco-innovation, supporting businesses and communities in shifting to more sustainable practices by establishing goals and supplying funding for green projects (Fetting & ESDN Office, 2020). Through policies like Energiewende, Germany has implemented how large-scale transitions to sustainable energy can be achieved through developed countries' supportive policy framework (Schweitzer et al., 2018).

This is a complex guide for the environmental sustainability approach and complicated policies that engage with regulatory procedures and frameworks, technological advancements, social relationships, and government and private relationships in developing countries (Haunschild et al., 2016).

Furthermore, developed countries' renewable energy policies and procedures have affected strong legal measures focused on reducing greenhouse gas emissions and improving energy efficiency. The Emissions trading systems, such as the European Union's Emissions Trading System (EU ETS), cap the total emissions authorised in a particular society, inspiring organisations to mitigate their carbon footprint. The tax incentives for renewable energy and carbon taxes in certain states have reinforced the push toward a greener economy in the United States (Hermann et al., 2015). These regulatory procedures and policies incentivise industries to develop and move to pure technologies, granting that economic growth does not come at the cost of environmental health.

The institutional quality and the cleaner are the other significant factors of the policies that ensure effectiveness and efficiency. Flexibility, the most potent and transparent institution, is essential for making and enforcing environmental regulations and procedures. Individualistic agencies like the U.S. Environmental Protected Agency (EPA) and the European Environment Agency (EEA) play critical roles in enforcing environmental rules and regulations, controlling pollution stats and ensuring compliance with emissions targets (Khazragui & Hudson, 2015). These world-famous organisations provide a stable regulatory environment that enables long–term strategic planning and fosters social trust in environmental initiatives.

The Public-Privet Partnerships (PPPs) effectively and efficiently encourage environmental sustainability. By doing so and linking with the government and private organisations, they may control the resources to develop large-scale green infrastructure projects. For example, the relationship between the government and industry has been crucial in developing electric vehicle infrastructure, including public transportation systems and charging stations powered by renewable energy. These relationships or partnerships also allowed the private sector to gain innovative solutions to the market, like smart grids, energy-efficient buildings and advanced recycling systems.

Also, private awareness and involvement are crucial and central to the long-term success of the environmental sustainability mechanism in developed countries. The government and the NGOs have invested heavily in educational campaigns and student awareness programs to raise awareness of environmental issues and inspire the public to adopt sustainable behaviours for the long- term.

Long-term sustainability policies in developed countries are designed to examine the link between environmental, economic, and social factors. Resource efficiency initiatives like reducing waste and promoting the reuse of materials focus on improving environmental outcomes and enhancing economic resilience (Greenhalgh & Fahy, 2015). When active, these models not only reduce the impact on the environment but also lead to the creation of a new and modern economic opportunity.

Another element is flexibility in environmental policies within developed countries. As these ecological challenges evolve, so do the policies and procedures that analyse them. The new changes thus ensure that the efforts towards sustainability stay on track as countries carry out measures to achieve long-term climate neutrality, resource efficiency, success effectiveness, and social equity.

Role of Intuition in Resource Management

The intuitions in innovative countries are promoting and encouraging resource-efficient and effective practices to reduce waste and protect natural ecosystems for the future. The Resource Management policies focus on minimising the environmental impact of production and consumption, increasing resource efficiency, and ensuring that natural resources are conserved for future generations. The Netherlands has set a national goal to be fully circular by 2050, meaning that products and materials will be repurposed rather than discarded, drastically reducing waste and preserving natural resources.

Institutions are significant when discussing resource management because they play a pivotal role and are unique when dealing with comprehensive environmental systems (Keeney 2012). It allows decision-makers to catch and draw on experience and implicit knowledge when faced with uncertain conditions and situations like fluctuating resource availability, scarcity, and unexpected ecological or environmental changes (Villa & McLeod, 2002).

Social Sustainability and Inclusivity Policymaking

Social Impact of Climate Policies

As eco-innovation drives and sources of new industries, governments in developed countries are working to ensure that green policies create jobs, improve public health and address social equity. For example, Sweden has implemented a "Smart Cities or developed countries" initiative that creates green jobs and sustainable urban planning to enhance environmental quality and social well-being (Nightingale & Scott, 2007). This initiative focuses on making cities liveable, reducing air pollution and ensuring that eco-innovation benefits are accessible to all citizens, not only those in higher income brackets. This approach ensures that communities, especially vulnerable groups, benefit from green transitions.

The social impact of climate policies and procedures can be learned as they influence the livelihoods, well-being and social equity of communities in the region. Nearly all the policies identifying climate-induced migration are always a result of extreme events. Increasing sea levels can strain public services and infrastructure related to society in receiving areas in the environment; more needed strategic planning and support to integrate displaced populations in a socially equitable way.

The social sustainability of climate policies is based on their ability to balance the needs of diverse social groups and ensure the vulnerable population. The low-income communities, indigenous populations and the women are not out of proportion to burdened. These groups are often the most implementing for climate change and are frequently overlooked in policy making and launching. Thus, it is essential for climate policies to prioritise social justice, ensuring that the advantages and disadvantages are equitably distributed (Nwokolo et al., 2023c). Altmetrics tracks the influence of research on public policy, helping to highlight how a particular knowledge or specific thongs can report, improve, and analyse social outcomes.

A deepened insight into how policies influence the social texture, from factors such as employment, resource base, and health to family and community resilience, would be crucial in developing climate. If put into place, this framework can help the public and private sectors implement policies that offer social benefits or higher advantages in the longer term, further equality and long-term survival, and resilience to climate-related vulnerabilities (Milat et al., 2015).

Economic Sustainability: Balancing Growth with Environmental Goals

Economic Incentive for Eco-Innovations

It is necessary to reach for climate neutrality, which also requires economic incentives to encourage industries to innovate sustainability. The Inflation Reduction Act supports renewable energy investment in the United States by providing significant tax credits to businesses and individuals who adopt green technologies. Similarly, Canada provides green tax credits for clean energy and green tax credits for clean energy projects to promote climate neutrality and green growth (Kapitonov, 2021).

Economic incentives for eco-innovations are vital for enhancing societies' financial sustainability. It helps harness economic growth without necessarily conflicting with environmental objectives. To this end, eco-innovations, defined as new, modified, or improved technologies, processes, and products that aim to minimise adverse environmental impacts, are key to transitioning to a green economy (Hermann et al., 2015).

Governments and International Organizations, however, have an equally important function in facilitating eco-innovations by creating markets for such products and services. For example, solar and wind technology advances have been considerable due to renewable energy incentives, enabling these technologies to become competitive with traditional fossil fuels (Nwokolo et al., 2023a; Nwokolo et al., 2023b). In addition, eco-labelling and certifications grant market-based incentives for industry organisations to align with established environmental standards, appealing to an increasingly eco-conscious consumer base.

By showcasing economic sustainability and environmental objectives through intelligent incentives, policymakers can stimulate a green economy conducive to growth and ecological harmony (PETIT 2004). Such steps not only handle pressing environmental issues but also lay down the necessary elements for sustainable development to ensure the economy remains viable and environmentally responsible.

The Link Between New Technological Innovations and the Environment

Previous studies have identified environmental innovations as one of the significant driving forces toward a green economy and change. Innovations or new creative things in environmentally related technologies play a critical and crucial role in reducing the ecological disadvantage in a manner that integrates innovation resources. These initiatives enhance regional competitiveness, encourage innovation and develop green economic systems (Xie et al., 2022). In addition, environment-based technology has become an essential tool for organisations to improve their market image, attain sustainable development, and adhere to environmental rules locally and internationally (Xu et al., 2020). The innovation offers practical solutions enabling the development of sustainable products, processes, and practices. They also allow companies to penetrate new markets and enhance competitiveness in the changeable economy (Ben Amara & Chen, 2022).

Technological innovation is one of the complex concepts that may significantly contribute to eco-technology development and environmental performance. Researchers have tried to find various proxies for measuring technological innovation and its impact on environmental quality. For example, R&D activities are considered a widely used indicator in assessing technological Development and its role in reducing CO_2 emissions, as found by Fernández-Fernández et al. (2018) and Petrović (2021). Similarly, energy efficiency is considered a vital factor indicating eco-technological innovation. Several research studies have shown its influence in reducing greenhouse gas emissions (Salman & Long, 2019). Patent development: Patent development is also an excellent standard measure, as seen by many scholars, for evaluating the relationship between technological improvement and environmental sustainability (Rehman & Raza, 2023). Moreover, FDI is a source of diffusing eco-innovations, while TFP captures the efficiency gains due to technological improvements.

Institutional Adapt to Smart Eco-Systematic Realities

Artificial intelligence will analyse the heavy environmental data sets, forecast the ecological effects, and use the resources for optimisation. At the same time, the Internet of Things (IoT) allows for ongoing energy use, soil health monitoring and water and air quality. When AI and IoT work together, they can produce policies that more precisely target sustainability goals while combining intuition with the correct way insights. IoT in places like Stockholm and Amsterdam maintains "smart grids" that increase energy efficiency and reduce carbon emissions. Such technological integration is motivated by European policy frameworks like the European Union's Digital Strategy for Green Transmission, which seeks to promote sustainable growth through digital innovation (Zhou, 2021).

Smart eco-cities combine advanced technologies such as Artificial Intelligence (AI) and the Artificial Intelligence of Things (IoT) with environmental sustainability principles to tackle urgent urban problems. (Ahvenniemi et al., 2017). This combination of both successful environmental management and waste reduction capabilities is the basis of day-to-day activity, which, at the same time, thanks to real-time data management, is the sorting out of ecological problems. Nonetheless, the move to smarter eco-cities is not without difficulties. Issues surrounding energy consumption, e-waste, and ethical dilemmas, such as privacy and data security, sustainable design principles and regulatory frameworks, are imperative. Although these obstacles exist, smarter eco-cities can change urban life and develop sustainable systems using modern technology (Bibri et al., 2024).

Data and Transparency in Climate Policy

Accountability and transparency are essential for climate programs to be effective. Boost public trust, advance global climate, and promote corporate compliance by collecting and disseminating environmental data. Systems for making ecological information available have been put in place in many developed countries. This will enable individuals and groups to hold governments responsible and actively participate in sustainable initiatives (Newell & Mulvaney, 2013). The public can monitor local pollution levels thanks to data on pollutant emissions from developed countries around Europe provided by the European Pollutant Release and Transfer Register (E-PRTR) (Noguera-Oviedo & Aga, 2016). Several digital technologies like satellite imaging and Internet of Things senses in countries make this data gathering easier. These data sources enable organisations to make their climate change policies.

Transparency in climate policy is key. It means clear and consistent reporting of climate data, policy outcomes and progress on emissions reductions. By being transparent, countries can show they are serious about their climate goals and hold themselves accountable to national and international audiences. The Paris Agreement has an Enhanced Transparency Framework (ETF), which requires countries to inform their carbon emissions and take climate action to build trust and cooperation worldwide.

A transparent climate policy system provides detailed feedback on progress so you can see gaps or barriers to success and address them. The role of technology in data collection and dissemination is also key, as digital tools and platforms allow for real-time monitoring and easy access to climate information, making the data global.

Neutrality in Global Corporations and the Role of Developed Nations

Climate neutrality can be defined as the state at which, within a period, the amount of greenhouse gases released into the atmosphere equals the amount taken out (Kartal et al., 2023). This is what has been described as a net-zero carbon footprint. Industries worldwide are critical in this process since they produce many carbon emissions, especially in manufacturing industries, which use energy and move goods (Adebayo et al., 2023). From the various perspectives of global corporations, developed countries have essential roles in enacting rules and regulations that incentivise companies to be greener.

Developed nations have primarily led the demand for a net zero future, as they have previously contributed most to greenhouse gas emissions. Well-established processes needed by developed countries the institutional capacity, financial resources, plus technology transfer and diffusion for scaling up sustainable development (Wang et al., 2024).

Also, developed countries extend beyond their borders, as highlighted by Hausfather and Peters. To meet and achieve the world climate goals, developed countries should give help with technology, skill growth, and money to poor lands. The study by Wu et al. (2024) emphasises that collaborative international policies, like technology transfer agreements and sustainable financing activities, can accelerate the global shift towards climate neutrality. By fostering partnerships and sharing best practices, developed countries can help bridge the gap between different regions' abilities to meet climate goals, ensuring a just and inclusive transition. Also, good teamwork worldwide needs more than just high goals in each place; it also helps in big group plans like the Paris deal. This deal keeps the earth's heat rise under 1.5 °C.

Challengers/Barriers to Implementing Sustainability Environmental Policies and Smart EcoSystematic Realities

Achieve climate neutrality via smart innovation in many ways. One main barrier is the climate's fragmented nature, which affects different nations' policies. Dellmuth and Gustafsson (2021) noted that intergovernmental organisations do not always cooperate, leading to failures in execution and conflicts between policies implemented at the global, national and local levels. However, technologies that would be crucial in facilitating net-zero emissions, such as renewable energy technologies and carbon capture, are still under development or are costly and not scalable. They are also constrained by economic challenges that many third-world countries face, so these regions do not have the finances to embrace sustainable technologies for a net zero economy (Joshi et al., 2024).

Some barriers also delay the effective implementation of sustainable environmental policies. Significant financial difficulty exists, especially in developing countries, where access to funding for climate adoption projects remains limited (Xu et al., 2020). The availability of investments in green technologies is restricted, slowing the transition to low-carbon economies. Barriers of the institutional kind, such as the Byzantine-fold delays and complexities that characterise responses to climate through instruments like the European Green Deal, show how slow-moving structures can hinder momentum around action on climate change.

Putting in place environmental sustainability policies and moving to innovative, eco-friendly systems has many significant challenges. Economically, the high upfront costs of new sustainable technologies like renewable energy and energy-saving infrastructure are an important issue, especially for small and medium enterprises (SMEs) that initially find it hard to pay, even though they can save money in the long run. Technologically, the limited options and ability to scale green technologies, such as renewable energy systems and storage, often make them more expensive than traditional solutions, which stops widespread use. Globally, uneven efforts among countries lead to different sustainability rules, with some areas falling behind due to political issues or lack of resources.

Strategies to Reduce the Challenges and Barriers

Global Cooperation: aligning national policies and international climate targets is essential. Countries can establish a global framework around their climate adaptation by promoting international treaties and collaboration. Investing in technological innovations is also necessary; public-private partnerships can enhance the lowered development and development of low-carbon technologies by cutting costs, increasing efficiency, and Corporate Social Responsibilities (CSR). Corporate governance and sustainability management encompass the relationship between business and society in making climate-oriented solutions. Also, ensuring a just transition is vital to safeguarding communities affected by the shift away from fossil fuels. Policies that include retraining industries (Dewasiri, 2024). Expanding financial mechanisms will guarantee resource support to climate-neutrality projects, especially for disadvantaged regions.

Governments and private companies should prioritise financing new ideas in renewable energies, energy-saving solutions, and ways of doing things sustainably. This will contribute to lowering the costs and, consequently, making the technologies more accessible to businesses and people. Collaboration between the public and private sectors can help build a helpful environment for sustainable growth. Strengthening and clarifying the rules and policies will further help businesses follow them and eventually encourage all the players within the industry to be more sustainable. Governments can also use such mechanisms as carbon taxes or trading systems to incentivise and promote closer cooperation between companies to reduce their carbon footprint. Educating people in all spheres-from businesses to simple citizens decreases resistance. It involves employee training, which may worry about losing their jobs, by introducing green technologies and providing new job opportunities in this very area.

Conclusion

An integrated approach addresses sustainability's interlinked environmental, social and economic realities and is needed to achieve climate neutrality. Developed countries are leading this endeavour with initiatives like Germany's Energiewende and the European Green Deal, which help with unbiased growth, increasing resource efficiency, renewable energy, and circular economy principles (Khan et al., 2024).

The policy drives in developing countries make a complex and multifaceted approach to environmental sustainability for future goals. By combining the rules and regulatory mechanisms, technological innovations, strong institutions, public–private collaborations and social engagement, these nations are making particular strides in examining environmental challenges while fostering long-term economic growth and social well-being. Using flexibility, adaptability and policies, developing countries are paving the way towards a more sustainable and equitable future of climate neutrality. To maximise resources, increasing energy efficiency and transparency in environmental data will support the sustainability of the countries. International collaboration is essential to meet climate targets, and developed countries should share financial and technological resources with imaging economics to promote fair transitions. These countries may move the world toward a more resilient, inclusive and carbon–neutral future by funding eco-innovations and trying sustainability to monetary rewards. Achieving climate neutrality through smart eco-innovation faces numerous barriers, as fragmented policies, financial constraints, and sociocultural challenges hamper this transformation. To overcome these challenges, global collaboration is required to mobilise resources to drive technological innovations through effective public–private partnerships. Inclusive strategies, including just transition, modernisation of regulatory systems and expanding financial support, should be developed by nations collectively to realise the goals of climate neutrality in an integrated manner, thus ensuring environmental sustainability and fostering socio-economic resilience for business and society in the transformative path to a sustainable future.

References

Adebayo, T. S., Ullah, S., Kartal, M. T., Ali, K., Pata, U. K., & Ağa, M. (2023). Endorsing sustainable development in BRICS: The role of technological innovation, renewable energy consumption, and natural resources in limiting carbon emission. *Science of the Total Environment, 859*, 160181. https://doi.org/10.1016/j.scitotenv.2023.160181

Ahmad, M., & Zheng, J. (2021). Do innovation in environmental-related technologies cyclically and asymmetrically affect environmental sustainability in BRICS nations? *Technology in Society, 67*(101746). https://doi.org/10.1016/j.techsoc.2021.101746

Ahvenniemi, H., Huovila, A., Pinto-Seppä, I., & Airaksinen, M. (2017). What are the differences between sustainable and smart cities? *Cities, 60*, 234–245. https://doi.org/10.1016/j.cities.2016.09.009

Bandara, C., Rathnasiri, M. S. H., Dewasiri, N. J., Chathuranga, B., & Singh, N. (2024). Greening the economy for a sustainable future: Opportunities and perceptions. Greening Our *Economy for a Sustainable Future*, 197–213. https://doi.org/10.1016/B978-0-443-23603-7.00016-9

Ben Amara, D., & Chen, H. (2022). Driving factors for eco-innovation orientation: Meeting sustainable growth in Tunisian agribusiness. *International Entrepreneurship and Management Journal, 18*(2), 713–732. https://doi.org/10.1007/s11365-021-00792-0

Bibri, S. E., Krogstie, J., Kaboli, A., & Alahi, A. (2024). Smarter eco-cities and their leading-edge artificial intelligence of things solutions for environmental sustainability: A comprehensive systematic review. *Environmental Science and Ecotechnology, 19*, 100330. https://doi.org/10.1016/j.ese.2024.100330

Bhushan, M., Singh, R. (2024). Talent management facilitates net-zero transition through employee green behavior. In R. Singh, S. Khan, A. Kumar, S. Luthra, & H. Chokshi (Eds.), *Net zero economy, Corporate social responsibility and sustainable value creation. csr, sustainability, ethics & governance*. Springer. https://doi.org/10.1007/978-3-031-55779-8_7

Chathuranga, B., Rathnasiri, M. S. H., Dewasiri, N. J., Jayarathne, P., Bandara, C., & Bakari, M. A. (2024). Direct and indirect effect on industries and services (A). *Greening Our Economy for a Sustainable Future*, 215–226. https://doi.org/10.1016/B978-0-443-23603-7.00017-0

Dellmuth, L. M., & Gustafsson, M. T. (2021). Global adaptation governance: How intergovernmental organisations mainstream climate change adaptation. *Climate Policy, 21*(7), 868–883. https://doi.org/10.1080/14693062.2021.1930072

Dewasiri, N. J. (2024). Sustainable strategies for net zero: An Asia Pacific perspective. In *Transition towards a sustainable future: Net zero policies and environmental sustainability* (pp. 3–25). Springer Nature. https://doi.org/10.1007/978-981-99-9911-2_1

Dewasiri, N.J., Gnanapala, W. K. A. C., Rathnasiri, M. S. H., Yadav, M., Tewari, L. M., & Mishra, U. (2024). The way forward for net zero: Aligning nationally determined contributions (NDCs) and holistic sustainable development in South Asia. In R. Singh, & D. Crowther (Eds.), *Transition towards a sustainable future. Approaches to global sustainability, markets, and governance*. Springer. https://doi.org/10.1007/978-981-97-5756-5_5

Fetting, C., & ESDN Office. (2020). *The European Green Deal*. ESDN Report, December 2020. ESDN Office. https://www.esdn.eu/fileadmin/ESDN_Reports/ESDN_Report_2_2020.pdf

Fernández Fernández, Y., Fernández López, M. A., & Olmedillas Blanco, B. (2018). Innovation for sustainability: The impact of R&D spending on CO_2 emissions. *Journal of Cleaner Production, 172*, 3459–3467. https://doi.org/10.1016/j.jclepro.2017.11.00112

Greenhalgh, T., & Fahy, N. (2015). Research impact in the community-based health sciences: An analysis of 162 case studies from the 2014 UK Research Excellence Framework. *BMC Medicine, 13*, 1–12. https://doi.org/10.1186/s12916-015-0353-6

Haunschild, R., Bornmann, L., & Marx, W. (2016). Climate change research given bibliometrics. *PLoS One, 11*(7), e0160393. https://doi.org/10.1371/journal.pone.0160393

Halperin, S. A., Ye, L., MacKinnon-Cameron, D., Smith, B., Cahn, P. E., Ruiz-Palacios, G. M., … & Zubkova, T. (2022). Final efficacy analysis, interim safety analysis, and immunogenicity of a single dose of recombinant novel coronavirus vaccine (adenovirus type 5 vector) in adults 18 years and older: An international, multicentre, randomised, double-blinded, placebo-controlled phase 3 trial. *The Lancet, 399*(10321), 237–248. https://doi.org/10.1016/S0140-6736(21)02885-0

Hermann, A. T., Pregernig, M., Hogl, K., & Bauer, A. (2015). Cultural imprints on scientific policy advice: Climate science–policy interactions within Austrian neo-corporatism. *Environmental Policy and Governance, 25*(5), 343–355. https://doi.org/10.1002/eet.1694

Isopahkala-Bouret, U., & Nori, H. (2021). On a level playing field? The effect of parents' education on master's degree graduates' labour market success in Finland. *Advances in Applied Sociology, 11*(12), 626–643. https://doi.org/10.4236/aasoci.2021.1112045

Joshi, A., Singh, R., Kumar, V., Rani, S. (2024). Net zero strategy: Exploring opportunities, confronting challenges for developing countries. In R. Singh, & D. Crowther (Eds.), *Transition towards a sustainable future. Approaches to global sustainability, markets, and governance*. Springer. https://doi.org/10.1007/978-981-97-5756-5_11

Kapitonov, I. A. (2021). Development of low-carbon economy as the base of sustainable improvement of energy security. *Environment, Development and Sustainability, 23*, 3077–3096. https://doi.org/10.1007/s10668-020-00751-w

Kartal, M. T., Ghosh, S., & Adebayo, T. S. (2023). Renewable energy effect on economy and environment: The case of G7 countries through novel bootstrap rolling window approach. *Renewable Energy, 216*, 119057. https://doi.org/10.1016/j.renene.2023.119057

Karunarathna, K. S. S. N., Dewasiri, N. J., Singh, R., & Rathnasiri, M. S. H. (2024). What does artificial intelligence–powered ChatGPT bring to academia? A review. In R. Singh, S. Khan, A.

Kumar, & V. Kumar (Eds.), *Artificial intelligence enabled management: An emerging economy perspective* (pp. 3–13). De Gruyter, https://doi.org/10.1515/9783111172408

Khan, S., Singh, R., Alnahas, J., Abbate, S., & Centobelli, P. (2024). Navigating the smart circular economy: A framework for manufacturing firms. *Journal of Cleaner Production, 480*, 144007. https://doi.org/10.1016/j.jclepro.2024.144007

Keeney, R. L. (2012). Value-focused brainstorming. *Decision Analysis, 9*(4), 303–313. https://doi.org/10.1287/deca.1120.0253

Keys, D. L. (2024). Getting to net zero by 2050. In *Using NEPA to combat global warming* (pp. 129–156). Springer Nature Switzerland. https://doi.org/10.1007/978-3-031-14555-3_5

Khazragui, H., & Hudson, J. (2015). Measuring the benefits of university research: Impact and the REF in the UK. *Research Evaluation, 24*(1), 51–62. https://doi.org/10.1093/reseval/rvu028

Kumar, K., Singh, R., Sharma, A., & Lehri, B. P. S. (2024). Climate change adaptation strategies for achieving net-zero economy. In R. Singh, D. Crowther (Eds.), T*ransition towards a sustainable future. Approaches to global sustainability, markets, and governance*. Springer. https://doi.org/10.1007/978-981-97-5756-5_8

Milat, A. J., Bauman, A., & Redman, S. (2015). Narrative review of models and success factors for scaling up public health interventions. *Implementation Science, 10*, 1–11. https://doi.org/10.1186/s13012-015-0301-6

Mishra, U., Rathnasiri, M. S. H., & Dewasiri, N. J. (2024). Net-zero economy and energy sustainability: A Nepalese perspective. In R. Singh, S. Khan, A. Kumar, S. Luthra, & H. Chokshi. (Eds.), *Net zero economy, corporate social responsibility and sustainable value creation. CSR, sustainability, ethics & governance*. Springer. https://doi.org/10.1007/978-3-031-55779-8_9

Newell, P., & Mulvaney, D. (2013). The political economy of the 'just transition.' *The Geographical Journal, 179*(2), 132–140. https://doi.org/10.1111/geoj.12008

Nightingale, P., & Scott, A. (2007). Peer review and the relevance gap: Ten suggestions for policy-makers. *Science and Public Policy, 34*(8), 543–553. https://doi.org/10.3152/030234207X244784

Noguera-Oviedo, K., & Aga, D. S. (2016). Chemical and biological assessment of endocrine disrupting chemicals in a full-scale dairy manure anaerobic digester with thermal pretreatment. *Science of the Total Environment, 550*, 827–834. https://doi.org/10.1016/j.scitotenv.2016.01.138

Nwokolo, S. C., Singh, R., Khan, S., Kumar, A., & Luthra, S. (2023a). Remedies to the challenges of renewable energy deployment in Africa. In *Africa's path to net-zero. CSR, sustainability, ethics & governance*. Springer. https://doi.org/10.1007/978-3-031-44514-9_3

Nwokolo, S. C., Singh, R., Khan, S., Kumar, A., Luthra, S. (2023b). Africa's awakening to climate action. In *Africa's path to net-zero. CSR, sustainability, ethics & governance*. Springer. https://doi.org/10.1007/978-3-031-44514-9_9

Nwokolo, S. C., Singh, R., Khan, S., Kumar, A., & Luthra, S. (2023c). Impacts of climate change in Africa. In *Africa's path to net-zero. CSR, sustainability, ethics & governance*. Springer. https://doi.org/10.1007/978-3-031-44514-9_7

Nwokolo, S. C., Singh, R., Khan, S., & Kumar, A. (2024). Threats to the rapidity of sustainability transitions posed by technological changes. In S. C. Nwokolo, R. Singh, S. Khan, & A. Kumar (Eds.), *Technological pathways for Africa's net-zero economy* (pp. 39–75). https://doi.org/10.1016/B978-0-443-31486-5.00002-1

Obobisa, E. S., Chen, H., & Mensah, I. A. (2022). The impact of green technological innovation and institutional quality on CO_2 emissions in African countries. *Technological Forecasting and Social Change, 180*, 121670. https://doi.org/10.1016/j.techfore.2022.121670

Petit, J. C. (2004). Why do we need fundamental research? *European Review, 12*(2), 191–207. https://doi.org/10.1017/S1062798704000173

Petrović, P. (2021). The impact of R&D expenditures on CO_2 emissions: Evidence from sixteen OECD countries. *Environmental Science and Pollution Research, 28*, 51149–51159. https://doi.org/10.1007/s11356-021-14288-51

Rathnasiri, M. S. H., Dewasiri, N. J., & Kumar, A. A. (2024). Policy framework and implementation strategies for Sri Lanka's Transition to a net-zero economy. In R. Singh, S. Khan, A. Kumar, S. Luthra, & H. Chokshi (Eds.), Ne*t zero economy, corporate social responsibility and sustainable value creation. csr, sustainability, ethics & gover*nance. Springer. https://doi.org/10.1007/978-3-031-55779-8_3

Rehman, F. U., Islam, M. M., & Raza, S. A. (2023). Does disaggregate energy consumption matter to export sophistication and diversification in OECD countries? A robust panel model analysis. *Renewable Energy, 206*, 274–284. https://doi.org/10.1016/j.renene.2023.06.115

Rogelj, J., Popp, A., Calvin, K. V., Luderer, G., Emmerling, J., Gernaat, D., … & Tavoni, M. (2018). Scenarios towards limiting global mean temperature increase below 1.5 C. *Nature Climate Change, 8*(4), 325–332. https://doi.org/10.1038/s41558-018-0091-3

Salman, M., Long, X., Dauda, L., & Mensah, C. N. (2019). The impact of institutional quality on economic growth and carbon emissions: Evidence from Indonesia, South Korea and Thailand. *Journal of Cleaner Production, 241*, 118331. https://doi.org/10.1016/j.jclepro.2019.118331

Schweitzer, H., Haucap, J., Kerber, W., & Welker, R. (2018). Modernising the law on abuse of market power: Report for the Federal Ministry for Economic Affairs and Energy (Germany). Available at SSRN 3250742. https://doi.org/10.2139/ssrn.3250742

Shabir, M., Hussain, I., Işık, Ö., Razzaq, K., & Mehroush, I. (2023). The role of innovation in environmental-related technologies and institutional quality to drive environmental sustainability. *Frontiers in Environmental Science, 11*, 1174827. https://doi.org/10.3389/fenvs.2023.1174827

Singh, R., Gupta, A., & Kandpal, V. (2024). ESG excellence in the digital world. In D. R. Ernesto, S. Gonzalez, V. Kandpal, P. K. Ozili, & P. Chatterjee (Eds.), *Fintech for ESG and the circular economy* (pp. 93–204). Scrivener Publishing LLC. https://doi.org/10.1002/9781394234035.ch10

Villa, F., & McLeod, H. E. L. E. N. A. (2002). Environmental vulnerability indicators for environmental planning and decision-making: Guidelines and applications. *Environmental Management, 29*, 335–348. https://doi.org/10.1007/s00267-001-0039-5

Vijerathne, D. T., Wahala, W., Rathnasiri, M. S. H., De Silva, V., & Dewasiri, N. J. (2024). Literature review on sustainability. In *Greening our economy for a sustainable future* (pp. 33–45). https://doi.org/10.1016/B978-0-443-23603-7.00004-2

Wang, C., Zhou, D., Guo, X., & Kayani, U. N. (2024). Role of natural resource rents, financial development and technological research in achieving sustainable development: A study of South Asian Countries. *Resources Policy, 89*, 104632. https://doi.org/10.1016/j.resourpol.2023.104632

Wu, L., Yi, F., Bu, Y., Lu, W., & Huang, Y. (2024). Toward scientific collaboration: A cost-benefit perspective. *Research Policy, 53*(2), 104943. https://doi.org/10.1016/j.respol.2024.104943

Xie, Y., Ou, P., Wang, X., Xu, Z., Li, Y. C., Wang, Z., … & Sargent, E. H. (2022). High carbon utilisation in CO_2 reduction to multi-carbon products in acidic media. *Nature Catalysis, 5*(6), 564–570. https://doi.org/10.1038/s41929-022-00757-1

Xu, X. W., Wu, X. X., Jiang, X. G., Xu, K. J., Ying, L. J., Ma, C. L., … & Li, L. J. (2020). Clinical findings in a group of patients infected with the 2019 novel coronavirus (SARS-Cov-2) outside of Wuhan, China: Retrospective case series. *BMJ, 368*. https://doi.org/10.1136/bmj.m606

Zhang, J., Terrones, M., Park, C. R., Mukherjee, R., Monthioux, M., Koratkar, N., … & Bianco, A. (2016). Carbon science in 2016: Status, challenges and perspectives. *Carbon, 98*, 708–732. https://doi.org/10.1016/j.carbon.2015.11.060

Chapter 13
Harnessing Emerging Technologies for Green Sustainability: Innovations, Challenges, and Future Pathways

Orhan Balcı, Dawoud Almohammad, Cumali Kılıç, and Rubee Singh

Abstract Emerging technologies have become pivotal in advancing green sustainability by fostering innovative solutions to environmental challenges while promoting economic growth. This chapter explores the transformative potential of technologies such as artificial intelligence (AI), blockchain, Internet of Things (IoT), and Industry 4.0 innovations in optimizing resource management, enhancing operational efficiency, and reducing environmental impacts. Also it highlights the critical role of sustainable energy technologies, digital circular economy practices, and financial innovations in driving systemic change. By examining the interplay between technological advancements, economic considerations, and policy frameworks, the chapter underscores the need for integrative approaches to achieve sustainability goals and also addresses limitations, including the digital divide, technological integration challenges, and lifecycle impacts, and identifies future research directions. These include equitable access to technologies, hybrid systems, and interdisciplinary strategies that align innovation with environmental and social objectives. To conclude, this chapter offers a comprehensive understanding of how emerging technologies can serve as catalysts for sustainable development, paving the way for a resilient and equitable future.

O. Balcı
Department of Business Administration, Faculty of Economics and Administrative Sciences, Karatekin University, Çankırı, Turkey
e-mail: orhanbalci@karatekin.edu.tr

D. Almohammad · C. Kılıç (✉)
Department of Business Administration, Faculty of Economics and Administrative Sciences, Artuklu University, Mardin, Turkey
e-mail: cumalikilic@artuklu.edu.tr

D. Almohammad
e-mail: dawoud79@artuklu.edu.tr

R. Singh
Institute of Business Management, GLA University, Mathura, India

Department of Management Studies, Kumaun University, Nainital, Uttarakhand, India

R. Singh and W. L. Filho (eds.), *Climate Neutrality Through Smart Eco-Innovation and Environmental Sustainability*, Climate Change Management,
https://doi.org/10.1007/978-3-031-83250-5_13

Keywords Emerging technologies · Green sustainability · Artificial intelligence · Digital circular economy · Sustainable development

Introduction

Emerging technologies play a crucial role in advancing green sustainability by fostering innovative solutions that address environmental challenges while simultaneously promoting economic growth (Nwokolo et al., 2023b). The integration of renewable energy technologies and sustainable engineering practices is pivotal in this context, enabling societies to transition toward environmentally conscious energy systems (Etukudoh et al., 2024). Technologies such as artificial intelligence (AI), 5G, internet of things (IoT), digital twins, and edge computing exhibit transformative potential by optimizing renewable energy systems and enhancing energy efficiency, which are critical for minimizing environmental impacts and achieving global sustainability goals (Pasupuleti, 2024). These rapidly evolving technologies, operating under the umbrella of Industry 4.0, not only facilitate the efficient management of energy resources but also adapt dynamically to fluctuating energy demands and supply conditions, ensuring a more resilient and sustainable energy infrastructure (Kılıç & Atilla, 2024a).

Moreover, the relationship between technological innovation and sustainable development underscores the profound impact of technology on green development (Yuan, 2024). Technological progress contributes significantly to creating sustainable urban environments that can support growing populations while minimizing ecological footprints (Ahmed et al., 2020). In this regard, sustainable energy technologies such as solar, wind, and biomass are indispensable for reducing dependency on fossil fuels. These technologies provide dual benefits: immediate reductions in energy consumption and long-term solutions to pressing environmental challenges (Wang et al., 2019). This dual focus aligns with the overarching objectives of sustainable development, which necessitate a harmonious balance among economic growth, environmental stewardship, and social equity (Ilieva, 2023; Nwokolo et al., 2023d).

From the foregoing, the convergence of emerging technologies, innovative practices, and sustainable policies serves as the foundation for fostering green sustainability. Leveraging advancements in AI, renewable energy systems, and financial innovations is crucial for establishing resilient frameworks that address current environmental challenges while paving the way for a sustainable future. This integrated approach ensures not only the efficient utilization of resources and the reduction of environmental impacts but also the creation of adaptive and inclusive systems that support long-term sustainability. Furthermore, fostering global collaboration, incentivizing technological adoption, and strengthening policy frameworks are vital to amplifying these efforts. As societies navigate the complexities of sustainable development, the synergy between technological innovation and environmental stewardship will play a decisive role in shaping a greener and more equitable future.

A Theoretical Approach on Emerging Technologies for Green Sustainability

Green Industry 4.0

Green Industry 4.0 represents a transformative approach to manufacturing that integrates sustainable practices with the advancements of the Fourth Industrial Revolution (Industry 4.0) (Bradu et al., 2022). This paradigm shift emphasizes the utilization of digital technologies, such as the IoT, big data, and AI, to enhance operational efficiency while minimizing environmental impact (Bibri et al., 2024). The integration of these technologies supports productivity and aligns with sustainability goals, making it a critical area of research and application in modern industrial practices (Bai et al., 2020).

The concept of Green Industry 4.0 is fundamentally rooted in the principles of sustainable manufacturing because the intersection of sustainable manufacturing and Industry 4.0 technologies is essential for developing a comprehensive understanding of how these innovations can contribute to environmental sustainability (Machado et al., 2019). Industry 4.0 with sustainable manufacturing practices, emphasizing the need for a cohesive framework that incorporates both technological advancements and sustainability principles (Kılıç, 2023).

Moreover, smart factories leverage advanced technologies to create more efficient production processes that reduce waste and energy consumption. The elements of Industry 4.0 play a pivotal role in promoting sustainable business performance among small and medium-sized enterprises (SMEs) (Haseeb et al., 2019). These technologies facilitate real-time monitoring and optimization of manufacturing processes, leading to improved resource management and reduced environmental footprints (Javaid et al., 2022). The implications of Industry 4.0 for sustainability extend beyond operational efficiency; they also encompass broader economic and social dimensions. Industry 4.0 technologies can help overcome technological challenges and enhance sustainable business performance (Reis et al., 2021), particularly in the context of SMEs and this issue highlights the potential for Industry 4.0 to drive not only environmental benefits but also economic growth and social responsibility (Asmolov & Ledentsov, 2022).

In addition to operational advancements, Green Industry 4.0 also necessitates a shift in corporate strategies towards digital transformation. The integration of digital tools into corporate strategies can accelerate green innovation, thereby enhancing sustainability metrics (Gan, 2023). This underscores the importance of aligning digital transformation efforts with sustainability objectives to achieve long-term environmental goals. By leveraging advancements in IoT, big data, and AI and other technologies, industries can enhance productivity while minimizing their environmental impact. The ongoing research and application of these technologies will be vital in shaping a sustainable future for manufacturing.

Blockchain for Transparent and Sustainable Supply Chains

Blockchain technology has emerged as a transformative force in enhancing transparency and sustainability within supply chains (Friedman & Ormiston, 2022). Its decentralized and immutable nature allows for real-time tracking of products, ensuring that every transaction is recorded and verifiable (Tijan et al., 2019). This capability is particularly crucial in today's globalized supply chains, where maintaining oversight and accountability has become increasingly challenging due to complexity and opacity (Searcy et al., 2022).

One of the primary benefits of blockchain in supply chain management is its ability to provide end-to-end traceability (Feng et al., 2020). Organizations are increasingly implementing blockchain to facilitate product lifecycle transparency and better control their environmental footprint (Hastig & Sodhi, 2020). This is particularly relevant in sectors such as agri-food, where the demand for improved supply chain visibility is driven by consumer preferences for sustainable and ethically sourced products (González-Mendes, 2024). The ability to trace products from origin to consumer not only enhances transparency but also fosters trust among stakeholders, which is essential for sustainable supply chain practices.

Moreover, the integration of blockchain with other technologies, such as IoT and smart grids, further amplifies its potential for sustainability. A blockchain-based smart grid application that enhances collaboration among stakeholders at all levels of the supply chain, thereby promoting sustainable practices (Du et al., 2021). This interconnectedness allows for more efficient resource management and reduces waste, contributing to overall sustainability goals and can eliminate intermediaries and streamline processes (Bischoff & Seuring, 2021). This necessitates a careful examination of the long-term implications of blockchain adoption on supply chain structures and dynamics.

Blockchain also offers significant opportunities for sustainable supply chain mapping, understanding the barriers to its implementation is equally important (Khan et al., 2022). Factors such as technological readiness, regulatory frameworks, and stakeholder collaboration play critical roles in determining the success of blockchain initiatives in supply chains (Balasubramanian et al., 2021). Addressing these barriers is essential for realizing the full potential of blockchain in promoting sustainability (Ronaghi & Mosakhani, 2022). Also, the introduction of blockchain in logistics can significantly enhance the sustainability of supply chains by providing clear visibility into the movement and handling of goods (Tijan et al., 2019). This transparency is vital for ensuring compliance with environmental regulations and ethical standards, thus fostering a more responsible supply chain ecosystem.

Furthermore, the application of blockchain in supply chains has been shown to improve sustainability performance across various dimensions, including environmental protection, social equity, and governance efficiency (Park & Li, 2021). This aligns with the growing recognition of the need for sustainable practices in supply chains, driven by both regulatory pressures and consumer demand (Saeed & Kersten,

2019). By enabling traceability, improving collaboration, and fostering accountability, blockchain can help organizations navigate the complexities of modern supply chains while adhering to sustainability principles. However, addressing the challenges associated with its implementation is crucial for unlocking its full potential.

Digital Circular Economy Technologies

Digital Circular Economy (DCE) Technologies represent a convergence of digital innovations and circular economy principles aimed at promoting sustainable practices across various industries (Singh et al., 2024). The circular economy paradigm emphasizes resource efficiency, waste reduction, and the continual use of materials, while digital technologies facilitate these objectives through enhanced data management, process optimization, and innovative business models (Khan et al., 2024; Kılıç & Atilla, 2024b). Automation, data analysis, and integration are key digital functions that can lead to the creation of circular economy strategies focused on extending product lifespans and optimizing material use (Lukić et al., 2023) and digital technologies are essential for operationalizing the transition to a circular economy, enhancing sustainability performance through improved resource recovery and waste management (Cagno et al., 2021; Nwokolo et al., 2024).

The integration of advanced digital technologies such as IoT, AI and blockchain is crucial for fostering a circular economy. The adoption of these technologies can significantly promote circular economy models by increasing energy efficiency, extending the useful life of products, and recovering value at the end of product life cycles (Cioffi et al., 2020). This is particularly relevant in sectors such as construction, where digital tools can facilitate the implementation of circular practices through better planning and resource allocation (Meng, 2023). Plus, digital technologies can facilitate strategies such as regenerating, narrowing, and closing resource loops in the built environment, thereby promoting sustainability (Aral, 2024). This is echoed by Çetin et al. (2021), who emphasize the importance of digital technologies in enabling circular economy practices specifically within the construction industry. The ability to track materials and optimize their use through digital platforms can lead to substantial reductions in waste and resource consumption (Maiurova et al., 2022).

Besides, digitization supports the evolution of circular economy business models by optimizing processes and enhancing innovation (Ţurcan, 2023) and can significantly contribute to energy conservation and efficiency improvements in production and manufacturing, thereby supporting circular economy goals (Xu, 2024). However, the transition to a digital circular economy is not without challenges. The integration of digital technologies must be carefully managed to ensure that they effectively contribute to the establishment of closed supply chains and the overall goals of the circular economy (Vetrova & Ivanova, 2021). Additionally, the potential for digital technologies to exacerbate existing inequalities must be also addressed (Lekan & Rogers, 2020).

To sum up, digital circular economy technologies play a pivotal role in advancing the principles of the circular economy by leveraging digital innovations to enhance resource efficiency, optimize processes, and foster sustainable practices. As industries increasingly adopt these technologies, the potential for achieving significant environmental and economic benefits becomes more pronounced, paving the way for a more sustainable future.

Sustainable Urban Design

Sustainable urban design is an essential framework that integrates environmental, social, and economic considerations into the planning and development of urban spaces (Kılıç, 2024). It aims to create cities that are not only livable and resilient but also capable of minimizing their ecological footprints while promoting social equity and economic vitality (Gözkaman, 2024). The principles of sustainable urban design are increasingly relevant in the context of rapid urbanization and climate change, necessitating innovative approaches to urban planning (Nwokolo et al., 2023c).

One of the foundational aspects of sustainable urban design is the strategic integration of various urban elements to enhance sustainability outcomes. Strategic urban design can facilitate impactful solutions to urban challenges by connecting the interests of public and private stakeholders (Bafarasat, 2023). This approach advocates for small-scale, multi-project initiatives that address specific sustainability issues, thereby fostering a culture of continuous improvement in urban environments (Louman et al., 2022). Moreover, data-driven technologies can be leveraged to address the challenges of urban sustainability to enhance urban management and planning (Bibri, 2021). This data-centric approach allows for more informed decision-making and resource allocation, ultimately leading to more sustainable urban outcomes.

Social sustainability is another critical dimension of sustainable urban design because social sustainability indicators can be applied to urban developments, highlighting the importance of community engagement and social equity in the planning process (Medved et al., 2020). The New Urbanism paradigm, which emphasizes walkable neighborhoods and mixed-use developments, aligns with these principles by fostering vibrant communities that prioritize social interaction and accessibility. Thus, the design of urban spaces must also consider ecological factors for harmonizing public open spaces, architectural forms, and ecological conditions in neighborhood planning to create sustainable urban environments (Choi, 2024). This holistic approach ensures that urban design not only meets human needs but also respects and enhances the natural environment.

Furthermore, the incorporation of green infrastructure, such as green roofs and sustainable drainage systems can reduce energy consumption and mitigate urban flooding, showcasing how such strategies can contribute to the overall sustainability of urban areas (Cascone, 2023). Similarly, the role of sustainable urban drainage systems in improving water quality and managing runoff are critical for maintaining urban ecosystems (Beecham & Razzaghmanesh, 2015). Compact cities, characterized by high density and mixed land use, can enhance sustainability by reducing reliance on automobiles and promoting public transportation (Tan & Lu, 2018). This aligns with the broader goals of sustainable urbanism, which advocate for efficient land use and reduced environmental impact and sustainable urban design encompasses a multifaceted approach that integrates social, environmental, and economic considerations into urban planning. By leveraging strategic design practices, data-driven technologies, and green infrastructure, cities can enhance their sustainability and resilience in the face of contemporary challenges.

Green Energy Innovations

Green energy innovations are pivotal in addressing the pressing challenges of climate change, energy security, and sustainable development (Nwokolo et al., 2023a). These innovations encompass a wide array of technologies and practices aimed at harnessing renewable energy sources, improving energy efficiency, and reducing carbon emissions (Yuan, 2024). The transition to green energy is not only a technological shift but also a socio-economic transformation that requires collaboration among various stakeholders, including governments, businesses, and communities (Kumar et al., 2024).

One of the most significant areas of green energy innovation is the development of green hydrogen (Dixit, 2023). This innovation has the potential to decarbonize multiple sectors, including transportation, industry, and energy production, thereby playing a crucial role in the global transition to sustainable energy systems and also green hydrogen can serve as a storage solution for excess renewable energy, contributing to grid stability and energy security (Elshafei & Mansour, 2023). In addition to hydrogen, the integration of renewable energy technologies in various sectors is essential for promoting sustainability such as liquefied natural gas (LNG) and hybrid systems that utilize solar and wind energy (Kizielewicz, 2021). This trend demonstrates the versatility of green energy innovations across different industries, highlighting their potential to reduce environmental impacts while maintaining operational efficiency (Joshi et al., 2024).

The economic aspects of green energy innovations are also critical because higher energy prices can incentivize the development of energy-efficient technologies (Ley et al., 2016) and also can foster a competitive market for sustainable energy solutions (Bhushan & Singh, 2024). In their precious work, Liu et al. (2020), explore how energy prices influence green energy innovation and indicate that market dynamics play a significant role in shaping the innovation landscape. Besides, there are various barriers to the development of green energy technologies, including technological, financial, and regulatory obstacles (Wang et al., 2023). For example, the role of policy and regulatory frameworks in facilitating green energy innovations such as the water resource regulations can promote corporate green energy-saving innovations by creating a conducive environment for sustainable practices (Kang, 2024).

Consequently, the impact of green energy innovations extends beyond environmental benefits; they can contribute to social and economic resilience (Arzo & Hong, 2024) and also the integration of renewable energy sources can create jobs, stimulate economic growth, and enhance energy independence (Batra, 2023). Thus, innovation in renewable energy technology is crucial for driving green growth and reducing energy security risks (Feng et al., 2022) and green energy innovations are essential for achieving a sustainable energy future. By leveraging advancements in technologies such as green hydrogen, renewable energy systems, and energy-efficient practices, societies can transition towards more sustainable and resilient systems. The interplay between economic incentives, policy frameworks, and technological advancements will be critical in shaping the future of green energy innovations.

Digital Twins for Sustainability

Digital twins are increasingly recognized as a transformative technology for promoting sustainability across various sectors, particularly in urban planning and development (Mazzetto, 2024). A digital twin is defined as a dynamic digital representation of a physical object or system that mirrors its states and behaviors throughout its lifecycle (Liu et al., 2021). This technology enables real-time monitoring, analysis, and simulation, providing valuable insights that can enhance decision-making processes related to sustainability (Kılıç & Atilla, 2024a).

Digital twin technologies can significantly enhance sustainability by enabling better resource management, optimizing energy use, and improving overall project efficiency (Zhang, 2024). By creating a virtual model of a building or infrastructure, stakeholders can simulate different scenarios, assess environmental impacts, and make informed decisions that align with sustainability goals (Lu et al., 2017). This capability is particularly crucial in the context of the construction industry's significant contribution to global carbon emissions (Jäkel, 2023). This approach underscores the potential of digital twins to provide ongoing insights into the performance and environmental impact of critical infrastructure, allowing for timely interventions and improvements.

In urban contexts, digital twins can be employed to enhance the quality of life for residents by integrating data from various sources, such as IoT devices and sensors. A digital twin architecture for urban ecosystems which collects and processes data on factors like carbon emissions, noise levels, and traffic patterns (Feng et al., 2022; Mukhacheva et al., 2022) enables city planners to make data-driven decisions that promote sustainability and improve urban living conditions.

Moreover, the integration of digital twins with machine learning and predictive analytics can further enhance urban resilience by the use of digital twin techniques combined with data analyses to forecast urban dynamics and improve public services (Gkontzis, 2024). This integration allows for proactive management of urban systems, enabling cities to adapt to changing conditions and mitigate potential sustainability challenges. However, issues related to data integration, quality, and computational complexity must be addressed to fully realize the potential of digital twins in urban environments (Mitkov, 2024) and the need for interdisciplinary collaboration among stakeholders is crucial to ensure that digital twin applications align with broader sustainability objectives (Tripathi et al., 2024).

In conclusion, digital twins represent a powerful tool for advancing sustainability in urban planning and infrastructure management. By providing real-time insights and facilitating data-driven decision-making, digital twins can help cities and organizations optimize resource use, reduce environmental impacts, and enhance the quality of life for residents. As the technology continues to evolve, its role in promoting sustainable practices is likely to expand, offering new opportunities for innovation and resilience in urban environments.

Artificial Intelligence for Green Decision-Making

Artificial Intelligence (AI) is increasingly recognized as a critical tool for enhancing green decision-making across various sectors (Yigitcanlar et al., 2021). By leveraging AI technologies, organizations can optimize resource use, reduce waste, and promote sustainable practices (Challoumis, 2024) and the integration of AI into decision-making processes not only enhances operational efficiency but also supports the broader goals of environmental sustainability (Bibri et al., 2024).

One of the key areas where AI contributes to green decision-making is in consumer behavior analysis (Chen, 2024) which allows businesses to gain insights into consumer preferences and behaviors, enabling them to tailor their strategies to promote eco-friendly products effectively (Savale et al., 2023). By understanding the motivations behind consumer choices, companies can enhance their offerings and foster environmentally responsible behaviors (White et al., 2019). In addition to consumer behavior, AI can enhance decision-making in various sectors, including urban planning and resource management. For instance, the development of AI-based decision-making frameworks can optimize energy efficiency and investment costs in building maintenance (Kim et al., 2018). By evaluating existing conditions and predicting future needs, AI can support sustainable innovation activities and inform regulatory policies (Challoumis, 2024).

However, the need for standardization in AI technologies is crucial to ensure their safe and effective use. Developing equilibrium-safe intelligent systems is essential for fostering trust and promoting the widespread adoption of AI in sustainability efforts (Russia, 2020). AI can facilitate the development of intelligent systems that support sustainable practices. Ghahremani-Nahr et al. (2021), propose a framework for a green supply chain based on the Artificial Intelligence of Things (AIoT), which combines AI with IoT technologies to improve environmental conditions. This integration allows for real-time monitoring and optimization of supply chain processes, leading to reduced waste and improved resource efficiency. Such systems can significantly enhance the sustainability of operations by enabling organizations to make informed decisions based on accurate data.

Addressing these challenges will require collaboration among stakeholders, including policymakers, businesses, and researchers. AI has the potential to significantly enhance green decision-making by providing insights into consumer behavior, optimizing resource management, and promoting sustainable practices. By leveraging AI technologies, organizations can make informed decisions that align with environmental sustainability goals, ultimately contributing to a more sustainable future.

Discussion

The rapid evolution of emerging technologies has fundamentally reshaped the pathways for achieving green sustainability. As highlighted in this chapter, innovations such as AI, blockchain, IoT, and other Industry 4.0 technologies have the potential to transform our life in different ways. However, their successful implementation requires a nuanced understanding of the opportunities and challenges these technologies present. While advancements in digital tools provide unparalleled capabilities for resource optimization, waste reduction, and efficiency enhancement, their deployment must align with broader sustainability goals to avoid unintended consequences, such as increased digital waste or exacerbation of social inequalities.

A critical theme emerging from this analysis is the interdependence of technological, economic, and policy dimensions. The role of financial innovation, as well as regulatory frameworks, is pivotal in supporting the transition to sustainable practices. For example, systemic innovations in energy and food production have shown how integrated approaches can address environmental challenges while fostering economic resilience. Yet, the scalability of such solutions remains a challenge, particularly in regions with limited technological infrastructure or regulatory support.

Also, the importance of collaboration among stakeholders, including governments, industries, and communities, to ensure that emerging technologies contribute meaningfully to green sustainability. This requires not only technological investments but also capacity building, public–private partnerships, and inclusive governance mechanisms that promote equitable access to sustainable benefits. Furthermore, the adoption of technologies such as digital twins and green energy innovations underscores the need for interdisciplinary approaches that bridge technical expertise with environmental and social considerations.

Ultimately, the integration of advanced technologies with sustainability principles is both an opportunity and a responsibility. As these technologies continue to evolve, fostering an adaptive, inclusive, and ethically grounded approach will be essential to maximizing their impact on sustainability goals. By addressing these challenges head-on, stakeholders can ensure that technological progress aligns with the broader vision of a sustainable and equitable future.

Concluding Remarks

The integration of emerging technologies, particularly within the framework of green sustainability, is indispensable for addressing the multifaceted challenges posed by environmental degradation and resource scarcity. This chapter has highlighted the transformative potential of digital innovations and Industry 4.0 technologies in driving sustainable practices. By optimizing resource management, reducing waste, and enhancing operational efficiency, these technologies align economic growth with environmental stewardship, showcasing the power of innovation to bridge the gap between ecological imperatives and economic imperatives.

Moreover, concepts such as the digital circular economy and sustainable urban design exemplify how digital tools can facilitate systemic changes, enabling industries and communities to transition toward sustainable models. The role of financial innovation and policy alignment further underscores the need for comprehensive approaches that integrate technological, economic, and regulatory dimensions. The potential for technologies like digital twins and green energy innovations to reimagine traditional systems also reflects the necessity of continued research and investment in this area.

The convergence of advanced technologies and sustainability principles offers a roadmap for creating resilient and adaptive systems that meet the demands of contemporary environmental challenges. However, achieving this vision requires collaboration across sectors, strategic policy support, and a commitment to embedding sustainability at the core of technological development. By doing so, societies can pave the way for a future that harmonizes innovation, economic growth, and environmental integrity, ensuring long-term benefits for both current and future generations.

Despite the transformative potential of emerging technologies in advancing green sustainability, there are several limitations that warrant attention. One significant limitation is the uneven access to these technologies across different regions and sectors. Developing countries and under-resourced industries often face barriers in adopting advanced tools such as AI, blockchain, and IoT due to high costs, lack of technical expertise, and inadequate infrastructure. This digital divide hinders the global scalability of sustainability solutions, creating disparities in environmental outcomes and economic benefits. Furthermore, the implementation of these technologies may inadvertently generate challenges such as increased electronic waste, energy consumption from digital operations, and cybersecurity vulnerabilities, which could offset their positive impacts on sustainability.

Another critical limitation lies in the integration of these technologies with existing systems and policies. Many organizations struggle with aligning new technologies with traditional practices, regulatory requirements, and stakeholder expectations. The absence of universal standards for the deployment of technologies such as digital twins or blockchain in sustainability contexts adds to this complexity. Moreover, the dynamic and fast-paced nature of technological advancements often outpaces the development of corresponding governance frameworks, creating gaps in accountability, ethical oversight, and environmental compliance.

Thus, future research is encouraged to address these limitations by exploring strategies to bridge the digital divide and ensure equitable access to emerging technologies. Studies could focus on scalable, cost-effective solutions tailored to the needs of developing regions and small to medium enterprises. Additionally, there is a need for interdisciplinary research that integrates technical innovation with socio-economic and policy analysis to create cohesive frameworks for sustainable development. Researchers may also investigate the lifecycle impacts of digital technologies, including their potential to generate waste or exacerbate energy consumption, and develop mitigation strategies that enhance their net-positive contributions to sustainability.

References

Ahmed, Z., Asghar, M. M., Malik, M. N., & Nawaz, K. (2020). Moving towards a sustainable environment: The dynamic linkage between natural resources, human capital, urbanization, economic growth, and ecological footprint in China. *Resources Policy, 67*, 101677. https://doi.org/10.1016/j.resourpol.2020.101677

Aral, D. (2024). Circular economy in the built environment: A systematic literature review on the role of digital technologies across regional contexts. *Iop Conference Series Earth and Environmental Science, 1363*(1), 012005. https://doi.org/10.1088/1755-1315/1363/1/012005

Arzo, S., & Hong, M. (2024). Resilient green infrastructure: Navigating environmental resistance for sustainable development, social mobility in climate change policy. *Heliyon, 10*(13). https://doi.org/10.1016/j.heliyon.2024.e33524

Asmolov, A., & Ledentsov, A. (2022). Industry modern: A solution for sustainable business performance's technology challenges. *Aptisi Transactions on Technopreneurship (Att)*, 4(3), 306–312. https://doi.org/10.34306/att.v4i3.283

Bafarasat, A. (2023). Strategic urban design for sustainable development: A framework for studio and practice. *Sustainable Development, 31*(3), 1861–1872. https://doi.org/10.1002/sd.2489

Bai, C., Dallasega, P., Orzes, G., & Sarkis, J. (2020). Industry 4.0 technologies assessment: A sustainability perspective. *International journal of production economics, 229*, 107776. https://doi.org/10.1016/j.ijpe.2020.107776

Balasubramanian, S., Shukla, V., Sethi, J. S., Islam, N., & Saloum, R. (2021). A readiness assessment framework for Blockchain adoption: A healthcare case study. *Technological Forecasting and Social Change, 165*, 120536. https://doi.org/10.1016/j.techfore.2020.120536

Batra, G. (2023). Renewable energy economics: achieving harmony between environmental protection and economic goals. *Social Science Chronicle, 2*(2), 1–32. https://doi.org/10.56106/ssc.2023.009

Beecham, S., & Razzaghmanesh, M. (2015). Water quality and quantity investigation of green roofs in a dry climate. *Water Research, 70*, 370–384. https://doi.org/10.1016/j.watres.2014.12.015

Bhushan, M., & Singh, R. (2024). Talent Management Facilitates Net-Zero Transition Through Employee Green Behavior. In Singh, R., Khan, S., Kumar, A., Luthra, S., & Chokshi, H. (Eds) *Net zero economy, corporate social responsibility and sustainable value creation. CSR, sustainability, ethics and governance.* Cham: Springer. https://doi.org/10.1007/978-3-031-55779-8_7

Bibri, S. E., Krogstie, J., Kaboli, A., & Alahi, A. (2024). Smarter eco-cities and their leading-edge artificial intelligence of things solutions for environmental sustainability: A comprehensive systematic review. *Environmental Science and Ecotechnology, 19*, 100330. https://doi.org/10.1016/j.ese.2024.100433

Bibri, S. (2021). The underlying components of data-driven smart sustainable cities of the future: a case study approach to an applied theoretical framework. *European Journal of Futures Research, 9*(1). https://doi.org/10.1186/s40309-021-00182-3

Bischoff, O., & Seuring, S. (2021). Opportunities and limitations of public blockchain-based supply chain traceability. *Modern Supply Chain Research and Applications, 3*(3), 226–243. https://doi.org/10.1108/mscra-07-2021-0014

Bradu, P., Biswas, A., Nair, C., Sreevalsakumar, S., Patil, M., Kannampuzha, S., & Gopalakrishnan, A. V. (2022). Recent advances in green technology and Industrial Revolution 4.0 for a sustainable future. *Environmental Science and Pollution Research International, 1*. https://doi.org/10.1007/s11356-022-20024-4

Cagno, E., Neri, A., Negri, M., Bassani, C., & Lampertico, T. (2021). The role of digital technologies in operationalizing the circular economy transition: A systematic literature review. *Applied Sciences, 11*(8), 3328. https://doi.org/10.3390/app11083328

Cascone, S. (2023). Green roof as a passive cooling technique for the mediterranean climate: an experimental study. *Rivista Tema, 09*(02). https://doi.org/10.30682/tema090006

Çetin, S., Wolf, C., & Bocken, N. (2021). Circular digital built environment: An emerging framework. *Sustainability, 13*(11), 6348. https://doi.org/10.3390/su13116348

Challoumis, C. (2024). Building a sustainable economy-how AI can optimize resource allocation. In: *XVI international scientific conference* (pp. 190–224), October.

Chen, C. (2024). Utilizing a hybrid approach to identify the importance of factors that influence consumer decision-making behavior in purchasing sustainable products. *Sustainability, 16*(11), 4432. https://doi.org/10.3390/su16114432

Choi, H. (2024). Reinventing sustainable neighborhood planning: A case study of le rheu, france. *Buildings, 14*(2), 536. https://doi.org/10.3390/buildings14020536

Cioffi, R., Travaglioni, M., Piscitelli, G., Petrillo, A., & Felice, F. (2020). Artificial intelligence and machine learning applications in smart production: Progress, trends, and directions. *Sustainability, 12*(2), 492. https://doi.org/10.3390/su12020492

Dixit, A. (2023). Green hydrogen for karnataka: regional solutions for a clean energy future. In *E3s web of conferences* (Vol. 455, p. 02020). https://doi.org/10.1051/e3sconf/202345502020

Du, X., Qi, Y., Chen, B., Shan, B., & Liu, X. (2021). The integration of blockchain technology and smart grid: Framework and application. *Mathematical Problems in Engineering, 2021*, 1–12. https://doi.org/10.1155/2021/9956385

Elshafei, A., & Mansour, R. (2023). Green hydrogen as a potential solution for reducing carbon emissions: A review. *Journal of Energy Research and Reviews, 13*(2), 1–10. https://doi.org/10.9734/jenrr/2023/v13i2257

Etukudoh, E. A., Fabuyide, A., Ibekwe, K. I., Sonko, S., & Ilojianya, V. I. (2024). Electrical engineering in renewable energy systems: a review of design and integration challenges. *Engineering Science & Technology Journal, 5*(1), 231–244. https://doi.org/10.51594/estj.v5i1.746

Feng, H., Wang, X., Duan, Y., Zhang, J., & Zhang, X. (2020). Applying blockchain technology to improve agri-food traceability: A review of development methods, benefits and challenges. *Journal of Cleaner Production, 260*, 121031. https://doi.org/10.1016/j.jclepro.2020.121031

Feng, J., Ma, L., Broyd, T., Chen, W., & Luo, H. (2022). Digital twin enabled sustainable urban road planning. *Sustainable Cities and Society, 78*, 103645. https://doi.org/10.1016/j.scs.2021.103645

Friedman, N., & Ormiston, J. (2022). Blockchain as a sustainability-oriented innovation?: Opportunities for and resistance to Blockchain technology as a driver of sustainability in global food supply chains. *Technological Forecasting and Social Change, 175*, 121403. https://doi.org/10.1016/j.techfore.2021.121403

Gan, Y. (2023). The impact of corporate digital transformation on green technological innovation in the context of "dual-carbon" goals. *AEDMR, 1*(1), 97. https://doi.org/10.61935/aedmr.1.1.2023.p97
Ghahremani-Nahr, J., Nozari, H., & Sadeghi, M. (2021). Green supply chain based on artificial intelligence of things (aiot). *International Journal of Innovation in Management Economics and Social Sciences, 1*(2), 56–63. https://doi.org/10.52547/ijimes.1.2.56
Gkontzis, A. (2024). Enhancing urban resilience: Smart city data analyses, forecasts, and digital twin techniques at the neighborhood level. *Future Internet, 16*(2), 47. https://doi.org/10.3390/fi16020047
González-Mendes, S. (2024). Discovering the conceptual building blocks of blockchain technology applications in the agri-food supply chain: A review and research agenda. *British Food Journal, 126*(13), 182–206. https://doi.org/10.1108/bfj-06-2023-0517
Gözkaman, A. (2024). Avrupa birliği, sürdürülebilir insani kalkınma ve insan hakları. *EURO Politika, 1*(20), 5–22.
Haseeb, M., Hussain, H., Ślusarczyk, B., & Jermsittiparsert, K. (2019). Industry 4.0: a solution towards technology challenges of sustainable business performance. *Social Sciences, 8*(5), 154. https://doi.org/10.3390/socsci8050154
Hastig, G., & Sodhi, M. (2020). Blockchain for supply chain traceability: Business requirements and critical success factors. *Production and Operations Management, 29*(4), 935–954. https://doi.org/10.1111/poms.13147
Ilieva, L. (2023). Role of technological innovation for sustainable management of tourism organizations. *Iop Conference Series Earth and Environmental Science, 1269*(1), 012038. https://doi.org/10.1088/1755-1315/1269/1/012038
Jäkel, J. (2023). Sustainability assessment of bridge structures in the operation phase based on a digital twin. *Ce/papers, 6*(5), 667–674. https://doi.org/10.1002/cepa.1988
Javaid, M., Haleem, A., Singh, R. P., Suman, R., & Gonzalez, E. S. (2022). Understanding the adoption of Industry 4.0 technologies in improving environmental sustainability. *Sustainable Operations and Computers, 3*, 203–217. https://doi.org/10.1016/j.susoc.2022.01.008
Joshi, A., Singh, R., Kumar, V., & Rani, S. (2024). Net zero strategy: exploring opportunities, confronting challenges for developing countries. In *Transition towards a sustainable future* (pp. 233–252). Singapore: Springer. https://doi.org/10.1007/978-981-97-5756-5_11
Kang, L. (2024). Can the water resource fee-to-tax reform promote the "three-wheel drive" of corporate green energy-saving innovations? quasi-natural experimental evidence from china. *Energies, 17*(12), 2866. https://doi.org/10.3390/en17122866
Khan, S., Mubarik, M., Kusi-Sarpong, S., Gupta, H., Zaman, S., & Mubarik, M. (2022). Blockchain technologies as enablers of supply chain mapping for sustainable supply chains. *Business Strategy and the Environment, 31*(8), 3742–3756. https://doi.org/10.1002/bse.3029
Khan, S., Singh, R., Alnahas, J., Abbate, S., & Centobelli, P. (2024). Navigating the smart circular economy: A framework for manufacturing firms. *Journal of Cleaner Production, 480*, 144007. https://doi.org/10.1016/j.jclepro.2024.144007
Kılıç, C., & Atilla, G. (2024a). The changing nature of the workforce: A qualitative exploration across selected countries. *International Journal of Eurasia Social Sciences/Uluslararasi Avrasya Sosyal Bilimler Dergisi, 15*(56), 507–536. https://doi.org/10.35826/ijoess.4449
Kılıç, C., & Atilla, G. (2024b). Industry 4.0 and sustainable business models: An intercontinental sample. *Business Strategy and the Environment, 33*(4), 3142–3166. https://doi.org/10.1002/bse.3634
Kılıç, C. (2023). Organizational stress and performance from the perspective of technological developments. *Journal of the Human and Social Science Researches, 12*(4), 2323–2343. https://doi.org/10.15869/itobiad.1308151
Kılıç C. (2024), Sustainable human development and human resource management strategies in building sustainable cities. *Kent Akademisi Dergisi 17(Sürdürülebilir İnsani Kalkınma ve Kent)*, 153–173. https://doi.org/10.35674/kent.1501835

Kim, J., Hadadi, O., Kim, H., & Kim, J. (2018). Development of a bim-based maintenance decision-making framework for the optimization between energy efficiency and investment costs. *Sustainability, 10*(7), 2480. https://doi.org/10.3390/su10072480

Kizielewicz, J. (2021). Eco-trends in energy solutions on cruise ships. *Energies, 14*(13), 3746. https://doi.org/10.3390/en14133746

Kumar, K., Singh, R., Sharma, A., & Lehri, B.P.S. (2024). Climate change adaptation strategies for achieving net-zero economy. In Singh, R., Crowther, D. (Eds.) *Transition towards a sustainable future. Approaches to global sustainability, markets, and governance*. Singapore: Springer. https://doi.org/10.1007/978-981-97-5756-5_8

Lekan, M., & Rogers, H. (2020). Digitally enabled diverse economies: Exploring socially inclusive access to the circular economy in the city. *Urban Geography, 41*(6), 898–901. https://doi.org/10.1080/02723638.2020.1796097

Ley, M., Stucki, T., & Woerter, M. (2016). The impact of energy prices on green innovation. *The Energy Journal, 37*(1), 41–76. https://doi.org/10.5547/01956574.37.1.mley

Liu, Y., Liu, S., Xu, X., & Failler, P. (2020). Does energy price induce china's green energy innovation? *Energies, 13*(15), 4034. https://doi.org/10.3390/en13154034

Liu, M., Fang, S., Dong, H., & Xu, C. (2021). Review of digital twin about concepts, technologies, and industrial applications. *Journal of Manufacturing Systems, 58*, 346–361. https://doi.org/10.1016/j.jmsy.2020.06.017

Louman, B., Girolami, E. D., Shames, S., Primo, L. G., Gitz, V., Scherr, S. J., & Brady, M. (2022). Access to landscape finance for small-scale producers and local communities: A literature review. *Land, 11*(9), 1444. https://doi.org/10.3390/land11091444

Lu, Y., Wu, Z., Chang, R., & Li, Y. (2017). Building Information Modeling (BIM) for green buildings: A critical review and future directions. *Automation in Construction, 83*, 134–148. https://doi.org/10.1016/j.autcon.2017.08.024

Lukić, J., Mirković, V., & Brkljač, M. (2023). Circular economy and banking industry in the digital age: case study of erste bank a.d. novi sad. *Ecologica, 30*(109), 157–164. https://doi.org/10.18485/ecologica.2023.30.109.22

Machado, C., Winroth, M., & Silva, E. (2019). Sustainable manufacturing in industry 4.0: an emerging research agenda. *International Journal of Production Research, 58*(5), 1462–1484. https://doi.org/10.1080/00207543.2019.1652777

Maiurova, A., Kurniawan, T. A., Kustikova, M., Bykovskaia, E., Othman, M. H. D., Singh, D., & Goh, H. H. (2022). Promoting digital transformation in waste collection service and waste recycling in Moscow (Russia): Applying a circular economy paradigm to mitigate climate change impacts on the environment. *Journal of Cleaner Production, 354*, 131604. https://doi.org/10.1016/j.jclepro.2022.131604

Mazzetto, S. (2024). A review of urban digital twins integration, challenges, and future directions in smart city development. *Sustainability, 16*(19), 8337. https://doi.org/10.3390/su16198337

Medved, P., Kim, J., & Uršič, M. (2020). The urban social sustainability paradigm in northeast asia and europe. *International Review for Spatial Planning and Sustainable Development, 8*(4), 16–37. https://doi.org/10.14246/irspsd.8.4_16

Meng, X. (2023). Integration of digital twin and circular economy in the construction industry. *Sustainability, 15*(17), 13186. https://doi.org/10.3390/su151713186

Mitkov, R. (2024). The role of computational fluid dynamics within city digital twins: opportunities and challenges. *Isprs Annals of the Photogrammetry Remote Sensing and Spatial Information Sciences, X-4/W4–2024*, 131–136. https://doi.org/10.5194/isprs-annals-x-4-w4-2024-131-2024

Mukhacheva, A., Ugryumova, M., Морозова, И., & Mukhachyev, M. (2022). Digital twins of the urban ecosystem to ensure the quality of life of the population. In *Proceedings of the international scientific and practical conference strategy of development of regional ecosystems "education-science-industry" (ISPCR 2021)*. https://doi.org/10.2991/aebmr.k.220208.047

Nwokolo, S. C., Singh, R., Khan, S., Kumar, A., & Luthra, S. (2023a). Africa's awakening to climate action. *In: Africa's path to net-zero: exploring scenarios for a sustainable energy transition* (pp. 299–310). Cham: Springer Nature Switzerland. https://doi.org/10.1007/978-3-031-44514-9_9

Nwokolo, S.C., Singh, R., Khan, S., Kumar, A., & Luthra, S. (2023b). Global ınvestment and development in Africa. In *Africa's path to net-zero. CSR, sustainability, ethics and governance.* Cham: Springer. https://doi.org/10.1007/978-3-031-44514-9_2

Nwokolo, S.C., Singh, R., Khan, S., Kumar, A., & Luthra, S. (2023c). Impacts of climate change in Africa. In *Africa's path to net-zero. CSR, sustainability, ethics and governance.* Cham: Springer. https://doi.org/10.1007/978-3-031-44514-9_7

Nwokolo, S.C., Singh, R., Khan, S., Kumar, A., & Luthra, S. (2023d). Remedies to the challenges of renewable energy deployment in Africa. In *Africa's path to net-zero. CSR, sustainability, ethics and governance.* Cham: Springer. https://doi.org/10.1007/978-3-031-44514-9_3

Nwokolo, S.C., Singh, R., Khan, S., & Kumar, A. (2024). Threats to the rapidity of sustainability transitions posed by technological changes. In *Technological pathways for Africa' s net-zero economy*. Elsevier. https://doi.org/10.1016/B978-0-443-31486-5.00002-1

Park, A., & Li, H. (2021). The effect of blockchain technology on supply chain sustainability performances. *Sustainability, 13*(4), 1726. https://doi.org/10.3390/su13041726

Pasupuleti, M. (2024). Sustainable engineering and renewable power: shaping the future of green technology. In *Conference proceedings* (pp. 298–315). https://doi.org/10.62311/nesx/97889

Reis, J. S. D. M., Espuny, M., Nunhes, T. V., Sampaio, N. A. D. S., Isaksson, R., Campos, F. C. D., & Oliveira, O. J. D. (2021). Striding towards sustainability: A framework to overcome challenges and explore opportunities through industry 4.0. *Sustainability, 13*(9), 5232. https://doi.org/10.3390/su13095232

Ronaghi, M. H., & Mosakhani, M. (2022). The effects of blockchain technology adoption on business ethics and social sustainability: Evidence from the Middle East. *Environment, Development and Sustainability, 24*(5), 6834–6859. https://doi.org/10.1007/s10668-021-01729-x

Russia, E. B. (2020). Standardization of artificial intelligence for the development and use of intelligent systems. *Advances in Wireless Communications and Networks, 6*(1), 1. https://doi.org/10.11648/j.awcn.20200601.11

Saeed, M. A., & Kersten, W. (2019). Drivers of sustainable supply chain management: Identification and classification. *Sustainability, 11*(4), 1137. https://doi.org/10.3390/su11041137

Savale, T. K., Byram Anand, P., Perumalla Varalaxmi, A. B., & Quaye, J. A. (2023). Green Marketing Strategies: Assessing Consumer Perception and Adoption of Eco-friendly Products. *Remittances Review, 8*(4).

Searcy, C., Castka, P., Mohr, J., & Fischer, S. (2022). Transformational transparency in supply chains: Leveraging technology to drive radical change. *California Management Review, 65*(1), 19–43. https://doi.org/10.1177/00081256221126204

Singh, R., Gupta, A., & Kandpal, V. (2024). ESG excellence in the digital world: circular economy E-business solutions. *Fintech for ESG and the Circular Economy, 2025*, 193–204. https://doi.org/10.1002/9781394234035.ch10

Tan, F., & Lu, Z. (2018). The impact of urban compactness on urban sustainable development in china: The case of nanjing. *Sustainable Development, 27*(3), 270–280. https://doi.org/10.1002/sd.1874

Tijan, E., Aksentijević, S., Ivanić, K., & Jardas, M. (2019). *Blockchain Technology Implementation in Logistics. Sustainability, 11*(4), 1185. https://doi.org/10.3390/su11041185

Tripathi, N., Hietala, H., Xu, Y., & Liyanage, R. (2024). Stakeholders collaborations, challenges and emerging concepts in digital twin ecosystems. *Information and Software Technology, 2024*, 107424. https://doi.org/10.1016/j.infsof.2024.107424

Ţurcan, I. (2023). Digitalization and its role in the development of circular economy business models., 103–109. In *Proceedings of the 5th economic ınternational conference (2nd-3rd november 2023).* https://doi.org/10.52326/csd2023.16

Vetrova, M., & Ivanova, D. (2021). Closed product life cycle as a basis of the circular economy. *Gatr Journal of Business and Economics Review, 5*(4), 36–50. https://doi.org/10.35609/jber.2021.5.4(4)

Wang, H., Sun, B., & Wang, P. (2019). Dominant technology identification model based on patent information toward sustainable energy development. *IEEE Access, 7*, 141374–141385. https://doi.org/10.1109/access.2019.2939381

Wang, D., Li, J., & Liu, Y. (2023). Evaluating barriers and strategies to green energy innovations for sustainable development: Developing resilient energy systems. *Frontiers in Energy Research, 11*, 1201692. https://doi.org/10.3389/fenrg.2023.1201692

White, K., Habib, R., & Hardisty, D. J. (2019). How to SHIFT consumer behaviors to be more sustainable: A literature review and guiding framework. *Journal of Marketing, 83*(3), 22–49. https://doi.org/10.1177/0022242919825649

Xu, J. (2024). Spatial spillover and threshold effects of digital rural development on agricultural circular economy growth. *Frontiers in Sustainable Food Systems, 8*, 1337637. https://doi.org/10.3389/fsufs.2024.1337637

Yigitcanlar, T., Mehmood, R., & Corchado, J. M. (2021). Green artificial intelligence: Towards an efficient, sustainable and equitable technology for smart cities and futures. *Sustainability, 13*(16), 8952. https://doi.org/10.3390/su13168952

Yuan, Q. (2024). Does technological innovation in national sustainable development agenda innovation demonstration zones promote green development?—the case from chengde city, china. *PLoS ONE, 19*(5), e0300315. https://doi.org/10.1371/journal.pone.0300315

Zhang, Z. (2024). A review of digital twin technologies for enhanced sustainability in the construction industry. *Buildings, 14*(4), 1113. https://doi.org/10.3390/buildings14041113

Chapter 14
Advancing Climate Neutrality Through Sustainable Human Systems: Transforming Transport, Energy, and Communication in South Asia

Narayanage Jayantha Dewasiri and Amit Joshi

Abstract This chapter examines the opportunities for attaining climate neutrality in South Asia concerning transforming transport, energy, and communication systems. The rationale is to raise awareness of such sectors as climate change solutions providers and address regional issues. In this context, adopting the Systematic Literature Review (SLR) approach, the chapter summarises and analyses the literature evidencing climate neutrality in South Asia regarding current emissions, emission sectors, and sustainable solutions. Mordant research outcomes indicated that incorporating renewable energy, transport, and communication green technologies could cut emissions, improve efficiency, and manage climate-smart outcomes. However, the efforts are constrained by economic constraints, political influences and infrastructure, which is comparatively archaic. The chapter underlines the four challenges that hinder these goals: multi-sectoral governance, technological advancement and regional collaboration. The outlook for policymakers, the corporate world, and societies in the South Asian region is summarised as follows: policy mixes should be integrated, clean technologies must be invested, and sustainability consciousness has to be promoted.

Keywords Climate neutrality · Sustainable transport · Renewable energy · Green communication technologies · South Asia

N. J. Dewasiri (✉)
Department of Accountancy and Finance, Faculty of Management Studies, Sabaragamuwa University of Sri Lanka, Belihuloya, Sri Lanka
e-mail: jayantha@mgt.sab.ac.lk

Qasim Ibrahim School of Business, Villa College, Malé, Maldives

A. Joshi
Department of Management Studies, Kumaun University, Nainital, Uttarakhand, India

R. Singh and W. L. Filho (eds.), *Climate Neutrality Through Smart Eco-Innovation and Environmental Sustainability*, Climate Change Management,
https://doi.org/10.1007/978-3-031-83250-5_14

Introduction

Carbon neutrality has become a global goal in combating climate change (Kumar et al., 2024). It aims to equilibrate emitted Greenhouse gas (GHG) emissions with their removal. This goal is critical to reducing the effects that global warming has on the climate system, such as global warming, sea level rise, extreme weather conditions, and species loss (Rathnasiri et al., 2024). Climate neutrality means a state where human system-related emissions have been substantially transformed. Transport, energy, and communication systems are some of the major areas of substantive systemic interventions possible.

South Asia comprises around one-quarter of the global population and has particular targets and opportunities for achieving climate neutrality (Dewasiri et al., 2024). Demographic pressure, increased urbanisation rates and economic growth have contributed to the enhanced energy demand and boosted voices for the carriage of emissions, strengthening the GHG emissions. However, there remains a high sensitivity towards climate change calamities, for instance, since the area's livelihoods rely on sectors sensitive to climate, such as agriculture and fisheries. However, South Asia also contains vast untapped reserves of renewable energy resources and is a technological hub (Mishra et al., 2024). It has promising opportunities for regional integration, making the region central to the global shift to sustainable development.

South Asia's Climate Profile and Human Systems

South Asia has comparatively very high levels of greenhouse gas emissions owing to its swift industrialisation, expanding population, and steadily rising energy requirements. The region jointly emits more than 10% of global greenhouse gases; major emitters include energy, transport, and agriculture (Dewasiri et al., 2024). India and Pakistan, for instance, are chasing the economic frontier while exhaling vast quantities of carbon dioxide by relying on fossil fuels to generate power and manufacture products. For example, the comparative missioners, including Nepal, Sri Lanka, and Bhutan, have tiny fractions of total emissions but receive the brunt of the adverse impacts of climate change variability, including weather incidences, floods, and droughts (Abas, 2017).

Transport, energy, and communication infrastructures and networks all relate to their South Asia footprint for human systems. Modern transportation networks, central to economic activity, rely on fossil fuels. Most public transport networks with ancient and inefficient vehicles significantly contribute to air pollution and emissions. Of all the developed energy systems, coal and oil are the most prominent in contributing to emissions, while the renewable energy systems, whose potentials are enormous, are under-harnessed (Mishra et al., 2024). However, communication systems are not direct emitters of greenhouse gases; however, these systems are crucial enablers of the kind of behaviours, resource management, and solutions that could help reduce emissions significantly. Relations between these systems demonstrate the interdependency of different aspects of sustainability, and further investigation of their multifaceted interactions is needed.

The transport, energy, and communication sectors are critical to South Asia's climate future (Raza et al., 2024). Traffic can also be efficiently de-carbonised through electric vehicles, efficient public transport and infrastructure for non-motorised transport. Using renewable energy sources like solar, wind, and hydropower can enable the energy source to be obtained from a source that is not necessarily intensive in carbon, thus cutting one of the significant emissions (Joshi et al., 2024). At the same time, employment of communication technologies will enable efficient use of resources, effective management of intelligent energy systems, and increased awareness of sustainable practices in energy management systems. Therefore, they comprise the fundamental structure for sustainable change and can allow South Asia to link economic progress to sustainability and climate adaptation (Singh et al., 2024).

Emphasis on sustainable transport, energy, and communication systems in South Asia is critical because these industries have significant potential for emissions and are also essential contributors to sustainable development. Energy transformation in transport with electrification, increasing efficiency with renewable energy, and optimising communication through digital technologies can jointly solve climate change and growth (Noussan & Tagliapietra, 2020). This chapter discusses the various transitional phases of these systems, providing knowledge about sustainable human systems for attaining climate neutrality in South Asia.

Methodology

This chapter uses a systematic literature review (SLR) following PRISMA guidelines and the POWER framework to analyse South Asia's transition to climate neutrality, focusing on transport, energy, and communication.

Systematic Literature Review (SLR)

The SLR methodology was employed to find, assess, and combine academic and industrial literature on the phenomenon, as emphasised by Vijerathne et al. (2024), Bandara et al. (2024), Chathuranga et al. (2024), Mishra et al. (2024), Karunarthne et al. (2024), and Rathnasiri et al. (2024). The databases, such as Science Direct and Google Scholar, were considered for the database search. Further, the researchers used keywords such as climate neutrality, sustainable transport, renewable energy and green communication technologies. All articles reviewed for recommendation were passed through the relevance filter that assesses the extent to which each article meets the objectives of the chapter, for instance, the identification of problems, solutions, and policy issues pertinent to the region. The SLR process helps reduce bias and make the results more reliable due to its inclusion of all the necessary studies.

PRISMA Framework

Reporting of the SLR followed the Preferred Reporting Items for Systematic Reviews and Meta-Analyses (PRISMA) framework used by Sarkis-Onofre et al. (2021). This method is popular in systematic reviews, given its clear structure and the evident and predefined approach to identifying literature and data extraction. The PRISMA flow diagram was utilised to outline the selection process flow from screening the studies to inclusion. This led to well-defined strategies of inclusion and exclusion of literature that had to be used such that the work only sourced the most suitable literature available. This approach defined a clear, methodological roadmap to evaluate transport, energy and communication infrastructure's role in climate neutrality and sustainability in South Asia.

POWER Framework

The POWER framework was employed to increase the methodological rigour of the review, as used by Rana et al. (2022). POWER is an acronym for Purpose, Output, Evaluation, Research, and Synthesis, all stages in the SLRs. Thus, this chapter aims to explain how transport, energy, and communication help achieve climate neutrality. The output here relates to the recommendations and vital notions from the review, including challenges and practices by the different sectors and sustainable practices and strategies. Cognitive analysis filtered the works, and only the most relevant and scientifically sound sources were considered. The research component provided the framework for directing the search strategy and data extraction. Synthesis entailed reiteratively synthesising data across the studies and providing a broad distillation of the current climate-neutral transformation in South Asia.

This chapter has thus applied the SLR methodology alongside PRISMA and the POWER framework to ensure that the paths to climate neutrality are well laid down in South Asia, with comprehensive, clear, and solution-oriented findings that are fit for policymakers, businesses, and the community. The final sample of this study consisted of thirty-nine (39) studies published in the recent decade.

Findings and Discussion

Transforming the Transport Sector

The transportation sector in South Asia thus is a significant emitter of GHG, which puts a large chunk of total emissions in this part of the world. The industry proliferates due to the growing population, urbanisation and increasing economic activities. However, this growth has come at a cost: poor road infrastructure, old-age vehicles, and poorly developed public transport systems are well known across the region. Large metro areas like Delhi, Dhaka and Karachi, particularly Delhi, continue to face acute problems like air pollution resulting from poor management of existing and increasing traffic congestion dominated by fossil fuel-based motorised transport (Uddin, 2022; Hassan et al., 2023). Rural areas face a different challenge: A lack of network capital hinders socioeconomic opportunities by restricting the flow to markets, health facilities and schooling, hence socially excluding people.

The change pathways must be transformative for South Asia to attain sustainable mobility. Metro rail, buses, and rapid transit systems can reduce emissions by absorbing less carbon per person when upgraded or expanded (Venter et al., 2018). The use of electrified transport, such as electric cars and electricity-operated buses, is an achievable solution towards de-carbonizing the mobility of people in cities when charged with renewable energy. Implementing non-motorized transport handling for walking and cycling may also lead to low-carbon transportation while improving the overall quality of life within the existing urban structure. These pathways lower emissions and pertain to problems such as air quality, health, and road safety.

Dynamics of this nature require support from policy frameworks and technology integration. Governments in the region have to formulate overt policies to make electric vehicles (EVs) the flagship technology, standardise charging solutions, and control vehicular emissions. Sustainable mobility must be implemented into urban planning as a part of transit-oriented development and through bright city designs that allow transport, housing, and workplaces to be in one place. Innovative mechanisation, such as ride-sharing firms and intelligent transportation systems, have the potential to increase productivity and decrease congestion (Li et al., 2022). Regional cooperation can also bring great value, especially in disseminating good practices and coordinating policies for international transport operations.

South Asia also provides several examples that establish sustainable transport is feasible. For instance, the Delhi Metro in India has provided an efficient urban transport system that dramatically reduces traffic jams and air pollution (Singh & Gupta, 2020). Likewise, given the development of environmentally friendly transport, Sri Lanka has launched programmes encouraging using electric three-wheelers and buses. Bhutan, a country with pronounced policies towards protecting the environment, has policies supporting the use of EVs and has minimised the use of fossil fuels. The following case studies depict the role of government and policies advocating for change, the Role of the private sector in creating sustainable transport systems and the Role of citizens and communities.

Transport challenges and opportunities in South Asia must be met to build an equitable and sustainable transport system that aligns with the region's climate neutrality agenda and enhances people's welfare (Nwokolo et al., 2023b; c).

Transitioning to Clean Energy Systems

Energy usage in South Asian countries has increased in the last decades as the regional economy grows, industrialises, and urbanises. Specifically, the shares of coal and oil in the region's total primary energy consumption remain significant, and these resources dominate the electricity generation and transport industries. Regarding the energy demand, South Asia has India, which contributes to the total energy demand by a considerable margin, given its population and industrial growth (Jain & Goswami, 2021). Although the region has recently accelerated efforts to extend electricity reach to previously unserved consumers, tens of millions remain without access to clean and reliable electricity, and many rely on biomass for cooking, which causes indoor air pollution. This results in high greenhouse gas (GHG) emissions and, hence, high risk to public health and the environment due to reliance on non-renewal energy sources.

South Asia, however, has many renewable energy resources available. The geographic plurality of the area provides ample opportunities, especially solar and wind power in the Sahara, the coastal zones, and the plateau of the Indian subcontinent, as well as potential hydropower in Nepalese and Bhutanese territories (Mishra et al., 2024). Traditional biomass energy production methods also looked promising when harnessed through efficient technology. If these renewable resources are correctly tapped, the carbon footprint of South Asia could be reduced, energy costs could be lowered, and energy security could be enhanced by minimising the reliance on imported fossil fuels.

For all the possibilities, connecting renewable energy to South Asia's power grid systems is difficult (Murshed, 2021). For instance, the power of the sun and wind is inconsistent; hence, there is a need to balance the supply with the help of proper energy-storing instruments. Furthermore, the current grid infrastructure, which has aged in many developed countries, is not competent in managing the variability and decentralisation of renewables. Limited financial resources for utilities for large capital-intensive renewable power projects, uncertainty in policies of different governments, and lack of knowledge in the application of renewable technologies also slow the pace of development (Mishra et al., 2024). Where energy is needed most, in developing rural communities or the hinterland, deploying renewable technologies such as solar mini-grid or wind turbines presents severe logistical problems.

Energy storage and decentralised energy solutions are possible solutions to these problems (Miglani et al., 2020). Lithium-ion and other advanced battery technologies can convert excess electricity produced in a peak generation period to electricity storage, enabling electricity distribution in the grid. Grid-connected systems involving roof-top solar power generation, off-grid systems such as standalone solar, and community mini-grid systems offer reliable electricity supply to remote regions with low transmission losses. These solutions enable local communities, address energy inequality, and relieve pressure on centralised power infrastructure.

Some South Asian countries have already started taking initiatives in renewable energy. For example, India has unveiled several strategic projects, including the National Solar Mission, which targets solar energy—28,000 MW of installed solar capacity by 2030 (Pandey et al., 2022). Hydro-power is a primary source of power in Bhutan that has been fully exploited, where the nation uses its power demands and sells excess energy to other countries. Rooftop solar programs have been launched in Sri Lanka to encourage the decentralisation of power generation (Jayasooriya et al., 2023). Pakistan has been working on wind farms in Sindh province (Asgar et al., 2022). The case studies presented here are a good representation of what can be achieved in the future adoption of renewable energy sources, setting the stage for further uptake across the region.

The move towards clean energy systems is vital to achieving the climate neutrality that South Asia aspires to. Integration challenges, the unused potential of renewable energy sources, and the capability of implementing innovative solutions will help any region create efficient strategies to develop a powerful and effective energy future (Nwokolo et al., 2023a).

Revolutionizing Communication for Sustainability

Information communication technologies have become somewhat effective facilitators in striving for climate neutrality, which is at the centre of promoting sustainable practices and efficient resource management (Slob & van Lieshout, 2013). Stakeholder engagement with new communication systems and technology transforms how governments, businesses, and society address environmental issues. Exploring communication for sustainability in South Asia reveals that where modernity revolutionises culture, technology for development holds excellent promise for the cohesiveness of development with climate change objectives.

Digital transformation and innovative technologies are central to the solutions needed for ongoing, systematic energy and transport management. Advanced intelligent transportation systems rely on communication networks and apply data analysis and real-time control of traffic flows, decreasing fuel consumption and improving the efficiency of public transportation. Based on Internet of Things (IoT) connections, remote and intelligent power networks enable direct participation of loads, incorporating renewable energies and minimisation of loss. Thus, the situation in South Asia, where the penetration of mobiles and broadband is opening up new possibilities to implement such solutions, provides the possibility for more effective delivery of sustainable systems between the urban and rural populations.

Electronic governance (e-governance) and digital education support the region's capability to adopt sustainable practices. E-Governance has provided citizens with an interface to access public services and deals transparently, reducing paperwork and minimising bureaucratic delay. For example, ticketing and energy billing, as implemented through digital means, help to minimise resource use and, at the same time, enhance convenience. Digital education activities helped raise awareness about climate change, especially climate-correct consumerism and farming techniques. Such platforms enable stakeholders to contribute and actively work collectively on climate issues.

Some South Asian programs can be cited to illustrate how green communication technologies and information and communication technology (ICT) are being used to promote the Sustainable Development Goals. In India, the Smart Cities Mission embraces ICT to make cities more resilient: intelligent street lighting, waste management applications and traffic solutions (Khan et al., 2024; Prasad & Alizadeh, 2020). Bhutan has formulated a framework for national e-governance, which minimises governmental bureaucracy and promotes green policies (Younus et al., 2023). Such innovations as real-time tracking technologies for public transport vehicles such as buses have made e-transport in Sri Lanka efficient and less reliant on polluting sources. These studies show how communication technologies create and enhance sustainability in several regional sectors.

A systematic change cannot result from merely introducing improvements to existing technologies that form communication systems but from actual changes in behaviour and systems. By adopting IT solutions in the energy and transport sectors and supporting e-governance and awareness, South Asia can leverage the potential of communication towards climate neutrality. The availability of ICT infrastructure in the region and the ever-increasing youth literacy in informational technology make the ground suitable for expanding these activities, hence establishing an innovative technology loop towards sustainable development.

Policy and Governance for Sustainable Human Systems

The foundation of sustainability transition is mainly anchored by policy and governance, especially in a complex and dynamic region like South Asia (Rathnasiri et al., 2024). Achieving sustainable human management also means cross-sectoral management and substantial cooperation between government entities, the private sector, and civil society. Such involvement confirms that policies are appropriate, just as they are sensitive to the unique characteristics of the socio-economic and natural conditions of the region. They must provide the necessary support to help the process prepare for innovations and investment. At the same time, environmental sustainability processes should fit with the general development frameworks.

Therefore, policy changes are essential for developing sustainable transport, energy and communication systems. The issues pertinent to transport include the policies that promote the use of electric cars, increase access to public transport, and place tight controls on emissions. Programs like India's faster adoption and manufacturing of electric vehicles and technologies are a few policies supporting the shift toward releasing electric mobility (Tarei et al., 2021). In energy, there are policies such as fiscal incentives for renewable energy projects, policy certainty in carbon reduction targets and policies in grid modernisation that facilitate the shift from fossil-based energy to clean energy. The case of Sri Lanka, through development policies in renewable rooftop solar and wind energy, manifests the need for target market policies to develop reasonable energy systems.

In communication, policies relating to the sustainable use of digital technologies, data sharing and technology usage can go a long way towards improving the outcomes of sustainability programs. For example, developing guidelines for green ICT technologies, promoting intelligent city initiatives, and providing research and development grants in communications technologies can catalyse sustainability (Bhushan & Singh, 2024). In addition, government consumer promotion agencies should pay significant attention to digital illiteracy and Fair distribution of devices in society to capture the benefits of these inventions.

This paper has argued that climate neutrality in South Asia requires regional cooperation, primarily because economic and ecological activities and impacts are interconnected. As with the shared resources, water and energy, cross-border systems like the River Basins and Energy Sharing require collective management. Regional cooperation can be beneficial, as the SAARC Energy Framework proved (Alam et al., 2015), BBIN (Bangladesh, Bhutan, India, Nepal) Motor Vehicles Agreement and others that regional partners can manage their problems with collaboration (Kumar, 2017). National activities need to be supported by international actions; existing and emerging synergy points like technology transfer capacity building or joint investments into the necessary infrastructure for renewable energy production can be enhanced (Nwokolo et al., 2024).

In addition, regional cooperation brings together the best talents and companies to work under one entity to avoid replication and promote innovation. For instance, flood management and disaster risk mitigation measures can be collectively adopted to build border security. International policymakers must, therefore, coordinate to synchronise the legal frameworks in creating a united litmus for a just transition on climate change and to increase the levels of trust in collective climate ambition towards attaining sustainable development in the region, in the case of South Asia.

Top management and governance mechanisms are crucial in redesigning human systems to become firm and climate-neutral. This means that through improving south-south and regional cooperation, enhancing policy practices, and promoting multi-sectoral planning approaches, South Asian leaders can steer their nations' regional development pathways to meet the ambitious climate targets without undue detriment to any socio-demographic group.

Barriers and Drivers of Change

The challenges of climate change are different at the South Asian level, meaning that many barriers to the transformative journey towards climate neutrality must be overcome (Dewasiri et al., 2024). Budgetary constraints are one of the most formidable constraints because most of the countries in the region are usually faced with issues such as inadequate funds, high levels of poverty, and many demands for resources as they address their developmental agendas. Sustainable infrastructure, renewable energy and clean technologies are sometimes capital intensive, thus forcing the national budgets. Several political barriers exist, such as policy restrictions, absence of policies or unequal implementation and interpretation, and organisational politics. Also, the infrastructure, including power, transport, and digital networks in the region, especially for rural facilities, remains a barrier to supporting the scalability of sustainable solutions.

Nonetheless, there are several push factors leading to sustainability. Technology is leading this shift, and solutions for humans to de-carbonise include renewable energy technologies, including wind and solar, electric vehicles, and intelligent communication systems. For instance, the cost of most renewable energy sources, including solar and wind power, has been reduced to compete with fossil energy sources. As with battery storage and grid management technology, traditional obstacles which stood in the way of renewable energy integration have been surmounted. With artificial intelligence and blockchain technologies, the IoT is changing the face of various energy and transport industries.

Funding mechanisms also create a significant imperative to change. Weather-related funds, including grants, credits, and equities, help fill the financing gap for sustainable projects worldwide. South Asia receives funding from institutions like the Green Climate Fund (GCF), the World Bank, and other institutions in renewable energy, infrastructure and climate change adaptation (Agarwal et al., 2021). Further, the emerging trend of blended capital, including public–private partnerships (PPPs) and other new funding mechanisms like green bonds and carbon trading, is a practical resource mobiliser for sustainability. These mechanisms open up the possibility for Government and business to work together to source large-scale transformations.

The social and cultural aspects of sustainability are as important as the technical and infrastructural ideas for a climate-neutral future of the region. South Asia consists of people with pluralism of culture, religion, and social structures with opportunities and barriers to sustainability. On the one hand, cultural values such as sustainable agriculture, organic farming, water-utilising technologies, and economic sustainability are desirable for reaching sustainable development goals. At the same time, convictions, including using biomass for cooking or lacking acceptance of modern energy technologies, hamper the advancement of efficient technologies. It is crucial to deepen public awareness strategies, involve the community, and promote education projects to meet these obstacles and provide a sustainable culture at various stages of society.

The intersections of barriers and drivers suggest a cautious possibility of the region attaining climate neutrality in the future. Despite current constraints and obstacles in economics, politics, infrastructure, and technological advancements, funds and attitude change support positive reform. The work carried out for this study shows that by addressing these barriers, the region has a clear pathway to achieving sustainable human systems and a climate-neutral future.

Conclusions

South Asia can and must reverse emissions no later than the second half of the century, addressing various challenges, but this task is also possible and critical. This chapter has centred its discussion on how the transport, energy and communication industries are essential in the drive for sustainability and the fight against climate change in the region. By analysing the climate profile for the present day and looking into the challenges faced at the sectoral level and the opportunities for transformation that have been discussed throughout, integrated climate solutions have been observed to offer a critical development path.

The presented climate structure of South Asia reveals that it has been rapidly developing densely populated areas that are falling into frontier environmental issues. These factors are enablers and challenges for achieving climate neutrality. The harmful reliance on fossil fuels and infrastructure and socio-economic inequalities complicate the journey, but there is tremendous possibility of change for the region. These sectors are critical to this change: transport, energy, and communication. Low carbon transportation, increased use of renewable energy, and green ICT can help decrease GHG emissions, enhance efficiency or enable climate change resilient development.

This chapter contains critical insights into sectoral integration, where the need for a more inclusive process is heavily underscored. Thus, the decarbonisation of the economy and transition to a climate-neutral future for South Asia is impossible to solve in isolation. Transportation must grow ecologically and withstand clean energy technology advancement, while transport and info-comms technology must embrace sustainable innovations. Similarly, the policy and governance architectures to enable these linkages and to ensure that sustainability becomes integrated within the fabric of the existing and emerging governance systems at the national and subnational levels require development.

Thus, economic, political, and infrastructural issues are still present. Given these challenges, the drivers of change—technological advancement, more international cooperation, and evolving cultural attitudes—provide the blueprint for surmounting these obstacles. Anticipating innovation, offering sufficient financial frameworks, and ensuring the participation of people within different niches of society are the primary ways for South Asia to realise its potential in protecting the environment and combating climate change.

Therefore, it is time to do something. The future can have sustainable human systems only if government, business, and communities pursue this goal together. Governments must promote policies relating to clean technologies, renewable energy, green urbanism and regional integration. This means consumers should engage and put their monies into green technologies and environmentally friendly activities in their companies. Communities also need to be educated and informed on how they can support environmental conservation initiatives. Climate neutrality is the end and the means to create a more sustainable, inclusive and prosperous South Asia. If South

Asian leaders act now, they can provide leadership to the rest of the world to ensure future generations can live sustainably on the planet.

Future Pathways and Recommendations

Climate-neutral South Asia is a global vision and demands a strategic and integrated action plan to plan for immediate and long-term imperatives for the region. International policy and action must develop progressive measures, processes, and enabling tools for a fair and equitable transition to a low-emission future and an economic and social growth vision. Given the new strategic priorities accentuating the need to align the economic development of the entire South Asia region with climate targets, the integrated approach to sustainability in sectors most vulnerable to climate change, as well as constructive international collaboration, the area is capable of creating a sustainable pathway to climate neutrality.

Strategic Priorities for South Asia

Hence, these South Asian governments must urgently and earnestly embark upon decarbonisation through ambitious policy templates and systemic and integrated strategies. There is a need to push for increased use of renewable energy sources, Source efficiency and sustainable transport. The primary goals for policymakers should be to set specific and quantifiable emissions reduction goals while guaranteeing their integration into a nation's development plans. In addition, aspects of managing cross-cutting relationships, like the relationship between energy demand and transport electrification, must be considered to develop an integrated and effective sustainability policy.

Integrating Sustainability Across Sectors

Moving to low-emission solutions can achieve sustainable solutions in the transport sector. These include increasing the availability and access to affordable and efficient public transport, adopting electric vehicles, and improving facilities promoting active modes of transport such as walking and cycling. Governments also have to consider other problems in urban planning, ensuring that cities never grow to the point that access to most resources is miles away.

Transforming a power tariff into renewable energy has been inevitable in the energy sector. South Asian countries can find rich resources in solar, wind and hydro energy, which can help meet increasing energy needs while remaining sustainable. Investments are needed to improve the energy grids for a new age, decentralise energy distribution, and develop battery technologies to address the change. The capacity for microgrids and solar rooftops at the community level should also be promoted to improve the availability of green power in rural communities.

Integrating green ICT practices in its operations is paramount, especially in the communication sector. New technology solutions must be implemented to improve the effectiveness of energy and transportation infrastructure, and e-governance and digital classrooms must be integrated to bring about large-scale sustainability. Governments should also respond to the imperative of innovation in technologies used in communication to foster sustainable ICT solutions like energy-efficient data centres or low-power Internet of Things technology solutions.

Role of International Collaboration and Innovation

South Asia, particularly, can only achieve climate neutrality much faster with international cooperation. Common problems such as air pollution, resource sharing, and energy infrastructure concerns across borders call for collective action. Regional actions complement national plans, such as jointly setting common sustainable regulation standards and collaborative renewable energy projects. In addition, South Asian countries need to join the relevant climate platforms to avail themselves of technology, Development capability, and bilateral/international financing.

Therefore, Technology will be a significant factor in achieving sustainable development in the South Asian region. Local and global leaders must work to encourage the development of advanced technologies in combating climate change consequences by impressing more innovations in colleges, business entities, and government. Grant funding, tax credits, and collaboration with the private sector can help build up and apply solutions appropriate to the area much faster as innovation is encouraged.

Call for Action

Entailing concepts for South Asia's climate-neutral change need a systematic framework and synthesis of policymaking, sectoral interfaces, and globally coordinated policy integrations. By integrating sustainability across transport, energy, and communication systems and encouraging and promoting innovation and partnership, the region can attain sustainable development and meet the antecedent goal of accomplishing climate change. Such endeavours hold potential for environmental protection and strengthen South Asian nations' international competitiveness in a changing climate future.

References

Abas, N., Kalair, A., Khan, N., & Kalair, A. R. (2017). Review of GHG emissions in Pakistan compared to SAARC countries. *Renewable and Sustainable Energy Reviews, 80*, 990–1016.

Agarwal, R., Balasundharam, V., Blagrave, P., Gudmundsson, R., & Mousa, R. (2021). Climate change in South Asia: Further need for mitigation and adaptation. Washington, D.C. International Monetary Fund.

Alam, A., Malik, I. A., Abdullah, A. B., Hassan, A., Awan, U., Ali, G., Zaman, K., & Naseem, I. (2015). Does financial development contribute to SAARC'S energy demand? From energy crisis to energy reforms. *Renewable and Sustainable Energy Reviews, 41*, 818–829.

Asghar, R., Ullah, Z., Azeem, B., Aslam, S., Hashmi, M. H., Rasool, E., Shaker, B., Anwar, M. J., & Mustafa, K. (2022). Wind energy potential in Pakistan: A feasibility study in Sindh Province. *Energies, 15*(22), 8333.

Bandara, C., Rathnasiri, M. S. H., Dewasiri, N. J., Chathuranga, B., & Singh, N. (2024). Greening the economy for a sustainable future: Opportunities and perceptions. *Greening Our Economy for a Sustainable Future, 197–213*,. https://doi.org/10.1016/B978-0-443-23603-7.00016-9

Bhushan, M., & Singh, R. (2024). Talent Management Facilitates Net-Zero Transition Through Employee Green Behavior. In: Singh, R., Khan, S., Kumar, A., Luthra, S., & Chokshi, H. (Eds) *Net zero economy, corporate social responsibility and sustainable value creation. CSR, sustainability, ethics and governance.* Cham: Springer. https://doi.org/10.1007/978-3-031-557 79-8_7

Chathuranga, B., Rathnasiri, M. S. H., Dewasiri, N. J., Jayarathne, P., Bandara, C., & Bakari, M. A. (2024). Direct and indirect effect on industries and services (A). *Greening Our Economy for a Sustainable Future,* (pp. 215–226). https://doi.org/10.1016/B978-0-443-23603-7.00017-0

Dewasiri, N. J., Gnanapala, W. K. A. C., Rathnasiri, M. S. H., Yadav, M., Tewari, L. M., & Mishra, U. (2024). The way forward for net zero: aligning nationally determined contributions (NDCs) and holistic sustainable development in South Asia. In Singh, R., & Crowther, D. (Eds.) *Transition towards a sustainable future. Approaches to global sustainability, markets, and governance.* Singapore: Springer. https://doi.org/10.1007/978-981-97-5756-5_5

Hassan, M. A., Mehmood, T., Liu, J., Luo, X., Li, X., Tanveer, M., Faheem, M., Shakoor, A., Dar, A. A., & Abid, M. (2023). A review of particulate pollution over Himalaya region: Characteristics and salient factors contributing ambient PM pollution. *Atmospheric Environment, 294*, 119472.

Jain, P., & Goswami, B. (2021). Energy efficiency in South Asia: Trends and determinants. *Energy, 221*, 119762.

Jayasooriya, V., Fernando, S., Silva, C., & Muthukumaran, S. (2023). Comparative analysis on the effectiveness of green roofs and photovoltaic panels as sustainable rooftop technologies. *Environmental Science and Pollution Research, 30*(44), 98977–98992.

Joshi, A., Singh, R., Kumar, V., & Rani, S. (2024). Net zero strategy: Exploring opportunities, confronting challenges for developing countries. In: Singh, R., Crowther, D. (Eds.) *Transition towards a sustainable future. Approaches to global sustainability, markets, and governance.* Singapore: Springer. https://doi.org/10.1007/978-981-97-5756-5_11

Karunarathna, K. S. S. N., Dewasiri, N. J., Singh, R., & Rathnasiri, M. S. H. (2024). What does artificial intelligence–powered ChatGPT bring to academia? a review. In Singh, R., Khan, S., Kumar, A., & Kumar, V. (Ed.) *Artificial intelligence enabled management: An emerging economy perspective* (pp. 3–13). De Gruyter: Berlin. https://doi.org/10.1515/9783111172408

Khan, S., Singh, R., Alnahas, J., Abbate, S., & Centobelli, P. (2024). Navigating the smart circular economy: A framework for manufacturing firms. *Journal of Cleaner Production, 480*, 144007. https://doi.org/10.1016/j.jclepro.2024.144007

Kumar, V. (2017). Sub-regionalism in South Asia: A case study of the Bangladesh–Bhutan–Nepal–India motor vehicles agreement. *Strategic Analysis, 41*(1), 1–13.

Kumar, K., Singh, R., Sharma, A., & Lehri, B. P. S. (2024). Climate change adaptation strategies for achieving net-zero economy. In: Singh, R., Crowther, D. (eds) *Transition Towards a Sustainable*

Future. Approaches to Global Sustainability, Markets, and Governance. Singapore: Springer. https://doi.org/10.1007/978-981-97-5756-5_8

Li, Z., Liang, C., Hong, Y., & Zhang, Z. (2022). How do on-demand ridesharing services affect traffic congestion? The moderating role of urban compactness. *Production and Operations Management, 31*(1), 239–258.

Miglani, A., Kumar, N., Chamola, V., & Zeadally, S. (2020). Blockchain for the Internet of Energy management: Review, solutions, and challenges. *Computer Communications, 151*, 395–418.

Mishra, U., Rathnasiri, M. S. H., & Dewasiri, N. J. (2024). Net-zero economy and energy sustainability: A nepalese perspective. In Singh, R., Khan, S., Kumar, A., Luthra, S., & Chokshi, H. (Eds.) *Net zero economy, corporate social responsibility and sustainable value creation. CSR, sustainability, ethics and governance.* Cham: Springer. https://doi.org/10.1007/978-3-031-55779-8_9

Murshed, M. (2021). Can regional trade integration facilitate renewable energy transition to ensure sustainability in South Asia? *Energy Reports, 7*, 808–821.

Noussan, M., & Tagliapietra, S. (2020). The effect of digitalisation in the energy consumption of passenger transport: An analysis of future scenarios for Europe. *Journal of Cleaner Production, 258*, 120926.

Nwokolo, S. C., Singh, R., Khan, S., Kumar, A., & Luthra, S. (2023a). Remedies to the challenges of renewable energy deployment in Africa. In *Africa's path to net-zero. CSR, sustainability, ethics and governance.* Cham: Springer. https://doi.org/10.1007/978-3-031-44514-9_3\

Nwokolo, S. C., Singh, R., Khan, S., Kumar, A., & Luthra, S. (2023b). Africa's awakening to climate action. In *Africa's path to net-zero. CSR, sustainability, ethics and governance.* Cham: Springer. https://doi.org/10.1007/978-3-031-44514-9_9

Nwokolo, S.C., Singh, R., Khan, S., Kumar, A., Luthra, S. (2023c). Impacts of Climate Change in Africa. In: Africa's Path to Net-Zero. CSR, Sustainability, Ethics & Governance. Cham: Springer. https://doi.org/10.1007/978-3-031-44514-9_7

Nwokolo, S. C., Singh, R., Khan, S., & Kumar, A. (2024). Threats to the rapidity of sustainability transitions posed by technological changes. In Nwokolo, S.C., Singh, R., Khan, S. & Kumar, A. (Eds.) *Technological pathways for Africa's net-zero economy* (pp. 39–75). https://doi.org/10.1016/B978-0-443-31486-5.00002-1

Pandey, A., Pandey, P., & Tumuluru, J. S. (2022). Solar energy production in India and commonly used technologies—An overview. *Energies, 15*(2), 500.

Prasad, D., & Alizadeh, T. (2020). What makes Indian cities smart? A policy analysis of smart cities mission. *Telematics and Informatics, 55*, 101466.

Rana, S., Sakshi, & Singh, J. (2022). Presenting the POWER framework of conducting literature review. In Rana, S., Sakshi & Singh, J. (Ed.) *Exploring the latest trends in management literature (review of management literature, Vol. 1)* (pp. 1–13). Leeds: Emerald Publishing Limited. https://doi.org/10.1108/S2754-586520220000001001

Rathnasiri, M. S. H., Dewasiri, N. J., & Kumar, A. A. (2024). Policy framework and implementation strategies for Sri Lanka's transition to a net-zero economy. In: Singh, R., Khan, S., Kumar, A., Luthra, S., & Chokshi, H. (Eds.) *Net zero economy, corporate social responsibility and sustainable value creation. CSR, sustainability, ethics and governance.* Cham: Springer. https://doi.org/10.1007/978-3-031-55779-8_3

Raza, A., Liaqat, M., Adnan, M., Iqbal, M. S., Jingzhao, L., & Ahmad, I. (2024). SAARC super smart grid: Navigating the future-unleashing the power of an energy-efficient integration of renewable energy resources in the SAARC region. *Computers and Electrical Engineering, 118*, 109405.

Sarkis-Onofre, R., Catalá-López, F., Aromataris, E., & Lockwood, C. (2021). How to properly use the PRISMA Statement. *Systematic Reviews, 10*, 1–3.

Singh, R., Gupta, A., & Kandpal, V. (2024). ESG excellence in the digital world. In Ernesto, D.R. Gonzalez, S., Kandpal, V., Ozili, P. K., & Chatterjee, P. (Eds.) *Fintech for ESG and the circular economy* (pp. 93–204). Beverly, MA: Scrivener Publishing LLC. https://doi.org/10.1002/9781394234035.ch10

Singh, M., & Gupta, S. (2020). Urban rail system for freight distribution in a mega city: A case study of Delhi, India. *Transportation Research Procedia, 48*, 452–466.

Slob, A., & van Lieshout, M. (2013). Information and Communication Technologies contribute to the transition towards a climate-neutral society. In *Global warming and social innovation* (pp. 143–159). Routledge.

Tarei, P. K., Chand, P., & Gupta, H. (2021). Barriers to the adoption of electric vehicles: Evidence from India. *Journal of Cleaner Production, 291*, 125847.

Uddin, W. (2022). Mobile and area sources of greenhouse gases and abatement strategies. In Lackner, M., Sajjadi, B., & Chen, W. Y. (Eds.) *Handbook of climate change mitigation and adaptation.* Cham: Springer. https://doi.org/10.1007/978-3-030-72579-2_23

Venter, C., Jennings, G., Hidalgo, D., & Valderrama Pineda, A. F. (2018). The equity impacts of bus rapid transit: A review of the evidence and implications for sustainable transport. *International Journal of Sustainable Transportation, 12*(2), 140–152.

Vijerathne, D. T., Wahala, W., Rathnasiri, M. S. H., De Silva, V., & Dewasiri, N. J. (2024). Literature review on sustainability. *Greening Our Economy for a Sustainable Future,* (pp. 33–45). https://doi.org/10.1016/B978-0-443-23603-7.00004-2

Younus, M., Pribadi, U., Nurmandi, A., & Rahmawati, I. Z. (2023). Comparative analysis of E-government development index: A case study of South Asian countries. *Transforming Government: People, Process and Policy, 17*(4), 552–574.

Part IV
Issues and Challenges in Climate Neutrality and Sustainability

Chapter 15
The Human Side of Sustainability: Behavioural Economics in Climate Action and Neutrality

Apurvaa Trivedi, Som Aditya Juyal, and Vinay Kandpal

Abstract Climate neutrality demands, therefore, not only a technological but also a change at the level of human behaviour. This chapter discusses the adaptation of behavioural economics in bringing about the desired change to promote sustainable practices and alleviate barriers to climate-neutral behaviour. Nudge theory, loss aversion, and social norms explain how cognitive bias, socioeconomic factors, and institutional constraints stand as barriers for sustainable behaviour. These challenges are counteracted by using the proposed INSPIRE framework, which combines strategies such as incentivization, nudges, feedback mechanisms, gamification, community-based interventions, reminders, and encouragement. Measurable impacts on carbon emissions have been seen in real-world examples such as programs for energy conservation, waste reduction, and adoption of renewable energy. Finally, the chapter addresses the policy implications of Behavioural Insights as a tool for integrating it with climate strategies, leveraging technologies like AI and IoT to scale interventions and promote cultural adaptability and inclusion. Methods of measuring intervention effects- challenges such as attribution and cultural differences are discussed along with recommendations for improved methodologies and multi-metric evaluation. Behavioural economics bridges the knowledge-action gap, offering effective tools for pro-environmental action. The book concludes with a call to policymakers, businesses, and communities to implement these insights to drive the transition to climate neutrality.

Keywords Climate neutrality · Sustainable behaviour · Sustainability frameworks · Behavioural economics · Human behaviour and sustainability · Environmental policy

---

A. Trivedi (✉) · S. A. Juyal
Himalayan School of Management Studies, Swami Rama Himalayan University, Jolly Grant, Dehradun, India
e-mail: apurvaa.trivedi@gmail.com

V. Kandpal
Department of Management Studies, Graphic Era Deemed to Be University Dehradun, Uttarakhand, India

R. Singh and W. L. Filho (eds.), *Climate Neutrality Through Smart Eco-Innovation and Environmental Sustainability*, Climate Change Management,
https://doi.org/10.1007/978-3-031-83250-5_15

Introduction

Contextualizing Climate Neutrality

Climate neutrality in pursuit of net-zero greenhouse gas emissions is essential in countering the impacts of climate change. According to the IPCC (2021), for the world to have any hope of holding the line at 1.5 °C above pre-industrial levels, it needs to get to net-zero by 2050. Otherwise, irreversibility could lead to effects such as loss of biodiversity, rising sea levels, and extreme weather events affecting vulnerable populations more, while also aggravating global inequalities in food security, health, and economic stability (Steffen et al., 2018).

The way toward achieving climate neutrality will necessarily be multi-stranded: it requires bringing together the innovation of technological advance with the uptake of renewable energies and the management of sustainable consumption. Yet more than that is needed in response to climate change, as most aspects of success rely on changing how individuals, communities, and institutions act and behave (Rockström et al., 2017). Climate-neutral strategy needs to appeal to all these behavioural features if they are to shape change in ways that last.

The Role of Human Behaviour in Climate Neutrality

Human behaviour is critical in climate action, which allows for practices such as reducing energy use, adopting renewables, and minimizing waste. However, despite the growing awareness of climate change, there are still significant barriers to action. These include cognitive biases, lack of resources, societal norms, and inadequate policy support for sustainable choices.

1. ***Cognitive Biases and Heuristics***: Bounded rationality, availability bias, and loss aversion often lead a person to focus on more immediate benefits rather than larger environmental gains (Kahneman & Tversky, 1979; Tversky & Kahneman, 1974).
2. ***Habitual Inertia***: Embedded routines and convenience often outweigh pro-environmental intentions, making it difficult to inculcate sustainable practices (Verplanken & Wood, 2006).
3. ***Underestimation of Impact***: People tend to view their contributions as small and negligible in comparison with global challenges, which lowers their motivation to act sustainability (Nyborg et al., 2016).

Social norms and community value systems have a high impact on environmental behaviour that leads to broader societal practices favouring sustainability and having a "spillover effect" to accelerate climate action (Van der Linden et al., 2015). Behavioural economics offers good tools to overcome barriers by having interventions that align with human psychology. Techniques like nudges, incentives, and personalized feedback make choices more accessible and appealing for sustainability and thereby foster pro-environmental behaviour (Thaler & Sunstein, 2008).

Objectives and Scope of the Chapter

This chapter discusses how behavioural economics can overcome psychological and social barriers to climate neutrality. The goals are:

1. Review key behavioural economics theories and frameworks relevant to climate action.
2. Analysing the INSPIRE framework, an integrative model affecting sustainable behaviours at both the individual and community levels.
3. Review case studies that highlight the effective implementations of behavioural interventions in promoting environmental behaviours.
4. To give policymakers, businesses, and institutions actionable insights on integrating behavioural strategies into climate policies.

The chapter includes comprehensive behavioural economic theories, practical strategies, and interdisciplinary literature from psychology, environmental science, and economics. It relies on secondary data and a rigorous review of academic and policy documents to establish a robust foundation, offering comprehensive insights for advancing climate action through behavioural interventions.

Theoretical Foundations

Overview of Behavioural Economics

Behavioural economics is the integration of psychology and economics to understand decision-making in the presence of constraints such as cognitive limitations, time pressures, and imperfect information. Unlike classical economics, it accepts that human decisions are not always optimal because of bounded rationality, heuristics, and cognitive biases (Simon, 1955). Bounded rationality suggests people "satisfice" by choosing satisfactory options rather than optimal ones (Simon, 1955). Heuristics, such as availability bias, result in the overestimation of dramatic events, such as extreme weather, and the underestimation of gradual issues, such as sea-level rise (Kahneman, 2011). These deviations are critical in climate-related decisions, which are characterized by complex trade-offs between immediate costs and long-term benefits (Tversky & Kahneman, 1974).

Key Behavioural Economic Theories Relevant to Climate Neutrality

- ***Nudge Theory:*** Nudge theory, as Thaler and Sunstein (2008) postulate, suggests that small changes in choice architecture can profoundly affect behaviour without limiting the freedom of choice. Nudges exploit cognitive biases such as inertia and status quo bias to promote desirable outcomes. Setting renewable energy as the default option has been shown to increase adoption (Pichert & Katsikopoulos, 2008).
- ***Loss Aversion and Prospect Theory***: According to the Prospect Theory of Kahneman and Tversky (1979), loss aversion is the notion that losses are felt more strongly than gains. This bias can be used in climate actions, where focusing on the financial losses of energy overuse can be a driving force for more sustainable behavior (Allcott & Mullainathan, 2010).
- **Social Norms Theory:** Social norms have strong effects on behaviour. Behaviour is likely to be embraced if an individual believes other people would embrace it also (Cialdini et al., 1990). Comparing energy consumption with that of neighbours lowers the energy usage (Schultz et al., 2007).
- ***Time Discounting***: Temporal discounting, the preference for immediate rewards over future gains, hinders long-term climate goals (Frederick et al., 2002). Interventions that visualize future consequences can reduce this bias, promoting environmental responsibility (Hardisty & Weber, 2009).

Application of Behavioural Economics to Environmental Challenges

Behavioural economics has been very key in the development of interventions promoting sustainable behaviour across different domains.

1. ***Energy Savings***: Real-time feedback systems, like smart meters, lowered household energy consumption by 5–15% through social comparisons and loss aversion (Fischer, 2008; Karlin et al., 2015).
2. ***Waste Management***: Behavioural nudges such as smaller trash cans and prominent recycling signs have greatly increased recycling rates and reduced contamination of waste (Goldstein et al., 2008).
3. ***Sustainable Consumption***: Such practices as carbon labelling of products and the default on reusing grocery orders placed online have made consumer choice move in favour of lower carbon consumption options (Thøgersen & Schrader, 2012).

These applications show the potential that behavioural insights can bring to the solution of environmental problems, by targeting the cognitive and motivational barriers which frequently impede climate-friendly behaviours. Addressing climate change requires mobilizing human capital for success in achieving sustainability goals (Kumar et al., 2024; Singh et al., 2024).

Behavioural Economics in Climate Neutrality

Behavioural Insights and Climate-Neutral Practices

Behavioural economics is important in bridging the intention-action gap in the area of environmental concern. More people are aware of the value of sustainable practices, but the psychological, social, and structural barriers tend to knock the reasoning. Thus, with regard to these barriers, it is possible to use behavioural insights in the form of simplified choice, enhancing motivation, and aligning individual actions with collective efforts toward long-term climate aims (Gifford, 2011). Human Resource can drive sustainability through net-zero initiatives and AI integration, aligning strategies with organizational goals (Singh et al., 2024).

Overcoming Cognitive Barriers

Behavioural economics recognizes the following cognitive biases that hinder sustainable decision-making:

1. ***Status Quo Bias***: People are prone to favor existing systems and avoid change, like staying with fossil fuel-based transportation instead of adopting electric vehicles (Samuelson & Zeckhauser, 1988).
2. ***Loss Aversion***: People overemphasize perceived losses, such as the upfront costs of installing solar panels, at the expense of long-term financial and environmental benefits (Kahneman & Tversky, 1979).
3. ***Temporal Discounting***: Costs of green alternatives that appear immediately in time discourage investments, even when future savings are substantial (Frederick et al., 2002).

Effective interventions can counteract these biases by changing the cost perception as losses averted or short-term benefits. For instance, by stressing immediate energy savings through appliances, individuals may make more environmentally friendly choices (Hardisty & Weber, 2009).

Strengthening Behavioural Interventions

Behavioural interventions can influence pro-environmental behaviours by altering the choice environment:

1. *Nudges*: Subtle design changes, such as renewable energy defaulting or recycling bin placement, nudge choices but do not limit freedom (Thaler & Sunstein, 2008).
2. *Incentives*: Financial and social rewards make sustainable practices more appealing and rewarding (Allcott & Greenstone, 2012).
3. *Feedback Mechanisms*: Real-time data on energy consumption or carbon footprints raise awareness and promote behavioural change (Fischer, 2008).

Real-World Applications and Evidence

Behavioural economics principles have been applied successfully to encourage climate-neutral practices across multiple sectors:

1. ***Energy Savings***: Households account for a large share of carbon emissions. Behavioural interventions such as comparative energy use feedback have shown measurable impact. For example, Opower's Home Energy Reports, which compared household energy use with neighborhood averages, had an average 2% reduction in energy consumption across millions of households, saving millions of tons of CO_2 emissions (Allcott, 2011).
2. ***Waste Management***: Behavioural nudges - putting out one small litter bin besides the usual large recycling bin - helps reduce landfill waste. In Britain, the "Love Food, Hate Waste" campaign ran a loss aversion programme on food wastage costs by highlighting financial as well as environmental costs of wasted food resulting in a reduction of household food waste of 21% (Quested et al., 2013).
3. ***Transportation:*** Behavioural interventions have been proven to be effective in promoting low-carbon transportation modes like cycling and public transit. A gamified app in the Netherlands rewarded users with redeemable points for cycling, increasing the number of bike commutes and reducing urban emissions (Hamari & Koivisto, 2015).
4. ***Sustainable Consumption***: Retailers have nudged consumers toward eco-friendly choices by labelling products with their carbon footprints. Thøgersen and Schrader (2012) found that such labelling significantly increased purchases of environmentally friendly products, demonstrating the impact of transparency on consumer behaviour.

Behavioural Economics and Climate Policy

Policymakers are integrating behavioural economics into climate policies to promote sustainability at scale:

1. ***Green Defaults***: Cities in Germany have implemented default renewable energy options, which have a take-up rate of over 90%, compared to 10–20% in opt-in models (Pichert & Katsikopoulos, 2008).

2. ***Carbon Pricing and Incentives***: Sweden's carbon tax aligns financial incentives with behavioural principles, reducing emissions by over 25% while supporting economic growth (Andersson, 2019).
3. ***Social Norm Campaigns***: Public campaigns focusing on collective action have driven behaviour change. Water conservation programs in California, where people were educated about what their peers were doing, produced measurable reductions in water use (Ferraro & Price, 2013).

The INSPIRE Framework for Sustainable Behaviour

Overview of the INSPIRE Framework

The INSPIRE framework is an integrative behavioural model designed to overcome psychological, social, and structural barriers to sustainable practices. It incorporates seven components, each targeting specific challenges to adopting climate-neutral behaviours:

1. ***Incentivize:*** Monetary and social benefits that promote responsible practices.
2. ***Nudge:*** Organisational adjustments in selection architecture to support eco-friendly choices.
3. ***Supply:*** Immediate, tailor-made information regarding environmental implications.
4. ***Play:*** Introducing a gamification perspective to boost participation and commitment.
5. ***Initiate:*** Grassroots-led activities that instigate action in the collective.
6. ***Remind:*** Helping pro-environmental behaviours stay salient by way of cues and reminders.
7. ***Encourage***: Long-term commitment building through intrinsic and extrinsic rewards.

The INSPIRE framework brings together seven behavioural strategies to overcome barriers to sustainable behaviour. These strategies apply the principles of behavioural economics to influence individual and collective decision-making toward climate neutrality. Below is an in-depth discussion of each component of the framework:

Insights into INSPIRE Components

Incentivize

Incentives play a very significant role in promoting sustainable practices. Financial incentives, including tax rebates and subsidies, eliminate economic barriers to green actions. For example, Germany's solar subsidy program increased the adoption of residential solar panels by covering installation costs (Rosenow et al., 2016). Social incentives, including public recognition, fulfill individuals' need for approval, thereby motivating them (Deci et al., 1999).

Nudge

Nudges subtly alter the decision-making environment to default to sustainable options. Thus, cities in Germany, with renewable energy set as the default choice of utility, achieved uptake rates of over 90% compared to less than 20% when having opt-in models (Pichert & Katsikopoulos, 2008). Analogously, digital receipts or reusable packaging reduce the complexity of 'green' choices.

Supply Feedback

Feedback mechanisms make individuals accountable by giving them information about their footprint on the environment. Opower's energy reports, comparing householder energy consumption to that of neighbors, resulted in 2% sustained energy savings per million users (Allcott, 2011). Smart meters amplify this effect with real-time feedback.

Play

Gamification engages individuals through interactive challenges, such as apps like "Eco Challenge," which reward eco-friendly actions. Dutch experiments with gamified transport apps significantly increased cycling and walking, demonstrating its effectiveness in low-carbon mobility (Hamari & Koivisto, 2015).

Initiate

Community-based programs, like California's "Cool Neighbours," utilize social norms for community action, reducing energy consumption by 12% (Abrahamse et al., 2005). Reminders include prompts near switches or mobile alerts to ensure behaviour is consistent (Goldstein et al., 2008).

Remind

Reminders, such as prompts near switches or mobile alerts, ensure behavioural consistency (Goldstein et al., 2008).

Encourage

Finally, intrinsic motivation and extrinsic rewards, such as green commuting incentives, ensure sustained environmental action and reduced emissions (Siero et al., 1996).

Barriers to Sustainable Behaviour and Climate Neutrality

Despite the general public's increased awareness regarding climate change, a set of psychological, socioeconomic, and institutional barriers prevents such an adoption of eco-friendly behaviours and technologies. Such an understanding is crucial to the design of effective interventions.

Psychological Barriers

Psychological factors often constrain individuals from adopting sustainable practices, even when they know about environmental benefits. Status quo bias, which is the tendency to hold onto familiar habits, prevents many from changing to eco-friendly alternatives such as electric vehicles despite the technological advancement (Samuelson & Zeckhauser, 1988). Loss aversion also prevents change since individuals fear losses more than they value equivalent future gains (Kahneman & Tversky, 1979). Similarly, present bias makes individuals more value current costs and efforts than future benefits and therefore discourages investment in sustainability (Frederick et al., 2002). Framing of environmental activities in terms of current benefits or avoided losses might mitigate these biases (Hardisty & Weber, 2009).

Socioeconomic Barriers

Socioeconomic constraints play a very critical role in the access of sustainable technologies and practices. Income inequality prevents poor families from adopting climate-neutral practices due to the high upfront costs of technologies such as renewable energy systems and electric vehicles (Rao & Min, 2018). Education gaps in environmental literacy restrict the adoption of sustainable behaviour, especially in less represented regions (Fraj & Martinez, 2007). Infrastructure-related issues, such as a lack of electric vehicle charging stations or dependable public transportation, create logistical barriers to implementing environmentally friendly practices, particularly in developing regions (Sovacool & Ryan, 2016).

Institutional and Market Barriers

Structural and institutional factors often compromise attempts toward climate neutrality. Market failures, such as the unavailability of inexpensive green technologies and the overrepresentation of fossil fuels in the market due to subsidies, are barriers to green innovations (Unruh, 2000). Regulatory challenges like conflicting environmental policies that favour industries also weaken the promotion of sustainable practices (Harrison, 2012). Behavioural lock-in, or the way in which society is anchored on the existing systems, is also a barrier against new technologies' acceptance in the market (Jaffe et al., 2005).

Overcoming Barriers

The challenges will be solved with a combination of behavioural, economic, and structural interventions. Environmental choices can be reframed in terms of avoided losses and long-term benefits to counteract psychological biases (Kahneman & Tversky, 1979). Subsidies for green technologies, such as Norway's electric vehicle incentives, can fill affordability gaps (Mersky et al., 2016). Campaigns that address diverse populations with environmental education improve environmental literacy and encourage pro-environmental behaviors (Stern, 2000). Strong regulatory frameworks, such as carbon taxes and green building mandates, can encourage green technologies (Stiglitz et al., 2017). Finally, investments in infrastructure, like renewable energy grids and public transportation systems, can overcome logistical barriers and facilitate the shift to sustainable practices (Sovacool & Ryan, 2016).

Policy Implications and Future Directions

Leveraging Behavioural Economics for Climate Policy

Behavioural economics gives policymakers insights into human psychology and decision-making, allowing them to design interventions that promote sustainable practices in an effective manner. In contrast to traditional mandates or penalties, behavioural strategies align with natural human tendencies, making eco-friendly actions more intuitive. Economic incentives such as carbon pricing, which Sweden has successfully implemented, have reduced emissions while boosting economic growth through loss aversion (Andersson, 2019). For instance, green subsidies for renewable energy systems and electric vehicles reduce cost barriers, thus leading to their adoption (Mersky et al., 2016).

Nudges, such as green defaults in renewable energy plans, increase adoption rates without constraining choice (Pichert & Katsikopoulos, 2008). Framing losses over gains increases participation in conservation (Hardisty & Weber, 2009). Public campaigns that focus on social norms, such as California's water conservation feedback, increase collective responsibility (Ferraro & Price, 2013).

Emerging Trends and Technologies

New technologies are increasingly opening doors to the extension of behavioural economics in climate policy. For instance, AI and big data analytics allow personalized interventions like suggesting energy-efficient practice to individuals according to their consumption patterns (Strengers, 2011). Smart city infrastructures, with air quality monitoring and real-time energy tracking as eco-feedback systems, inspire people to lead sustainable lifestyles by making them aware of their effects in real time (Trivedi & Trivedi, 2024). Gamification and digital platforms, through apps rewarding good eco-friendly actions with points, badges, or discounts, engage users in sustainability. These strategies have proved effective in promoting low-carbon mobility and waste reduction among the tech-savvy younger populations (Hamari & Koivisto, 2015).

Addressing Challenges in Policy Implementation

Despite the promise of behavioural economics in climate policy, a number of challenges persist. Ethical objections to the diminution of individual liberty by nudges have emerged, with critics arguing that nudges should be made transparent and opt-outable in order to preserve choice (Sunstein, 2015). Cultural diversity is another challenge, whereby policies that work in one region may not work elsewhere, and therefore need adaptation to local contexts (Stern, 2000). Moreover, interventions that work in the smallest, most homogeneous context cannot be scaled up, when the population is vast, diverse, and thus creates a problem. And secondly, behavioural lock-in prevents change by offering traditional practices against new ones and sustainable alternatives (Bicchieri, 2017; Jaffe et al., 2005).

Recommendations for Policymakers

Behavioural interventions can be strengthened in their policies if the latter allows equal access to such alternatives for everyone irrespective of the income levels. Public consultations and community and community leadership involvement improve intervention fit into regional and cultural circumstances hence enhancing their effectiveness as documented by Ferraro and Price (2013). Infrastructure investments such as investment in public transport networks and renewable energy supply, for instance, help smoothen logistics (Trivedi & Trivedi, 2024). Monitoring systems, from longitudinal studies with real-time data, can track impacts of policies and guide improvement (Strengers, 2011). Inter-sectoral collaboration as well as new technologies, such as AI and IoT, can further drive scalable, personalized solutions for climate neutrality

(Hamari & Koivisto, 2015). Climate change impacts business models: companies must integrate climate strategies for a net-zero economy (Kumar et al., 2024).

Measuring Impact of Behavioural Interventions

Metrics for Measuring Impact

For climate neutrality behavioural interventions, good metrics are needed that assess both behavioural changes and environmental impacts. Some major indicators include the reduction in carbon emissions through increased utilization of renewable energy or reductions in energy consumption (Gillingham et al., 2009). The rates at which green practices-such as recycling or transition to green energy-are adopted indicate the level of sustainable behaviours (Fischer, 2008). Cost-effectiveness evaluates the environmental benefits relative to intervention costs (Allcott & Mullainathan, 2010). Social norm changes, particularly public views on sustainability, help gauge societal impact (Schultz et al., 2007). Lastly, assessing the sustainability of behaviour changes ensures long-term impact (Verplanken & Wood, 2006).

Methodologies for Impact Assessment

The impacts of behavioural intervention are measured through different methods. Randomized Controlled Trials (RCTs) are considered the gold standard, where participants are randomly assigned to intervention or control groups, often showing reductions in household energy usage (Allcott, 2011). When RCTs are not feasible, quasi-experimental designs like differences-in-differences analysis compare treated and untreated groups over time (Ferraro & Price, 2013). These tools range from smart meters to applications that track different behaviours like energy consumption and waste management (Fraj & Martinez, 2007). Attitude and behaviour changes are tracked with longitudinal surveys, whereas system dynamics and agent-based models make predictions about long-term impacts of interventions (Poteete et al., 2010).

Case Studies of Measured Impact

Case studies offer concrete examples of successful behavioural interventions. In the US, the Opower energy feedback program showed the effect of comparative feedback: savings of 1.4–3.3% were sustained across millions of households (Allcott, 2011). In the UK, the 5p tax on single-use plastic bags led to an 85% reduction in use within the first year: loss aversion was quite effective in public campaigns (Poortinga et al., 2016). Subsidies for solar panel installations increased the adoption of renewable energy by 25% in five years, indicating that financial incentives are quite persuasive (Rosenow et al., 2016).

Challenges in Impact Measurement

Measuring the effectiveness of behavioural interventions is beset by several challenges. Problems of attribution arise because effects of a particular intervention may be confused with more pervasive changes in society or the economy (Gillingham et al., 2009). The temporal constraint is often so that the study is carried out over a short term, failing to capture long-term maintenance of behavioural changes, thus requiring longer follow-up periods (Verplanken & Wood, 2006). Cultural variations across regions also influence the effectiveness of interventions, making generalizability of the findings challenging (Stern, 2000).

Recommendations for Improved Measurement

Incorporation of both quantitative and qualitative measures would help in enhancing the measurement of impact. Incorporation of carbon savings data along with public attitudes information would provide a holistic picture of the intervention effect. It would also be crucial to do longitudinal studies on how these behavioural changes persist long into the future (Schultz et al., 2007). Leveraging technologies like AI and IoT-enabled sensors for real-time data collection can provide actionable insights for refining interventions (Strengers, 2011). Finally, conducting cross-cultural studies across diverse socioeconomic and cultural contexts will help identify universally effective strategies while allowing for regional adaptations (Stern, 2000).

Conclusion and Recommendations

Key Insights

This chapter will demonstrate that it is behavioural economics that will provide the stimulus for pro-environmental behaviours necessary to achieve climate neutrality. This will enable the demonstration of how cognitive biases, social norms, and structural barriers affect sustainable practice (Stern, 2000). Interventions such as nudges, financial incentives, and feedback mechanisms have proven effective in reducing emissions and enhancing long-term sustainability (Allcott, 2011). The INSPIRE model offers an all-around approach towards the overcoming of the barriers that lead to sustainable behaviour. It is crucial in measuring such interventions to take into consideration metrics like carbon reductions, rates of adopting behaviour, and changes in social norms (Andersson, 2019; Schultz et al., 2007).

Practical Recommendations

Interventions should be all-inclusive of socioeconomic groups and tailored to local cultural contexts to increase the behavioural strategies for climate neutrality (Ferraro & Price, 2013). Policymakers should incorporate behavioural insights into environmental policies such as green defaults and carbon pricing to encourage sustainable practices (Andersson, 2019). Technology such as AI and IoT can amplify interventions through real-time data and gamification (Hamari & Koivisto, 2015). Monitoring with longitudinal studies and adaptive feedback loops ensures the ongoing change of behaviour (Strengers, 2011). Finally, cooperation between governments, businesses, and civil society has to be established for adaptation of successful interventions into other regions (Bicchieri, 2017).

Conclusion and Future Research Directions

The potential for growth in the application of climate neutrality by behavioural economics is huge and even greater with the new technologies, such as blockchain, used to track carbon (Trivedi & Trivedi, 2024). Future studies should examine cultural differences so that universally applicable frameworks may be designed (Stern, 2000), and predictive models developed to analyse the long-term effects of behavioural strategies on carbon goals (Poteete et al., 2010). By confronting psychological, social, and structural barriers, behavioural economics can fill the gap between awareness and action, allowing for a culture of sustainability that aligns individual and collective behaviour with environmental goals (Schultz et al., 2007).

Fig. 15.1 INSPIRE framework for sustainable behaviour. *Source: User Generated*

References

Abrahamse, W., Steg, L., Vlek, C., & Rothengatter, T. (2005). A review of intervention studies aimed at household energy conservation. *Journal of Environmental Psychology, 25*(3), 273–291. https://doi.org/10.1016/j.jenvp.2005.08.002

Allcott, H. (2011). Social norms and energy conservation. *Journal of Public Economics, 95*(9–10), 1082–1095. https://doi.org/10.1016/j.jpubeco.2011.03.003

Allcott, H., & Greenstone, M. (2012). Is there an energy-efficiency gap? *Journal of Economic Perspectives, 26*(1), 3–28. https://doi.org/10.1257/jep.26.1.3

Allcott, H., & Mullainathan, S. (2010). Behaviour and energy policy. *Science, 327*(5970), 1204–1205. https://doi.org/10.1126/science.1180775

Andersson, J. J. (2019). Carbon taxes and CO_2 emissions: Sweden as a case study. *American Economic Journal: Economic Policy, 11*(4), 1–30. https://doi.org/10.1257/pol.20170144

Bicchieri, C. (2017). *Norms in the wild: How to diagnose, measure, and change social norms.* Oxford University Press.

Cialdini, R. B., Reno, R. R., & Kallgren, C. A. (1990). A focus theory of normative conduct: Recycling the concept of norms to reduce littering in public places. *Journal of Personality and Social Psychology, 58*(6), 1015–1026. https://doi.org/10.1037/0022-3514.58.6.1015

Deci, E. L., Koestner, R., & Ryan, R. M. (1999). A meta-analytic review of experiments examining the effects of extrinsic rewards on intrinsic motivation. *Psychological Bulletin, 125*(6), 627–668. https://doi.org/10.1037/0033-2909.125.6.627

Ferraro, P. J., & Price, M. K. (2013). Using nonpecuniary strategies to influence behaviour: Evidence from a large-scale field experiment. *Review of Economics and Statistics, 95*(1), 64–73. https://doi.org/10.1162/REST_a_00344

Fischer, C. (2008). Feedback on household electricity consumption: A tool for saving energy? *Energy Efficiency, 1*(1), 79–104. https://doi.org/10.1007/s12053-008-9009-7

Fraj, E., & Martinez, E. (2007). Ecological consumer behaviour: An empirical analysis. *International Journal of Consumer Studies, 31*(1), 26–33. https://doi.org/10.1111/j.1470-6431.2006.00565.x

Frederick, S., Loewenstein, G., & O'Donoghue, T. (2002). Time discounting and time preference: A critical review. *Journal of Economic Literature, 40*(2), 351–401. https://doi.org/10.1257/002205102320161311

Gifford, R. (2011). The dragons of inaction: Psychological barriers that limit climate change mitigation and adaptation. *American Psychologist, 66*(4), 290–302. https://doi.org/10.1037/a0023566

Gillingham, K., Newell, R. G., & Palmer, K. (2009). Energy efficiency economics and policy. *Annual Review of Resource Economics, 1*(1), 597–620. https://doi.org/10.1146/annurev.resource.102308.124234

Goldstein, N. J., Cialdini, R. B., & Griskevicius, V. (2008). A room with a viewpoint: Using social norms to motivate environmental conservation in hotels. *Journal of Consumer Research, 35*(3), 472–482. https://doi.org/10.1086/586910

Hamari, J., & Koivisto, J. (2015). Why do people use gamification services? *International Journal of Information Management, 35*(4), 419–431. https://doi.org/10.1016/j.ijinfomgt.2015.04.006

Hardisty, D. J., & Weber, E. U. (2009). Discounting future green: Money versus the environment. *Journal of Experimental Psychology: General, 138*(3), 329–340. https://doi.org/10.1037/a0016433

Harrison, K. (2012). The comparative politics of carbon taxation. *Annual Review of Law and Social Science, 8*(1), 217–238. https://doi.org/10.1146/annurev-lawsocsci-102811-173828

Intergovernmental Panel on Climate Change (IPCC). (2021). *Climate change 2021: The physical science basis. Contribution of working group i to the sixth assessment report of the intergovernmental panel on climate change.* Cambridge University Press. https://doi.org/10.1017/9781009157896

Jaffe, A. B., Newell, R. G., & Stavins, R. N. (2005). A tale of two market failures: Technology and environmental policy. *Ecological Economics, 54*(2–3), 164–174. https://doi.org/10.1016/j.ecolecon.2004.12.027

Kahneman, D. (2011). *Thinking, fast and slow.* Farrar, Straus, and Giroux.

Kahneman, D., & Tversky, A. (1979). Prospect theory: An analysis of decision under risk. *Econometrica, 47*(2), 263–291. https://doi.org/10.2307/1914185

Karlin, B., Zinger, J. F., & Ford, R. (2015). The effects of feedback on energy conservation: A meta-analysis. *Psychological Bulletin, 141*(6), 1205–1227. https://doi.org/10.1037/a0039650

Kumar, K., Singh, R., Sharma, A., & Lehri, B. P. S. (2024). Climate change adaptation strategies for achieving net-zero economy. In: Singh, R., & Crowther, D. (Eds.) *Transition towards a sustainable future. Approaches to global sustainability, markets, and governance.* Springer, Singapore. https://doi.org/10.1007/978-981-97-5756-5_8\

Mersky, A. C., Sprei, F., Samaras, C., & Qian, Z. S. (2016). Effectiveness of incentives on electric vehicle adoption in Norway. *Transportation Research Part d: Transport and Environment, 46*, 56–68. https://doi.org/10.1016/j.trd.2016.03.011

Nyborg, K., Anderies, J. M., Dannenberg, A., Lindahl, T., Schill, C., Schlüter, M., & de Zeeuw, A. (2016). Social norms as solutions. *Science, 354*(6308), 42–43. https://doi.org/10.1126/science.aaf8317

Pichert, D., & Katsikopoulos, K. V. (2008). Green defaults: Information presentation and pro-environmental behaviour. *Journal of Environmental Psychology, 28*(1), 63–73. https://doi.org/10.1016/j.jenvp.2007.09.004

Poortinga, W., Whitmarsh, L., & Suffolk, C. (2016). The introduction of a single-use carrier bag charge in Wales: Attitude change and behavioural spillover effects. *Journal of Environmental Psychology, 47*, 126–135. https://doi.org/10.1016/j.jenvp.2016.05.008

Poteete, A. R., Janssen, M. A., & Ostrom, E. (2010). *Working together: Collective action, the commons, and multiple methods in practice.* Princeton University Press.

Quested, T. E., Marsh, E., Stunell, D., & Parry, A. D. (2013). Spaghetti soup: The complex world of food waste behaviours. *Resources, Conservation and Recycling, 79*, 43–51. https://doi.org/10.1016/j.resconrec.2013.04.011

Rao, N. D., & Min, J. (2018). Decent living standards: Material prerequisites for human well-being. *Social Indicators Research, 138*(1), 225–244. https://doi.org/10.1007/s11205-017-1650-0

Rockström, J., Gaffney, O., Rogelj, J., Meinshausen, M., Nakicenovic, N., & Schellnhuber, H. J. (2017). A roadmap for rapid decarbonization. *Science, 355*(6331), 1269–1271. https://doi.org/10.1126/science.aah3443

Rosenow, J., Platt, R., & Flanagan, B. (2016). Fuel poverty and energy efficiency obligations: A critical assessment of the supplier obligation in the UK. *Energy Policy, 62*(1), 989–998. https://doi.org/10.1016/j.enpol.2013.06.007

Samuelson, W., & Zeckhauser, R. (1988). Status quo bias in decision making. *Journal of Risk and Uncertainty 1*, 7–59 . https://doi.org/10.1007/BF00055564

Schultz, P. W., Nolan, J. M., Cialdini, R. B., Goldstein, N. J., & Griskevicius, V. (2007). The constructive, destructive, and reconstructive power of social norms. *Psychological Science, 18*(5), 429–434. https://doi.org/10.1111/j.1467-9280.2007.01917.x

Siero, F. W., Bakker, A. B., Dekker, G. B., & van Den Burg, M. T. C. (1996). Changing organizational energy consumption behaviour through comparative feedback. *Journal of Environmental Psychology, 16*(3), 235–246. https://doi.org/10.1006/jevp.1996.0019

Simon, H. A. (1955). A behavioural model of rational choice. *The Quarterly Journal of Economics, 69*(1), 99–118. https://doi.org/10.2307/1884852

Singh, R., Ravesangar, K., Narayanan, S., & Singh, K. K. H. (2024). HRM Practices in Creating Net Zero: Supporting Organizational Sustainability. In: Singh, R., Khan, S., Kumar, A., Luthra, S., Chokshi, H. (Eds.) *Net zero economy, corporate social responsibility and sustainable value creation. CSR, sustainability, ethics and governance.* Cham: Springer. https://doi.org/10.1007/978-3-031-55779-8_1

Singh, R., Crowther, D., & Ravesangar, K. (2024). harnessing human capital for a sustainable future: HR's pivotal role toward net zero policies. In Singh, R., & Crowther, D. (Eds.) *Transition towards a sustainable future. Approaches to global sustainability, markets, and governance.* Singapore: Springer. https://doi.org/10.1007/978-981-97-5756-5_2

Sovacool, B. K., & Ryan, S. E. (2016). The geography of energy and climate transitions. *Nature Climate Change, 6*(9), 847–856. https://doi.org/10.1038/nclimate3057

Steffen, W., Rockström, J., Richardson, K., Lenton, T. M., Folke, C., Liverman, D., & Schellnhuber, H. J. (2018). Trajectories of the Earth system in the Anthropocene. *Proceedings of the National Academy of Sciences, 115*(33), 8252–8259. https://doi.org/10.1073/pnas.1810141115

Stern, P. C. (2000). Toward a coherent theory of environmentally significant behaviour. *Journal of Social Issues, 56*(3), 407–424. https://doi.org/10.1111/0022-4537.00175

Stiglitz, J. E., Stern, N., Duan, M., Edenhofer, O., Giraud, G., Heal, G., & Winkler, H. (2017). Report of the high-level commission on carbon prices. *Carbon Pricing Leadership Coalition.*

Strengers, Y. (2011). Designing eco-feedback systems for everyday life. *Interactions, 18*(1), 42–45. https://doi.org/10.1145/1897239.1897250

Sunstein, C. R. (2015). *Choosing not to choose: Understanding the value of choice.* Oxford University Press.

Thaler, R. H., & Sunstein, C. R. (2008). *Nudge: Improving decisions about health, wealth, and happiness.* Yale University Press.

Thøgersen, J., & Schrader, U. (2012). From knowledge to action—New paths towards sustainable consumption. *Journal of Consumer Policy, 35*(1), 1–6. https://doi.org/10.1007/s10603-012-9180-7

Trivedi, A., & Trivedi, N. (2024). Integrating circular economy in smart cities: Challenges and pathways to sustainable urban development. In V. Kandpal, E. D. Santibanez-Gonzalez, P. Chatterjee, & M. K. Nallapaneni (Eds.), *Smart cities and circular economy* (pp. 71–82). Emerald Publishing Limited. https://doi.org/10.1108/978-1-83797-957-820241006

Tversky, A., & Kahneman, D. (1974). Judgment under uncertainty: Heuristics and biases. *Science, 185*(4157), 1124–1131. https://doi.org/10.1126/science.185.4157.1124

Unruh, G. C. (2000). Understanding carbon lock-in. *Energy Policy, 28*(12), 817–830. https://doi.org/10.1016/S0301-4215(00)00070-7

Van der Linden, S., Maibach, E., & Leiserowitz, A. (2015). Improving public engagement with climate change: Five "best practice" insights from psychological science. *Perspectives on Psychological Science, 10*(6), 758–763. https://doi.org/10.1177/1745691615598516

Verplanken, B., & Wood, W. (2006). Interventions to break and create consumer habits. *Journal of Public Policy & Marketing, 25*(1), 90–103. https://doi.org/10.1509/jppm.25.1.90

Chapter 16
On the Way Towards Climate Neutrality: The Positives and Negatives of Climate Policy Changes in Australia

Wei Qian and Ming Xue

Abstract This study examines the impacts of climate policy changes on business carbon performance in Australia as the country strives for climate neutrality. In this study, we analyse the carbon efficiency data of National Greenhouse and Energy Reporting (NGER) registrants—the largest emitters in Australia—from 2009 to 2020 to examine the carbon efficiency results of these top polluters against the significant climate policy changes over the decade, identifying patterns, challenges, and opportunities. The findings show that although over half of Australia's top emitters improved their carbon efficiency, only a quarter were strong carbon-efficiency performers (i.e., improved both carbon and economic performance). Carbon efficiency developments were predominantly driven by economic performance, with few companies solely focusing on environmental performance ('Green Stars'). However, companies also increasingly avoided being perceived as 'dirty' or as 'Dirty Cash Cows'. Carbon performance improvement was more prominent among companies operating in less environmentally sensitive industries compared to highly sensitive industries. As economic performance rather than environmental improvement remained the main driver for carbon efficiency development, the carbon tax lowered carbon efficiency outcomes due to increased economic and compliance costs, while its repeal led to a slight improvement in carbon efficiency. However, the failure of the National Energy Guarantee and ongoing policy uncertainty regarding climate change significantly reduced carbon efficiency outcomes in more recent years.

Keywords Climate neutrality · Carbon efficiency · Eco-efficiency · Climate change policy · Australia

W. Qian (✉) · M. Xue
UniSA Business, University of South Australia, Adelaide, Australia
e-mail: wei.qian@unisa.edu.au

M. Xue
e-mail: ming.xue@mymail.unisa.edu.au

R. Singh and W. L. Filho (eds.), *Climate Neutrality Through Smart Eco-Innovation and Environmental Sustainability*, Climate Change Management,
https://doi.org/10.1007/978-3-031-83250-5_16

Introduction

Climate change presents a significant social, political and technological challenge for governments worldwide (Joshi et al., 2024; Nwokolo et al., 2023a, 2024). As the largest developed country in the Pacific region, Australia is highly sensitive to climate variation because it is surrounded by many developing countries and small islands that are sensitive to rising temperatures. Numerous calls have been made in the past decade for Australia to lead the region in tackling climate challenges and reduce carbon emissions to achieve climate neutrality. However, among the OECD countries, Australia has the highest per capita emissions and is a leading emitter globally (Qian et al., 2018).

Since 2007, the climate issue has dominated political debates in Australia. Following the Australian Labour Party's success in the 2007 election, Australia intensified its commitment to combat carbon pollution, introducing the National Greenhouse and Energy Reporting Act 2007 (NGER Act). This legislation mandates corporations exceeding certain emission thresholds to report their greenhouse gas emissions to the Clean Energy Regulator. Since 2009, this national framework has collected the carbon emission and energy data of heavy polluters and disseminated this information for public scrutiny. In the following decade, Australia experienced incredible carbon policy changes to address climate challenges. The government attempted to implement the Emissions Trading Scheme (ETS), which ultimately failed in 2009–2011. It established the carbon tax in 2012 but abolished it just two years later. It embarked on National Energy Guarantee (NEG) initiatives in 2017, but they collapsed after two years of debates. All of these attempts have had profound impacts on the business management of carbon performance and efficiency.

Despite policy uncertainties, the business sector understands that active responses or proactive actions regarding climate change are crucial for obtaining reputational dividends from investors and seeking additional equity financing (Kumar et al., 2024; Qian et al., 2020). Business managers are increasingly acknowledging the impact of environmental performance on share prices, leading them to place greater emphasis on corporate sustainability and achieving eco-efficiency (i.e., both economic and environmental efficiency) or a 'win–win' resolution (Qian et al., 2018; Zhou et al., 2022). Nevertheless, businesses also understand that environmental improvement comes at a price (Nwokolo et al., 2023b; Nwokolo et al., 2023d). Carbon pollution reduction requires substantial investment that would receive financial returns only in the long term. For example, high-energy-intensive companies may be willing to transition to cleaner energy producers to contribute to climate neutrality, but the closure of facilities such as coal plants not only involves significant financial costs but could also create massive social and environmental controversies. More often, public companies—especially environmentally sensitive companies—may suffer significantly in market valuation due to carbon policy changes (Chapple et al., 2013; Cooper et al., 2018), as announcements of carbon emission reductions are sometimes associated with substantially negative market reactions (Cooper et al., 2018).

Despite growing interest in Australia's constant and often chaotic policy changes regarding carbon emission reduction (Crowley, 2017; Firsova et al., 2012), little research has examined how Australian companies have reacted to such changes over time. Most studies focus on earlier discussions of the Carbon Pollution Reduction Scheme (CPRS, i.e., the ETS) or carbon tax, or they merely assess the relationship between carbon performance and financial performance. For instance, Ramiah et al. (2013) identified a significant negative impact of the CPRS on the alternative energy sector. Meanwhile, Qian et al. (2020) found that companies with better carbon performance achieved greater market returns during the repeal of the carbon tax but not during the ETS and carbon tax changes, indicating that political uncertainties limit the value-enhancing potential of carbon performance improvements. However, research has yet to fully explore how companies adapt their eco-efficiency performance in response to different policy changes over time.

Using the carbon efficiency typologies developed by Qian et al. (2018), this chapter examines how Australia's top emitting companies adjusted their carbon efficiency in response to various policies over a 12-year period. We conducted a systematic comparison and analysis against four major carbon policy changes to assess corporate carbon performance over the period and its variation among industries. Our results contribute greater understanding of the impact of carbon policy changes on business carbon behaviour – notably, the efficiency levels of carbon management. More importantly, as carefully designed government regulations and policies for environmental protection can stimulate corporate environmental innovation and risk management and motivate businesses to seek opportunities that allow them to stay profitable while being environmentally responsible (Birchall et al., 2013), an improved understanding of corporate responses to various carbon polices provides regulatory suggestions and implications for the future design of carbon strategy.

A Snapshot of Australia's Decade-Long Carbon Policy Changes

Since the implementation of the NGER Act 2007, the Australian government has actively combated climate change, aiming to limit corporate carbon emissions through various regulations. This section reviews Australian carbon-related regulations from 2009 to 2020, providing a detailed timeline.

The ETS Initiative (2009–2011)

The Rudd government's CPRS faced heavy criticism from the Opposition, the Greens, and some energy-intensive companies. After months of negotiations, the government announced a modified CPRS in May 2009, which passed in the House of Representatives (HoR) in June but faced intense debate in the Senate. Opposition from both the Liberal Party and the Greens led to the bill's failure in the Senate in August.

Most disagreements centred on the energy generation sector's future liabilities. The government increased compensation for energy-intensive industries like coal mining and reintroduced the CPRS in October 2009. Although it passed in the HoR, Senate debates escalated in November. A series of amendments were proposed to gain support, but Malcolm Turnbull's ousting as Liberal Party leader on 1 December and Tony Abbott's subsequent opposition led to the bill's second failure in the Senate, causing political turmoil and a double dissolution election over the ETS. Although the Rudd government reintroduced the CPRS bill in February 2010, again passing in the HoR, it was unlikely to succeed due to declining public support and internal political instability. Kevin Rudd was ousted as Prime Minister in April 2010, and the ongoing political uncertainty led to the CPRS bill being shelved until it lapsed on 28 September 2010.

The Carbon Tax Bill (2012–2014)

The elected Gillard government postponed the CPRS implementation in favour of a less aggressive climate change policy, the carbon tax. Unlike the direct 'cap and trade' carbon pricing scheme, a carbon tax is easier to manage and provides price/cost certainty for industries. This proposal was more acceptable to the Greens and independents after the ETS political battles. In February 2011, the Gillard government outlined plans for a carbon tax, aiming for a path to the ETS by 2015. By July 2011, the government had announced carbon pricing details and a $15 billion compensation package for high-energy industries and households.

Although debate intensified in September 2011, the Clean Energy Bill (the carbon tax bill) was introduced to Parliament on 23 September. It passed in the HoR in October with the support of the Greens and independents. Following further details on the bill, it passed in the Senate on 8 November 2011—the first climate change legislation victory in Australian history. The legislation took effect from 1 July 2012.

The Repeal of the Carbon Tax (2015–2017)

Carbon pricing for major polluters faced significant challenges. The Opposition relentlessly attacked the carbon tax, citing its negative impact on the national economy, which was exacerbated by the early 2013 collapse of carbon prices in the EU trading market. Tensions escalated after Gillard lost the Labour Party leadership to Rudd in June 2013, and Rudd subsequently failed to win the federal election. In September, the Liberal and National coalition claimed victory, and Tony Abbott was sworn in as Prime Minister.

During the election campaign, Abbott had promised to repeal the carbon tax. His government quickly worked to fulfill this promise, introducing the Carbon Tax Repeal Bill in October 2013, which passed in the House of Representatives in November but was defeated in the Senate in March 2014. A revised bill was introduced in June 2014 but initially failed due to opposition from Clive Palmer of the United Australia Party.

With continued volatility in the new Senate, the Abbott government worked to secure Palmer's support, which proved to be a critical turning point. The bill was reintroduced in July 2014, passing the HoR without issue. Palmer supported the government due to key amendments made to the bill. Although the debate continued for two days, the bill was ultimately passed by the Senate on 17 July 2014, taking effect on 1 July 2014, to abolish the carbon tax.

The National Energy Guarantee and Onwards (2018–2020)

After the repeal of the carbon tax, Australia experienced a carbon policy void for several years. Although a 26–28% emissions reduction target by 2030 was set in 2015 (increased to 43% in 2022) and the Paris Agreement was ratified in 2016, no specific actions were taken. In late 2017, the Turnbull government proposed the National Energy Guarantee (NEG), aiming to reduce energy prices, enhance market reliability, and lower carbon emissions in the electricity sector. However, the proposal to remove renewable energy subsidies drew objections from both inner and opposition parties.

The NEG draft was released in June 2018 and gained majority support after significant changes, including adjusted emissions reduction requirements. Despite this, internal party rebellions and two leadership challenges hindered its passage, culminating in Turnbull's ousting in August 2018. Consequently, no significant legislative actions on climate change occurred until Australia recently committed to a net zero target by 2050.

Business Responses to 'Green' Policies

Company managers have increasingly understood that their success or struggles in managing carbon efficiency and performance are influenced by and sometimes result in changes in government policies (Schaltegger & Burritt, 2018; Nwokolo et al., 2023c). Previous studies have explored the strategies used by company managers to navigate carbon performance and efficiency changes in response to 'green' policy pressures (Qian et al., 2018; Schaltegger & Burritt, 2000). Specifically, Qian et al. (2018) revealed that companies respond to carbon policy changes in eight different ways, which constitute eight carbon efficiency portfolio categories (see Table 16.1).

According to Qian et al. (2018), 'Golden Stars' and 'Green Stars' are active respondents who succeed in achieving strong eco-efficiency improvements,

Table 16.1 Corporate eco-efficiency portfolio

Portfolio	Eco-efficiency indicators	Direction and strength of change	Description
Efficient categories			
- Strong eco-efficiency development			
Golden Stars Green Stars	$Econ\uparrow > Envir\uparrow$	++	Economically strong EE improvement
	$Envir\uparrow > Econ\uparrow$	++	Environmentally strong EE improvement
- Weak eco-efficiency development			
Greedy Cash Cows Green Question Marks	$\lvert Econ\uparrow\rvert > \lvert Envir\downarrow\rvert$	+	Economically favourable weak EE improvement
	$\lvert Econ\downarrow\rvert < \lvert Envir\uparrow\rvert$	+	Environmentally favourable weak EE improvement
Inefficient categories			
- Minor eco-efficiency deterioration			
Dirty Cash Cows	$\lvert Econ\uparrow\rvert < \lvert Envir\downarrow\rvert$	–	Economically favourable EE decrease
Red Question Marks	$\lvert Econ\downarrow\rvert > \lvert Envir\uparrow\rvert$	–	Environmentally favourable EE decrease
- Significant eco-efficiency deterioration			
Dirty Dogs Stressed Dogs	$\lvert Econ\downarrow\rvert < \lvert Envir\downarrow\rvert$	–	Destructive EE improvement
	$\lvert Econ\downarrow\rvert > \lvert Envir\downarrow\rvert$	–	Destructive EE decrease

(*Source*: adapted from Qian et al., 2018).

enhancing both carbon and economic performance (Schaltegger & Burritt, 2000). The difference between them is that Golden Stars achieve economically strong eco-efficiency, where their economic performance increases exceed the increase in environmental performance, whereas Green Stars manage to increase more environmental than economic performance. 'Greedy Cash Cows' and 'Green Question Marks', while above the eco-efficiency hurdle line (i.e., achieving positive eco-efficiency development), are weak eco-efficiency performers, achieving either economic or environmental improvements. Greedy Cash Cows are economically favourable weak eco-efficient performers, as they still attempt to limit environmental deterioration when pursuing profit maximisation. In contrast, Green Question Marks focus mainly on environmental aspects without achieving efficient resource use or economic value creation.

In the inefficient categories, 'Dirty Cash Cows' and 'Red Question Marks' represent companies experiencing minor eco-efficiency deterioration. Dirty Cash Cows aim predominantly for economic growth regardless of environmental damages, whereas Red Question Marks mainly experience financial distress despite some environmental improvement. Finally, 'Dirty Dogs' and 'Stressed Dogs' are two groups of destructive eco-efficient performers. They grapple with both environmental and economic pressures, and the decrease in both types of performance is likely to take a toll on their ability to maintain environmental compliance or generate sufficient economic returns to stay in the market.

There has been an increasing acceptance in business practice and academic research of achieving eco-efficiency or a 'win–win' as a resolution for corporations to take adequate climate responsibility. Although the rewards of improved environmental performance from such investments may only materialise in the long term (Berrone & Gomez-Mejia, 2009), and it may take an even longer period to translate better environmental management into direct financial value (Horváthová, 2010), many researchers supporting a business rationale for sustainability and eco-efficiency consider that a positive industry and financial market response is achievable (Wahba, 2008). Environmental policies may be able to incentivise companies to invest in production redesign, new equipment, and employee coordination, contributing to positive financial returns. Therefore, it is likely that more companies would actively respond to policy changes such as the proposed ETS and carbon tax, becoming strong carbon efficiency performers. Conversely, the failure of these policies could lead to reduced investment in carbon efficiency improvements.

However, the equity market may negatively price legislative changes for carbon reduction (Chapple et al., 2013; Clarkson et al., 2015), as policy changes may impose compliance costs and hinder productivity (Walley & Whitehead, 1994). Public companies rely heavily on shareholders for diverse investments. If the market perceives that stricter regulations will increase costs and necessitate significant investments—often without short-term financial returns—this could hinder anticipated improvements from 'green' policies (Walley & Whitehead, 1994). Indeed, monetary mechanisms to address environmental issues may send negative signals to the market. Proposed ETS and carbon pricing could raise costs, which companies might pass on to customers, leading to unfavourable responses and negatively impacting carbon efficiency results; meanwhile, repealing the carbon tax and failing the NEG might improve efficiency.

Frequent changes in environmental policies are likely to complicate business responses, as perceived uncertainty may further negatively impact economic value creation. Previous literature has mainly examined the impacts of individual policies such as the NGER Act 2007 (Qian et al., 2018) or the carbon tax (Kumarasiri & Lodhia, 2017), but it is essential to understand whether the significant policy changes over a decade have effectively altered business behaviours. Thus, this study investigates how top polluting companies in Australia have responded and adjusted their carbon efficiency portfolios from 2009 to 2020.

Research Methods

Sample

The data used for this study were drawn from the Australian National Greenhouse and Energy Reporting (NGER) database and Company 360 Select database, with any missing data supplemented by information obtained from the companies' annual reports. Under the Australian NGER Act 2007, business entities that exceed specified thresholds of greenhouse gas emissions must register and report emissions and energy consumption information to the government for public consumption. The NGER data have been published on an annual basis since 2009. This study focuses on NGER-registered business entities, excluding non-corporate entities, with data spanning 12 years from 2009 to 2020. Due to the variation in the companies included each year and missing data in the NGER dataset, we ensure consistency in the companies included across different years and facilitate the comparison of changes over time by selecting companies that contain at least six (out of 12) consecutive years of environmental and financial data in the sample. After cleaning and sorting the data, the final study sample includes 181 companies that have reported more than six years of data over the 12-year study period.

Based on the Global Industry Classification Standard, we categorised the sample companies into 10 major industry sectors: Materials, Utilities, Energy, and Industrials are considered high environmental sensitivity industries, while sectors such as Consumer Staples, Consumer Discretionary, Financials, Health Care, Telecommunication Service, and Information Technology are classified as low environmental sensitivity industries. Table 16.2 presents the industry profile of these companies.

Companies in Materials are the dominant NGER reporters, making up nearly 30% of the total sample. This is followed by Industrials, which constitutes 20% of the sample. These two sectors combined represent nearly half of the total sample

Table 16.2 Industry profile of sample companies

Industry sector	No	%
Materials	50	28.6
Industrials	35	19.3
Consumer Staples	23	12.7
Utilities	19	10.5
Consumer Discretionary	16	8.8
Financials	14	7.7
Energy	15	8.3
Health Care	6	3.3
Telecommunication Service	2	1.1
Information Technology	1	0.6
Total	181	100

companies. The numbers of companies in other highly environmentally sensitive industries, including Utilities (10.5%) and Energy (8.3%), are relatively smaller and similar to those in low environmentally sensitive industries, such as Consumer Staples (12.7%), Consumer Discretionary (8.8%), and Financials (7.7%). Health Care, Telecommunication Service, and Information Technology only account for a small proportion of the NGER companies.

Carbon Efficiency Measurement and Classification

According to the eco-efficiency portfolio (Table 16.1), carbon efficiency compares the changes of carbon performance with the economic return of a company. We used total carbon emissions (i.e., the aggregate of Scope 1 and Scope 2 emissions reported in NGER data[1]) to measure corporate carbon performance. As an environmentally responsible company needs to be responsible for reducing both Scope 1 (direct) and Scope 2 (indirect) emissions, total carbon emissions can better reflect the overall carbon performance than the single measure of emissions. However, it is important to note that total emissions were calculated to measure carbon pollution levels. Therefore, a negative change in total emissions implies an improvement in corporate carbon performance in this study.

We used earnings before interest and tax (EBIT) as the measure of corporate economic performance. EBIT can effectively reflect a company's capacity to utilise its assets for generating revenue and profit, regardless of how those assets are financed (interest-bearing) or taxed. This measurement has been widely employed in prior studies to assess corporate economic performance (Qian et al., 2018). After collecting each company's carbon emission and economic performance data, we calculated the percentage changes of carbon and economic performance from a base year to a specified benchmarking year and then compared the change directions and percentages to determine each company's carbon efficiency development category.[2] We then coded the companies based on the typological classifications defined in Table 16.1.

[1] According to the NGER Act 2007, Scope 1 emissions are greenhouse gases released into the atmosphere as a direct result of a business activity or series of business activities; Scope 2 emissions are greenhouse gases released from energy consumption, such as electricity, heating, cooling, or steam that is consumed by the companies.

[2] As noted, the directions of carbon emission changes were reversed because emissions reduction indicated carbon performance improvement.

Results

Overview of Carbon Efficiency Changes from 2009 to 2020

We first analyse the companies falling within each carbon efficiency category from 2009 to 2020. The results are presented in Table 16.3.

The results show that within 53.2% of the efficient companies, only 27.1% of Australian high emitters achieved strong carbon efficiency improvements from 2009 to 2020, thereby making a 'win–win' case for sustainability. Among these companies, about 22.8% were Golden Stars, achieving economically strong carbon efficiency improvements. Only a small number were Green Stars (4.3%). This finding suggests that while these companies were able to achieve a 'win–win' result for sustainability, most companies focused more on economic development rather than environmental improvement.

We observed a slightly different composition for companies making weak carbon efficiency developments. Similar numbers of companies could trade off economic (environmental) improvement and environmental (economic) costs, making them

Table 16.3 Distribution of companies in each eco-efficiency category

Eco-efficiency category	All companies	High environmental sensitivity companies	Low environmental sensitivity companies
Efficient categories			
Strong carbon efficiency development			
Golden Stars	22.8%	18.1%	34.6%
Green Stars	4.3%	6.1%	0.0%
Total	27.1%	24.2%	34.6%
Weak carbon efficiency development			
Greedy Cash Cows	10.9%	13.6%	3.9%
Green Question Marks	15.2%	7.6%	34.6%
Total	26.1%	21.2%	38.5%
Inefficient categories			
Minor carbon efficiency deterioration			
Dirty Cash Cows	2.2%	3.0%	0.0%
Red Question Marks	26.1%	27.3%	23.0%
Total	28.3%	30.3%	23.0%
Significant carbon efficiency deterioration			
Dirty Dogs	4.4%	6.1%	0.0%
Stressed Dogs	14.1%	18.2%	3.9%
Total	18.5%	24.3%	3.9%
N	92	66	26

stay above the eco-efficiency hurdle line. About 10.9% of companies were Greedy Cash Cows, while 15.2% were Green Question Marks. Together, the weak carbon efficiency performers accounted for about one quarter of total companies, a proportion similar to that of strong carbon efficiency achievers.

Within the inefficient categories, 28.3% of companies experienced minor carbon efficiency deterioration, while 18.5% suffered significant carbon efficiency deterioration. The significantly larger number of Red Question Marks (26.1%) compared to Dirty Cash Cows (2.2%) and of Stressed Dogs (14.1%) compared to Dirty Dogs (4.4%) again indicates that economic performance was still the barometer of carbon efficiency results. Very few companies wanted to be perceived as 'dirty', and they still attempted to limit their carbon footprint even when faced with financial challenges or struggling to balance economic and environmental performance.

In high environmental sensitivity industries, about 45.4% improved carbon efficiency over the examined decade. Among these, 24.2% achieved strong improvements, while 21.2% were weak performers. Notably, all Green Stars came from high-sensitivity sectors. Despite a lack of legitimacy pressure to enhance carbon efficiency, the top performers in these industries outshone those in less sensitive sectors, where no Green Stars were found. Their strategies prioritised environmental performance to maintain their reputation and stakeholder support (Cowan & Deegan, 2011; Qian et al., 2018). Conversely, weak performers in these industries appeared 'greedier', with more Greedy Cash Cows (13.6%) than Green Question Marks (7.6%), indicating a stronger focus on economic rather than environmental improvement.

Companies in less environmentally sensitive industries exhibited more positive trends, with 73.1% improving carbon efficiency. Nearly 34.6% were Golden Stars, while none were Green Stars. Another 34.6% were Green Question Marks, suggesting a prioritisation of carbon performance over economic development. This trend was consistent among those experiencing carbon efficiency deterioration; none were classified as Dirty Cash Cows or Dirty Dogs.

Impacts of Carbon Policy Changes on Carbon Efficiency Portfolio

Carbon Efficiency Development in Different Stages of Regulatory Changes

Table 16.4 presents companies' carbon efficiency development during the four stages of carbon regulatory changes from 2009 to 2020.

Stage 1: The ETS Initiative (2009–2011)

We observed an encouraging development of carbon efficiency among the sample during the ETS initiative. Overall, 66.4% of companies achieved carbon efficiency

Table 16.4 Carbon efficiency development from 2009 to 2020

	Stage 1: The ETS initiative (2009–2011)	Stage 2: The carbon tax bill (2012–2014)	Stage 3: The repeal of the carbon tax (2015–2017)	Stage 4: The NEG and onwards (2018–2020)
	%	% (% change)	% (% change)	% (% change)
Panel A: Efficient categories				
Strong carbon efficiency development				
Golden Stars	28.0	24.1 (-3.9)	25.3 (+1.2)	15.2 (-10.1)
Green Stars	5.6	6.2 (+0.6)	6.5 (+0.3)	2.9 (-3.6)
Subtotal	33.6	30.3 (-3.3)	31.8 (+1.5)	18.1 (-13.7)
Weak carbon efficiency development				
Greedy Cash Cows	29.0	19.1 (-9.9)	18.8 (-0.3)	15.2 (-3.6)
Green Question Marks	3.8	3.4 (-0.4)	7.7 (+4.3)	5.1 (-2.6)
Subtotal	32.8	22.5 (-10.3)	26.5 (+4.0)	20.3 (-6.2)
Total	66.4	52.8 (-13.6)	58.3 (+5.5)	38.4 (-19.9)
Panel B: Inefficient categories				
Minor carbon efficiency deterioration				
Dirty Cash Cows	4.7	3.4 (-1.3)	4.7 (+1.3)	4.3 (-0.4)
Red Question Marks	19.6	27.5 (+7.9)	20.6 (-6.9)	38.4 (+17.8)
Subtotal	24.3	30.9 (+6.6)	25.3 (-5.6)	42.7 (+17.4)
Significant carbon efficiency deterioration				
Dirty Dogs	0.9	4.5 (+3.6)	3.5 (-1.0)	1.5 (-2.0)
Stressed Dogs	8.4	11.8 (+3.4)	12.9 (+1.1)	17.4 (+4.5)
Subtotal	9.3	16.3 (+7.0)	16.4 (+0.1)	18.9 (+2.5)
Total	33.6	47.2 (+13.6)	41.7 (-5.5)	61.6 (+19.9)
N	107	178	170	138

improvements, with nearly 33.6% achieving strong improvements and 32.8% classified as weak carbon efficiency performers. A closer look at the companies that made the positive carbon efficiency development reveals that the majority were Greedy Cash Cows (29.0%) and Golden Stars (28.0%). This result implies that although companies began to address carbon issues after the NGER Act 2007, the increase of carbon efficiency was mainly driven by maximising economic outputs as a response to the potential ETS changes: the business-as-usual economic-focused response was the main strategy.

We also observed that Red Question Marks (19.6%) accounted for a significant proportion of all companies compared to other inefficient companies during the ETS initiative. This finding implies that companies were mostly willing to make some improvement in environmental performance, demonstrating their active responses to the potential regulatory changes rather than being seen as only pursuing profit maximisation. Companies struggling with positive carbon efficiency outcomes understood that they had to signal some actions because market analysts started to scrutinise environmental policy and performance announcement implications (Guenster et al., 2011), and investors and the capital market became increasingly sensitive to how companies respond to changes in environmental regulations and policies since the ETS move (Molina-Azorín et al., 2009; Ramiah et al., 2013).

Stage 2: The Carbon Tax Bill (2012–2014)

Following the passage of the carbon tax bill, the total number of strong carbon efficiency achievers increased considerably, although their overall proportion (30.3%) remained similar to that in Stage 1 (33.6%). There was a slight increase of Green Stars (+0.6%) as a result of the regulatory drive. However, the economically driven strong carbon efficiency achievers (Golden Stars) slipped by 3.9%. The weak carbon efficiency performers significantly decreased by 10.3%, from 32.8% in Stage 1 to 22.5% in Stage 2. Overall, the carbon efficient companies (the combination of strong and weak carbon efficiency performers) dropped to 52.8%, which was 13.6% lower than Stage 1 (66.4%). These results suggest that carbon efficiency developments slowed after the carbon tax implementation. The direct increase in economic and compliance costs caused the equity market to react negatively to legislative changes aimed at reducing carbon emissions (Chapple et al., 2013; Clarkson et al., 2015). If economic performance was still the main driver for carbon efficiency, it was not surprising that the direct increase in economic costs as a result of the carbon tax led to an immediate reduction in carbon efficiency levels.

During this stage, about 47.2% of companies experienced deteriorating economic performance, a rise of 14.5% from Stage 1. The proportion of Red Question Marks grew by 7.9% to 27.5%, making it the largest category of the total sample. The proportion of Dirty Dogs and Stressed Dogs also exhibited a small increase. These results imply that despite an increasing number of companies experiencing financial pressure during the carbon tax period, more companies aimed to achieve environmentally favourable outcomes to signal environmental compliance and improvement.

Stage 3: The Repeal of the Carbon Tax (2015–2017)

The overall carbon efficiency results in Stage 3 were similar to those in Stage 2, with a slight increase (+5.5%) of companies achieving positive eco-efficiency outcomes. After the repeal of the carbon tax, 31.8% of the companies managed to achieve strong carbon efficiency development, representing a 1.5% increase from the previous period. This is likely attributed to the reduced financial burden imposed on companies in Stage 3. However, the pattern of Golden Stars and Green Stars remained similar to previous stages, with Golden Stars (25.3%) dominating the strong eco-efficiency achievers.

Additionally, we found that the proportion of companies falling into the Green Question Marks category increased by 4.3%, while the proportion of Red Question Marks dropped by 6.9%. These results imply that financially distressed companies became more willing to address carbon issues than before. The proportions of other efficiency categories did not present any significant change; the sample companies seemed to continue managing carbon efficiency in a similar way as in Stage 2 despite the weaker regulatory environment.

Stage 4: The NEG and Onwards (2018–2020)

Carbon efficiency developments in 2018–2020 showed considerable inconsistencies. Both strong and weak carbon efficiency performers fell significantly, with strong performers dropping to 18.1%, a decrease of 13.7% from Stage 3. Golden Stars were nearly halved, and the total number of companies improving their carbon efficiency dropped to 38.4%, nearly 20% lower than the previous stage. These results may suggest that years of policy and regulatory uncertainties took a toll on carbon efficiency improvement, as research has confirmed (Bui & Fowler, 2019). In contrast, the proportion of Red Question Marks increased by 17.8%, becoming the dominant category again (38.4%). The deterioration of economic performance seemed to be a major hindrance to carbon efficiency development, despite these companies' efforts in improving environmental performance.

Comparison of High and Low Environmental Sensitivity Industries

Corporate reactions to legislative changes for carbon reduction are unlikely to be homogenous across industry sectors (Clarkson et al., 2015) Highly sensitive sectors such as energy and mining face greater exposure to regulation and political costs (Qian et al., 2018). Heavily polluting industries often risk a breach of public legitimacy, as they are perceived to bear the most costs and responsibility for improving environmental performance (Cho & Patten, 2007). Chapple et al. (2013) found that high-energy-intensive firms were penalised by the equity market during the ETS initiative, with market value reductions of 7–10%. Firms in high environmental sensitivity industries showed consistently negative abnormal returns after green policy announcements in Australia, with little evidence that investors rewarded clean energy companies (Ramiah et al., 2013). Thus, highly sensitive industries may suffer the most and exhibit more negative reactions during each episode of carbon policy changes.

Table 16.5 reports the carbon efficiency development among high and low environmental sensitivity industries during the four stages of carbon regulatory changes.

The comparison of carbon efficiency development between high and low environmental sensitivity industries reveals several notable similarities and differences. In highly sensitive industries, carbon efficiency trends fluctuated significantly among company categories. During the initial regulatory stage (2009–2011), a considerable proportion of companies – particularly Golden Stars and Red Question Marks – achieved efficiency improvements, though the latter struggled in later stages. The introduction of the carbon tax (2012–2014) had a pronounced impact on these industries, with approximately 47.4% achieving improvements, an 18% decrease from the previous period. The tax heavily penalised highly sensitive industries, leading to a

Table 16.5 Comparison between high and low environmental sensitivity industries

	Stage 1: The ETS initiative (2009–2011)		Stage 2: The carbon tax bill (2012–2014)		Stage 3: The repeal of the carbon tax (2015–2017)		Stage 4: The NEG and onwards (2018–2020)	
	High	Low	High	Low	High	Low	High	Low
	%	%	% (% change)	% (% change)	% (% change)	% (% change)	% (% change)	% (% change)
Panel A: Efficient categories								
Strong carbon efficiency development								
Golden Stars	24.0	37.5	23.3 (-0.7)	25.8 (-11.7)	25.0 (+1.7)	25.9 (+0.1)	11.0 (-14.0)	23.4 (-2.5)
Green Stars	6.7	3.1	6.0 (-0.7)	6.5 (+3.4)	5.3 (-0.7)	8.6 (+2.1)	1.1 (-4.2)	6.4 (-2.2)
Subtotal	30.7	40.6	29.3 (-1.4)	32.3 (-8.3)	30.3 (+1.0)	34.5 (+2.2)	12.1 (-18.2)	29.8 (-4.7)
Weak carbon efficiency development								
Greedy Cash Cows	30.7	25.0	16.4 (-14.3)	24.2 (-0.8)	19.6 (+3.2)	17.2 (-7.0)	15.4 (-4.2)	14.9 (-2.3)
Green Question Marks	4.0	3.1	1.7 (-2.3)	6.4 (+3.3)	4.5 (+2.8)	13.8 (+7.4)	3.3 (-1.2)	8.5 (-5.3)
Subtotal	34.7	28.1	18.1 (-16.6)	30.6 (+2.5)	24.1 (+6.0)	31.0 (+0.4)	18.7 (-5.4)	23.4 (-7.6)
Total	65.4	68.7	47.4 (-18.0)	62.9 (-5.8)	54.4 (+7.0)	65.5 (+2.6)	30.8 (-23.6)	53.2 (-12.3)
Panel B: Inefficient categories								
Minor carbon efficiency deterioration								
Dirty Cash Cows	2.7	9.4	4.3 (+1.6)	1.6 (-7.8)	6.3 (+2.0)	1.7 (+0.1)	6.6 (+0.3)	0.0 (-1.7)
Red Question Marks	21.3	15.6	30.2 (+8.9)	22.6 (+7.0)	20.5 (-9.7)	20.7 (-1.9)	38.4 (+17.9)	38.3 (+17.6)
Subtotal	24.0	25.0	34.5 (+10.5)	24.2 (-0.8)	26.8 (-7.7)	22.4 (-1.8)	45.0 (+18.2)	38.3 (+15.9)
Significant carbon efficiency deterioration								
Dirty Dogs	1.3	0.0	5.2 (+3.9)	3.2 (+3.2)	3.6 (-1.6)	3.5 (+0.3)	1.1 (-2.5)	2.1 (-1.4)
Stressed Dogs	9.3	6.3	12.9 (+3.6)	9.7 (+3.4)	15.2 (+2.3)	8.6 (-1.1)	23.1 (+7.9)	6.4 (-2.2)
Subtotal	10.6	6.3	18.1 (+7.5)	12.9 (+6.6)	18.8 (+0.7)	12.1 (-0.8)	24.2 (+5.4)	8.5 (3.6)

(continued)

Table 16.5 (continued)

	Stage 1: The ETS initiative (2009–2011)		Stage 2: The carbon tax bill (2012–2014)		Stage 3: The repeal of the carbon tax (2015–2017)		Stage 4: The NEG and onwards (2018–2020)	
	High	Low	High	Low	High	Low	High	Low
Total	34.6	31.3	52.6 (+18.0)	37.1 (+5.8)	45.6 (-7.0)	34.5 (-2.6)	69.2 (+23.6)	46.8 (+12.3)
N	75	32	116	62	112	58	91	47

decrease in Greedy Cash Cows and an increase in Red Question Marks. The repeal of the carbon tax (2015–2017) yielded mixed results, with improvements mainly among Golden Stars, while more Red Question Marks (-9.7%) shifted towards Greedy Cash Cows or Dirty Cash Cows. This finding suggests that the repeal negatively impacted carbon performance, as financial relief reverted heavy polluters back to being cash cows. During the NEG discussion stage (2018–2020), highly sensitive industries experienced an overall decline in efficiency, marked by increases in Red Question Marks (+17.9%) and Stressed Dogs (+7.9%).

In contrast, low environmental sensitivity industries demonstrated a more stable carbon efficiency trend. During the earliest period, these industries achieved higher improvement rates, with approximately 68.7% improving efficiency compared to 65.4% in highly sensitive industries. The impact of the carbon tax was less pronounced, with fewer companies declining in efficiency, although Green Question Marks increased. Unlike the significant reduction among highly sensitive firms, the proportion of carbon efficiency performers in low-sensitivity industries only decreased by 5.8% during the carbon tax period. The repeal of the carbon tax led to a decrease in Greedy Cash Cows and an increase in Green Question Marks, suggesting that companies in low-sensitivity industries began prioritising carbon management over financial concerns. In the NEG stage, low-sensitivity sectors also faced carbon efficiency challenges, with a significant rise in Red Question Marks, although most companies maintained or improved their performance. Overall, low-sensitivity industries demonstrated better carbon efficiency outcomes than high-sensitivity industries, reinforcing the finding that they managed carbon efficiency more effectively.

The findings highlight that Australia's frequent changes in carbon policy have impacted corporate efforts to improve carbon efficiency. This instability has resulted in inconsistent carbon-efficiency development among top carbon emitters, who often prioritise economic performance over environmental performance. Despite some companies showing improvements, particularly during periods of more consistent regulatory signals, the lack of long-term, stable policy has limited sustained environmental progress. To achieve significant and lasting improvements in carbon efficiency, Australia may need to adopt a more predictable regulatory approach that encourages companies to integrate sustainability into their core strategies without the disruption of frequent policy changes.

Conclusions

The findings show that over half (53.2%) of these emitters improved their carbon efficiency over the decade examined, though only 27.1% achieved strong performance in both carbon and economic outcomes. The analysis reveals a stronger economic motivation behind these improvements, with fewer companies focused solely on environmental gains. Economic drivers were especially pronounced among companies achieving higher efficiency, while others struggled with both financial and environmental objectives. Some companies, aiming to avoid a 'dirty' image, prioritised economic performance over environmental progress, though most still committed to reducing carbon footprints towards carbon neutrality.

Across the four stages of policy shifts, carbon efficiency proved to be dynamic, reflecting regulatory and economic conditions. Early regulatory movement saw over two thirds of companies enhancing their carbon efficiency, but momentum weakened when carbon policy seemed uncertain, especially during the NEG discussion. The ETS proposal initially drove efficiency through economic gains, whereas the carbon tax introduced compliance costs that slowed progress. After the tax's repeal, companies maintained their carbon efficiency levels, but the NEG's failure led to a 15% decline in high-performing companies, illustrating the negative impact of prolonged regulatory uncertainty.

Sectoral comparisons show that companies in low environmental sensitivity industries managed carbon efficiency more stably than those in high-sensitivity sectors, consistent with prior research (Chapple et al., 2013; Qian et al., 2018). The carbon tax penalised high-sensitivity sectors, reducing Greedy Cash Cows and increasing Red Question Marks. In contrast, low-sensitivity sectors experienced less severe impacts but saw a rise in Green Question Marks. During the NEG discussion, high-sensitivity industries faced declines, while low-sensitivity companies largely maintained or improved their performance.

This study clarifies a decade of carbon performance changes, achievements, and challenges, offering insights for future strategy. Despite progress toward sustainability, most companies prioritised economic growth over environmental improvement. If financial performance remains the focus, carbon efficiency and environmental innovation will stay limited. A shift toward balanced economic and environmental goals is needed to boost investment in solutions that support both growth and sustainability. The findings of this study also clearly indicate a lack of legitimacy pressure, especially among highly sensitive and polluting companies as they continued to perform poorly without fully aligning with societal expectations to gain acceptance or approval from stakeholders, despite being subject to much greater public scrutiny compared to less sensitive companies over time. Therefore, this highlights the need for stronger, consistent regulatory frameworks to drive deeper commitments from these major polluters.

There are several limitations to this study. We focused on the Australian context and included only top carbon emitters registered with the NGER scheme. Therefore, caution must be exercised in generalising the study findings. Additionally, we only examined the companies' carbon efficiency changes within the time frame studied. We do not consider other factors that may impact corporate carbon efficiency development, such as company size, industry, and cash flow management. Future studies could broaden the scope by including a larger number of companies in the sample or by extending the investigation of carbon efficiency developments to encompass other countries and regions. These approaches would yield a more comprehensive understanding of corporate carbon efficiency development. Moreover, employing alternative research methods in future studies could enhance control over factors that might influence corporate carbon efficiency development, thus leading to more robust and reliable results.

References

Al-Najjar, B., & Anfimiadou, A. (2012). Environmental policies and firm value. *Business Strategy and the Environment, 21*(1), 49–59. https://doi.org/10.1002/bse.713

Berrone, P., & Gomez-Mejia, L. R. (2009). Environmental performance and executive compensation: An integrated agency-institutional perspective. *Academy of Management Journal, 52*(1), 103–126. https://doi.org/10.5465/amj.2009.36461950

Birchall, S. J., Ball, A., Mason, I., & Milne, M. J. (2013). Managing carbon in times of political change: The rise and fall of the New Zealand Carbon Neutral Public Service program. *Australasian Journal of Environmental Management, 20*(1), 63–78. https://doi.org/10.1080/14486563.2012.720455

Bui, B., & Fowler, C. J. (2019). Strategic responses to changing climate change policies: The role played by carbon accounting. *Australian Accounting Review, 29*(2), 360–375. https://doi.org/10.1111/auar.12213

Chapple, L., Clarkson, P. M., & Gold, D. L. (2013). The cost of carbon: Capital market effects of the proposed emission trading scheme (ETS). *Abacus, 49*(1), 1–33. https://doi.org/10.1111/abac.12006

Cho, C. H., & Patten, D. M. (2007). The role of environmental disclosures as tools of legitimacy: A research note. *Accounting, Organizations and Society, 32*(7–8), 639–647. https://doi.org/10.1016/j.aos.2006.09.009

Clarkson, P. M., Li, Y., Pinnuck, M., & Richardson, G. D. (2015). The valuation relevance of greenhouse gas emissions under the European Union carbon emissions trading scheme. *European Accounting Review, 24*(3), 551–580. https://doi.org/10.1080/09638180.2014.927782

Cooper, S. A., Raman, K. K., & Yin, J. (2018). Halo effect or fallen angel effect? Firm value consequences of greenhouse gas emissions and reputation for corporate social responsibility. *Journal of Accounting and Public Policy, 37*(3), 226–240. https://doi.org/10.1016/j.jaccpubpol.2018.04.003

Cowan, S., & Deegan, C. (2011). Corporate disclosure reactions to Australia's first national emission reporting scheme. *Accounting & Finance, 51*(2), 409–436. https://doi.org/10.1111/j.1467-629X.2010.00361.x

Crowley, K. (2017). Up and down with climate politics 2013–2016: The repeal of carbon pricing in Australia. *WIREs Climate Change, 8*(3), https://doi.org/10.1002/wcc.458

Firsova, A., Strezov, V., & Taplin, R. (2012). After 20 years of creating Australian climate policy: Was the proposed carbon pollution reduction scheme a change in direction? *Australasian Journal of Environmental Management, 19*(1), 21–34. https://doi.org/10.1080/14486563.2011.646751

Guenster, N., Bauer, R., Derwall, J., & Koedijk, K. (2011). The economic value of corporate eco-efficiency. *European Financial Management, 17*(4), 679–704. https://doi.org/10.1111/j.1468-036X.2009.00532.x

Horváthová, E. (2010). Does environmental performance affect financial performance? *A Meta-Analysis. Ecological Economics, 70*(1), 52–59. https://doi.org/10.1016/j.ecolecon.2010.04.004
Joshi, A., Singh, R., Kumar, V., Rani, S. (2024). *Net Zero Strategy: Exploring Opportunities, Confronting Challenges for Developing Countries*. In: Singh, R., Crowther, D. (eds) Transition Towards a Sustainable Future. Approaches to Global Sustainability, Markets, and Governance. Springer, Singapore. https://doi.org/10.1007/978-981-97-5756-5_11
Kumar, K., Singh, R., Sharma, A., Lehri, B.P.S. (2024). Climate change adaptation strategies for achieving Net-Zero economy. In: Singh, R., Crowther, D. (eds.) Transition Towards a Sustainable Future. Approaches to Global Sustainability, Markets, and Governance. Springer, Singapore. https://doi.org/10.1007/978-981-97-5756-5_8
Nwokolo, S.C., Singh, R., Khan, S., Kumar, A., Luthra, S. (2023a). *Impacts of Climate Change in Africa. In: Africa's Path to Net-Zero*. CSR, Sustainability, Ethics & Governance. Springer, Cham. https://doi.org/10.1007/978-3-031-44514-9_7
Nwokolo, S.C., Singh, R., Khan, S., Kumar, A., Luthra, S. (2023b). *Global Investment and Development in Africa*. In: Africa's Path to Net-Zero. CSR, Sustainability, Ethics & Governance. Springer, Cham. https://doi.org/10.1007/978-3-031-44514-9_2
Nwokolo, S.C., Singh, R., Khan, S., Kumar, A., Luthra, S. (2023c). *Africa's Awakening to Climate Action. In: Africa's Path to Net-Zero*. CSR, Sustainability, Ethics & Governance. Springer, Cham. https://doi.org/10.1007/978-3-031-44514-9_9
Nwokolo, S.C., Singh, R., Khan, S., Kumar, A., Luthra, S. (2023d). *Remedies to the Challenges of Renewable Energy Deployment in Africa*. In: Africa's Path to Net-Zero. CSR, Sustainability, Ethics & Governance. Springer, Cham. https://doi.org/10.1007/978-3-031-44514-9_3
Nwokolo, S.C., Singh, R., Khan, S., Kumar, A. (2024). Threats to the rapidity of sustainability transitions posed by technological changes. In: Technological Pathways for Africa' s Net-Zero Economy. https://doi.org/10.1016/B978-0-443-31486-5.00002-1
Qian, W., & Schaltegger, S. (2017). Revisiting carbon disclosure and performance: Legitimacy and management views. *The British Accounting Review, 49*(4), 365–379. https://doi.org/10.1016/j.bar.2017.05.005
Qian, W., Kaur, A., & Schaltegger, S. (2018). Managing eco-efficiency development for sustainability: an investigation of top carbon polluters in Australia, in *Accounting for Sustainability: Asia Pacific Perspectives*: 103–124. https://link.springer.com/chapter/https://doi.org/10.1007/978-3-319-70899-7_5
Qian, W., Suryani, A., and Xing, K. (2020). Does carbon performance matter to market returns during climate policy changes? Evidence from Australia. *Journal of Cleaner Production, 259*, https://doi.org/10.1016/j.jclepro.2020.121040
Ramiah, V., Martin, B., & Moosa, I. (2013). How does the stock market react to the announcement of green policies? *Journal of Banking & Finance, 37*(5), 1747–1758. https://doi.org/10.1016/j.jbankfin.2013.01.012
Schaltegger, S., & Burritt, R. (2000). *Contemporary environmental accounting: Issues, concepts and practice*. Greenleaf Publishing.
Schaltegger, S., & Burritt, R. (2018). Business cases and corporate engagement with sustainability: Differentiating ethical motivations. *Journal of Business Ethics, 147*, 241–259. https://doi.org/10.1007/s10551-015-2938-0
Wahba, H. (2008). Does the market value corporate environmental responsibility? An empirical examination. *Corporate Social Responsibility and Environmental Management, 15*(2), 89–99. https://doi.org/10.1002/csr.153
Walley, N., & Whitehead, B. (1994). It's not easy being green. *Harvard Business Review, 72*(3), 46–52. https://hbr.org/1994/05/its-not-easy-being-green
Zhou, Y., Luo, L., & Shen, H. (2022). Community pressure, regulatory pressure and corporate environmental performance. *Australian Journal of Management, 47*(2), 368–392. https://doi.org/10.1177/03128962211017172

Chapter 17
Climate Change and the Path to Climate Neutrality: Understanding Causes, Impacts, and Action Strategies

Rubee Singh, Seema Rani, and Walter Leal Filho

Abstract Climate neutrality (CN) is a target to reduce the greenhouse gas emissions to zero for combating the worldwide issue of climate change (CC). CC is a term that describes the alterations in climate and weather patterns that are occurring all over the world as because of the global warming of the atmosphere, which is produced by greenhouse gases. The effects of climate change, which include severe temperatures, intense storms, droughts, and floods, are being felt universally over the globe. This chapter makes an attempt to offer a descriptive picture on the basics of CC by reviewing the existing literature. The authors minutely presents the causes and impacts of CC in order bring a clearer perspective on the topic to broaden our knowledge on the same. With the intention of providing readers, who are less familiar with climate change, with an introduction that is not technical in nature, the significance of the 1.5 °C and 2 °C targets is explained, as well as the reasons why the process of controlling climate change must be expedited. An explanation of the potential effects that CC along with a briefing on climate action is given. The strategies for CC mitigation and adaption are also mentioned to enhance the understanding on the topic. Finally, implications are offered for policy as well as practice along with study's limitations and scope for further research.

Keywords Climate action · Climate change · Climate neutrality · Net zero · Sustainability

---

R. Singh
Institute of Business Management, GLA University, Mathura, Uttar Pradesh, India

Department of Management Studies, Kumaun University, Nainital, Uttarakhand, India

S. Rani (✉)
Department of Commerce, Kurukshetra University, Haryana, India
e-mail: rajpootseema11@gmail.com

W. L. Filho
Faculty of Life Sciences, Humburg University of Applied Sciences, Humburg, Germany
e-mail: walter.leal2@haw-hamburg.de

R. Singh and W. L. Filho (eds.), *Climate Neutrality Through Smart Eco-Innovation and Environmental Sustainability*, Climate Change Management,
https://doi.org/10.1007/978-3-031-83250-5_17

Introduction

Significant changes, including actual and expected climatic changes and global warming, have occurred over the past 65 years in the present century globally. Climate change (CC) is a worldwide issue affecting numerous aspects across different fields such as environment, society, and economy (Feliciano et al., 2022; Raihan, 2023). It entails increased temperatures in many regions along with extreme weather conditions such as heavy rains, droughts, and floods (Schuurmans, 2021). The industrial transformations coupled with technical innovations significantly exacerbated the issue of Earth's climate (Leppänen et al., 2014). It is implausible to ascertain the precise impacts of CC on a sectoral level for immediate treatment and appropriate measures to enhance the likelihood of mitigating its catastrophic effects (Izaguirre et al., 2021), as evidenced by the rising acknowledgment and incorporation of CC for designing policies at national as well as local levels (Fawzy et al., 2020). CC is labelled as drastic changes in precipitation and temperature for a long term, together with various factors such as pressure and humidity present in the atmosphere. Furthermore, erratic meteorological conditions, the melting glaciers, and the resultant rise in sea levels are prominent consequences of CC across the globe (Lipczynska-Kochany, 2018).

Before the industrial revolution, the primary sources of greenhouse gases (GHGs) such as carbon dioxide (CO^2), methane (CH^4), and nitrous oxide (N^2O) released into the atmosphere were mainly the remained the natural events including wildfires and volcanic outbreaks (Sovacool et al., 2021). The United Nations Framework Convention on Climate Change (UNFCCC) established an accord to address CC and enhance the necessary funding and actions for a sustainable future with low or zero carbon, at the Paris Conference on December 12, 2015 which is commonly referred as Paris Agreement (Kumar et al., 2024). The agreement builds upon the Convention with an attempt to unite all states for a collective effort to implement rigorous strategies to mitigate CC and counter all its effects, accompanied by enhanced financial support for developing countries (Abbass et al., 2022). This pointed towards a historical moment in the worldwide battle against CC. The agreement primarily focused on enhancing the response of al member countries to the threats posed by CC by maintaining the escalation in temperature largely below 2 °C relative to pre-industrial levels and striving to limit it to 1.5 °C for this century (Sharma et al., 2020). Moreover, it also targets to enhance governments' ability to address the impacts of CC and synchronize financial movements with low GHG emissions and paths which are climate-resilient (Shahbaz et al., 2019).

To attain these ambitious objectives, sufficient economic means needs to be organised and allocated, alongside the establishment of a new technological framework, enabling the developing nations to operate within their own priorities. The agreement additionally offers a clear and easier framework for measures and assistance. All members are mandated to exert maximum effort under "nationally determined contributions" (NDCs) and to keep enhancing their initiatives for the future time period (Balsalobre-Lorente et al., 2020). It encompasses requirements for all participating countries to consistently publicize their emissions and actions taken in this direction. It also specifies that collective advancement towards the agreement's objective will be assessed after each five year for evaluating the effectiveness of plans and guide the subsequent actions for further improvement. The agreement was opened to all countries in 2016 on 22nd of April coinciding with Earth Day, at Headquarters of UN in New York for signing in order to become the member. In the same year, on 4th of November, it became effective after a month when the twofold threshold was achieved where it was ratified by 55 countries representing around 55% of global emissions). Since then, additional countries have become members of the pact, resulting in 195 members at present as per the UNFCCC website. An action plan was established for delineating methods, processes, and ideas for the comprehensive implementation (Murshed et al., 2021).

The strategies for mitigating and adapting CC for reducing emissions encompasses establishing long-term objectives temperature control, as this would mitigate risks and impacts of CC in the most effective ways along with acknowledging the necessity of minimizing the global emissions. Every action here needs to be taken with an understanding that this may require more time for developing nations; and lastly, implementing rapid reductions thereafter, guided by the top accessible systematic methods for achieving a harmony amongst emissions and removals (Kyriakopoulos & Sebos, 2023; Nwokolo, 2023a). Conversely, certain adaptation techniques include enhancing the capacity of societies for addressing the impacts of CC and perpetuating and augmenting international backing for the adaption of developing nations. Nonetheless, manmade activities are presently considered the primary contributors to CC (Murshed et al., 2022). In addition to this, other human-induced activities encompass intensive agricultural practices, characterized by significant reliance on technology, fuel-based automation, and incineration of agrarian remainders, combustion of fossil fuels, desertification, and both national and domestic transport divisions (Huang et al., 2016; Nwokolo et al., 2024). Accordingly, these human deeds result in climate catastrophes, impairing the infrastructure, social health, and overall output.

Heightened demands for energy has escalated GHG levels in relation to rising temperatures, as the in most of the developing countries energy generation is undertaken from fossil fuels (Balsalobre-Lorente et al., 2022). Present chapter seek out to offer a comprehensive analysis of CC, its causes, and its detrimental effects on critical sectors, including temperature, oceans, agriculture, biodiversity, and health, while proposing practical preventive measures and mitigation strategies as viable alternatives to withstand the impacts of CC. The main objective is to assess the current state of research on "Global Climate Change" through a comprehensive evaluation of published studies. This study starts with a literature review section, detailing the basic concepts of climate neutrality and climate change along with its causes and impacts. The climate action is also discussed under this section. Then a research methodology section is given followed by implications sections followed by conclusion of the study that discusses future research avenues as well as limitations.

Literature Review

Climate Change and Climate Neutrality (CC and CN)

CC is characterized as the alteration in normal patterns of weather primarily resulting from emissions of GHGs. These emissions results into entrapment of heat within the atmosphere of our planet, serving as the major catalyst for global warming. The natural systems and anthropogenic activities are the main sources of such emissions. The natural causes encompass everything from volcanic eruptions, wildfires, to earthquakes and changes in the earth's orbit (Yue & Gao, 2018), whereas human causes primarily pertain to production of energy, high rate of industrialization, and activities associated with the use of land (Raihan, 2023). An analysis of related to the emissions of GHGs in atmosphere due to aforementioned causes concluded that the Earth has a natural ability which works as a balancing entity, whereas the human induced emissions exert added pressure on this ability of the planet (Liang et al., 2019).

The concept of CN describe the goal of achieving net zero GHG emissions through creating a balance between the emissions to make them equal to, or less than, those are removed (Kyriakopoulos & Sebos, 2023). Additionally, climate neutrality takes into account the regional or local bio-geo-physical effects of human activities, such as changes in surface albedo or local climate. In the most fundamental sense, it means that we mitigate our emissions through climate action in order to guarantee that we have no overall impact on the climate system. As part of this, the UNFCCC launched "Climate Neutral Now" in 2015 with the intention of encouraging stakeholders all over the world to strive towards goals of net zero and a world that is climate neutral (Wei et al., 2022). The goal of Climate Neutral Now is to establish a world that is "climate neutral" by the year 2050; however, the focus of this initiative is on the necessity of taking action right now in order to achieve this goal.

Causes of Climate Change: Natural Causes vs. Human Causes

Prior to the Industrial Era, before eighteenth century, natural forces consistently impact the climate and account for temperature variations. Yet, recent climatic changes should not be completely linked to the natural phenomena. Nowadays, anthropogenic actions have extensively released GHGs including CH^4 and CO^2 in the environment which causes CC.

Natural Causes

Variations in solar radiation and explosions of volcanos have a direct effect on the Earth's temperature. Alterations in the axial spin and orbit of earth have significantly influenced historical climatic patterns. The orbit's variations greatly impact the sunlight quantity in the Northern Hemisphere during summer which is viewed as the principal cause behind historical ice age cycles, characterized by extended durations of cold temperatures and shorter interglacial intervals of comparatively warmer temperatures (National Academy of Sciences, 2020). In case of coolest phase of the past icy epoch, the standard over-all temperature was about 11°F lower than the present day. Similarly, for the apex period of interglacial time, the temperature was no more than 2°F higher compared to present (Fahey et al., 2017). Variations in the quantity of can solar light influence its magnitude as well that reaches the surface of land. Although such alterations may affect the temperature, no impact of these variations is witnessed during the past few decades. The satellites are quantifying the energy received by the Earth from the Sun since 1978 (Liang et al., 2019). The data from these satellites have not indicated any augmentation in the sun's output, despite the rise in world surface temperatures. The quantity of sunlight received or reflected by the globe is contingent upon the surface and atmosphere of our planet. The forest, seas, and land, generally absorb a greater amount of sunlight because they are dark. While the objects and surfaces with light hues, such as vapours and ice, typically reflect sunlight. Approximately two third of the sunlight reaching the surface is absorbed by the dark objects and hence it can be said that the natural alterations in the surface of our planet has let to CC. (Fahey et al., 2017).

Volcanoes also influence climate, and historical volcanic eruptions emitted CO^2 in substantial volumes. Certain explosions of volcanos might eject elements in the stratosphere, where these may replicate ample sunlight back in the universe that causes an effect of cooling down the surface not for a shorter period but for a longer period of many years (Hagen & Azevedo, 2023). Volcanic particulates arising from a singular outburst have no long-term climatic alterations as they are very short-lived than GHGs. Moreover, anthropogenic activities release many folds of the amount of CO^2 formed by volcanoes annually (Gerlach, 2011). Throughout the past several hundred thousand years, CO^2 concentrations fluctuated in conjunction with glacial cycles. CO^2 levels were elevated during warm interglacial eras. During chilly glacial epochs, carbon dioxide concentrations were decreased. The thermal alteration of the land and oceans may induce modifications in the natural foundations of GHGs, in this manner changing their atmospheric concentrations (Fawzy et al., 2020). The fluctuations in these concentrations is considered as a beneficial environmental response, intensifying the changes in temperature induced by prolonged alterations in the Earth's orbit.23.

2 Human Causes

Human activity has indisputably elevated the temperatures of the atmosphere, ocean, and land. Researchers collected a comprehensive record of the climate through an examination of the several pointers, including sea level, glaciers, forests, and ice sheets along with investigating alterations in the Earth's orbital dynamics (Wuebbles et al., 2017). All the statistics in this regard shows that changes occur in the climate fluctuates over long time period; nevertheless, the rising temperature level since the mid-twentieth century are not considered. It is highly probable (> 95%) that primary cause of this increased heat is linked with human actions (IPCC, 2013a). Human actions have significantly exacerbated the CC through of GHGs emissions as well as alteration on reflectivity or absorption of solar energy.

Concentrations of principal GHGs have risen since the Industrial Revolution because of the anthropogenic activity. The quantities of GHGs in the atmosphere are currently higher compared to a specific point of time in the previous 8 lakh years (Fahey et al., 2017). The augmentation of GHG discharges led to intensified greenhouse effect, as a consequence of heightened temperature of the planet. The combustion of fossil fuels has a major role in the CC in contrast to any other human related acts. Over 30 billion tons of CO^2 is emanated annually in the environment (Hayhoe et al., 2018). The level of CO^2 present in the atmosphere is nearly 40 percent higher than pre-industrial times, increased roughly to 414 parts per million (ppm) from 280 ppm within the earlier two centuries (National Oceanic and Atmospheric Administration, 2021). In the same way, human activities elevated methane concentrations over 2.5 times, to 1,867 ppb rising from around 722 ppb (Sahu, 2023). Likewise, around 20 percent increase is observed in N^2O since then, with a notable acceleration to 332 ppb from 270 ppb (National Oceanic and Atmospheric Administration, 2021).

Further, activities like farming, construction of roads, and tree planting has the potential to alter the reflectance of the surface, resulting in localized cooling or warming. This phenomenon occurs in high temperature landmasses that exhibit higher temperatures as compared to their nearby less populous regions (Liu et al., 2021). Although the process of cutting forests and making them available for farming may enhance the Earth's albedo by substituting dark surfaces to lighter agricultural ones, leading to a minor cooling effect (Fahey, 2017). The release of tiny particles, termed aerosols, may help in reflecting solar energy. Various categories of air contaminants engage in atmospheric chemical interactions to form aerosols. In summary, anthropogenic aerosols influence the heat level and making the planet a bit cooler.

Impacts of Climate Change

CC is occurring as the normal temperature of our planet has risen approximately 1.7°F in the past few decades (National Academy of Sciences, 2020). Variations in the temperature even of one or two degrees might induce hazardous alterations in overall weather conditions. The tangible, obvious alterations are called CC impacts, as they represent the evident effects of extreme changes in the ecological environment on the planet. Several regions have undergone changes in the rainfall levels, giving rise to the occurrences of heavy rainfall, floods, and droughts, or alongside severe heatwaves (CRED, 2019). The glaciers and oceans have also undergone alterations as increasing acidity and temperature level are observed in the oceans, while melting glaciers are leading to increased sea levels. All such changes turn out to be increasingly evident in the forthcoming times, they will undoubtedly pose problems to the atmosphere and society at large.

As GHG levels rise, the atmospheric temperature rise alongside. The decade from 2011 to 2020 is the hottest recorded period (World Meteorological Organization, 2023). After 1980s, each following decade has demonstrated higher temperatures than it was before. Almost all terrestrial regions are experiencing intensified heat waves and increased number of hot days. All these has led to exacerbated heat-related illnesses and complicate open-air work. Wildfires ignite more readily and propagate more swiftly under elevated temperatures. Arctic temperatures is rising at twice the rate of the global average (CRED, 2019). Destructive storms have intensified and increased in frequency throughout numerous regions. Increasing temperatures lead to heightened evaporation of moisture, intensifying intense rainfall and flooding, hence resulting in more devastating storms (IPCC, 2013b). The warming ocean impact the intensity and frequency of tropical storms as typhoons, hurricanes, and cyclones derive energy from warm waters of the ocean's surface. Such storms frequently devastate residences and communities, resulting in fatalities and significant economic damage.

CC is altering water supply, rendering it scarcer in numerous regions. The issue of water scarcity is intensified due to global warming in countries already under water stress it increases the likelihood of droughts which greatly impacts the agricultural yields and then there is ecological droughts amplifying the vulnerability of biomes (Fuss, 2018). Droughts leads to desertification, further reducing the fertilized land for available for agriculture. A large share of human population is confronting the peril of insufficient water availability on a consistent basis. As oceans are absorbing the extreme heat, it has multiplied the ocean warming rate which has significantly accelerated over the past two decades at all depths (Venegas et al., 2023). As the temperature of ocean increases, its quantity compliments due to the thermal spreading out of water. The melting of ice sheets results in elevated sea levels, endangering coastal and island populations. Furthermore, the ocean sequesters CO^2 and its discharge into the environment is prevented. Increased CO^2 levels result in greater ocean acidity, jeopardizing marine organisms and coral reefs (Shi & Li, 2024).

CC threatens the survival of terrestrial and marine organisms. The risks escalate with rising temperatures. Due to CC the globe is seeing a highest loss of biodiversity in history which is 1,000 times higher from all the previous era (Ripple et al., 2023). One million species are at the edge of extinction during the forthcoming decades. Forest fires, severe weather, and exotic pests and illnesses constitute numerous dangers associated with climate change. Certain species will be capable of relocating and surviving, while others will not. CC is the foremost health hazard confronting humanity. CC is adversely affecting almost every aspect related to society and mankind including hunger and malnutrition to the mental health stresses and ailments due to extreme weather events as well as forced displacement from the regions where individuals cannot survive due to changed environmental conditions. Annually, environmental causes result in the deaths of millions of individuals (Sharma et al., 2023). Altered meteorological patterns are proliferating diseases, while extreme weather occurrences elevate mortality rates and challenge healthcare systems' capacity to respond.

Climate Action: Strategies for Mitigation and Adaption

Climate action constitutes a collaborative endeavor to deal with the CC and its repercussions, which encompasses Mitigation which means decreasing GHG emissions that add to CC and adaptation which means readying for and modifying in reaction to the present and impending results of CC (Nwokolo et al., 2023a). Mitigation measures aim to diminish both the beneficial and detrimental effects of CC, therefore alleviating the general demands of accomodation. On the other side, adaptation is inherently discriminatory and leverage beneficial effects while mitigating adverse ones (Klein et al., 2007). Although mitigation and adaptation both aim at combating CC, the intentions of each differs. The former tackles the fundamental origins of CC, specifically the buildup of GHGs, while the latter focuses on managing the consequences of CC. Nevertheless, adaptation alone cannot fully alleviate all adverse effects, highlighting the importance of mitigation strategies in reducing alterations in the climate (Locatelli, 2011). The dimensions of both these initiatives vary, with adaptation benefits predominantly local and mitigation benefits global, even though CC is a global concern.

The literature discusses three primary approaches to climate change mitigation. Initially, traditional mitigation strategies include measures for decarbonization and methods that diminish the emissions of CO^2, including carbon capture and storage, use of nuclear energy and renewable sources of energy, efficiency improvements, and fuel substitution (Bataille et al., 2018). The majority of these methods are tried and tested with a tolerable level of risk involved (Bustreo et al., 2019; Victor et al., 2018). The next approach comprises a novel array of methodologies that are introduced recently and referred as CO^2 removal methods or negative emissions technologies because these strategies may be utilized to absorb and seize CO^2 (Ricke et al., 2017). The principal ones are associated with the use of bioenergy for capturing and storing carbon either directly from air or soil or biochar. It also includes alkalinity enhancement and fertilization in ocean, along with preserving forests and planting new trees. The restoration and construction of wetlands are also a part of this approach. Alternative negative emissions utilization and storage methods are like use of biomass in construction and carbonation of minerals (Goglio et al., 2020; Yan et al., 2019).

A third one is more focused on transforming the radiation stability of the planed through regulating terrestrial and solar radiation. These techniques are referred as "radiative forcing geoengineering technologies" aiming at lowering and stabilizing temperature. It is attained without modifying atmospheric concentrations of GHGs, in contrast to aforementioned measures. The principal methods under this categories encompass space-based reflectors, marine cloud brightening, management of assorted radiation, and stratospheric aerosol injection. All these strategies remain hypothetical or in preliminary trial phases, encompassing significant uncertainty and risk for implementing in practical and therefore these methods are absent from policy agendas (Lawrence et al., 2018; Lockley et al., 2019).

Moreover, adaptation and mitigation differ with time dimensions and the economic sectors involved. Previous research has highlighted the significance of recognizing synergies and adjustments in ecological systems, biomass, and use of land as opposed to societal, economic, and technical frameworks. It is equally essential to assess the other dimensions of these synergies such as time period and area while considering the simulated consequences of foreign commodities. Across sectors different strategies are being adopted to cope with the changing climate (Raihan, 2023). A broad spectrum of agricultural activities entails the utilization of energy water, land, and other resources. Initiatives for climate-resilience in the agriculture sector for conserving the soil and saving it from being desert through improved irrigation, and various farming methodologies to enhance sustainablity, commonly termed "climate-smart agriculture" (Fawzy et al., 2020).

Public transportation can indirectly benefit communities by enhancing resilience to adverse health effects, promoting environmental value and community health, and concurrently decreasing expenses related to transportation and domiciliary healthcare (Kyriakopoulos & Sebos, 2023). Various measures and green infrastructures, including vegetated building roofs grasslands, and, parks serve as substitutes for impermeable planes; each offers advantages of diminished energy consumption and improved weather resilience. The green infrastructures can promote management of rainwater, combating the alleviated temperature level of urban areas, and improve quality of air (Fawzy et al., 2020). Decentralized systems based on renewable resources can provide mitigation advantages and enhance adaption capabilities, hence reducing strain on water-scarce areas (Nwokolo et al., 2023b). Conversely, centralized systems may suffer considerable power loss during high temperatures and severe storms, which coincide with greatest demands for energy.

Research Methodology

This chapter is completely based on published research work. To broaden the understanding of the topic, we discussed basic concept of climate change and climate neutrality. The reasons behind CC are thoroughly deliberated together with the nature induced as well as human caused. Further, the impact of CC on several aspects including atmosphere and health of mankind are explained minutely. Finally, the climate action is described along with the approaches to alleviate the CC and implications are offered for the same.

Implications

On the basis of this review we offer a few considerations for combating climate change. First for public policy such as there is a need for international measures that are more strict and enforced in order to combat concerns related to CC. as per the international frameworks such as the Paris Agreement, it may be necessary for governments to enact more stringent regulations regarding carbon emissions, establish systems for pricing carbon, and enforce more aggressive emission reduction aims (Sovaool et al., 2021). There is a necessity of policy coordination among governments in order to ensure that climate efforts occurring in one location do not impede the progress of other locations. Further, the decisions should not focus solely on adaptation or mitigation but a combination of both strategies. Enhanced communication strategies that make the science of climate change more understandable and practical for a wider audience may be required as a result of this. For the purpose of fosteing a coordinated effort in the fight against CC, this may involve initiatives implemented at the grassroots level, campaigning for legislation, and partnerships with the commercial sector.

Furthermore, numerous advantages of adaptation can be identified for a concise development of urban areas that prioritizes improved approachability, association, and diverse land use. This will not only mitigates exposure to threats like flooding and wildfires, but also safeguards precious ecologies which includes forestry and marshlands. The natural resources can provide critical amenities for mitigating risks associated with extreme heat and floods. A variety of strategies exist for modifying crop and livestock production systems in answer to CC which can yield socio-economic benefits through the creation of new and improved agricultural prospects, particularly for farmers with small landholdings and restricted water supply and synthetic fertilizers as means of adjusting to CC (Abbas et al., 2022).

There are indirect benefits which can be obtained by increased workplace productivity, facilitated by well-structured transport system and the promotion of commercial events near pedestrian-friendly areas. Alongside these initiatives, other significant actions related to the means of transport should be taken that aid in achieving mitigation objectives while enhancing adaptation capabilities encompass enhancing standards for vehicle efficiency, electrification of transport, and promoting carpooling actions. These initiatives would foster resilience in economic and environmental terms by enhancing the capacity to withstand energy shocks through energy efficiency measures.

Green Infrastructure initiatives have garnered significant attention across multiple sectors but are not very common and hence need policy support for mitigating the climate risk. Carbon-sink devices with extended longevity and superior insulating characteristics encompass green roofs and facades need to be promoted. A prolonged life expectancy is crucial for mitigation, given the substantial discharges of GHGs associated with construction supplies. The reductions in these amounts might be derived by cooling effects which decrease the use of infrastructure indirectly. The primary advantages of green infrastructure are food security, managed heating system, proper management of water, social as well as economic resilience, along with improved environmental conditions (Fawzy et al., 2020). The green building initiatives hold great potential that advocate for incorporating of such designs and energy-saving technologies that can help mitigate energy related issues, as well as to diminish anthropogenic CO^2 emissions. Consequently, an intelligent climate strategy that simultaneously tackles mitigation and adaptation within integrated land-use, water, and energy arrangements will enhance chances for recognizing combined effect.

There is a requirement to decentralize and diversify system of supplying energy to mitigate threats and guarantee if a few elements do not work properly than it would not lead to the failure of complete system. Integrating a shadow pricing for CC externalities in substantial schemes for developing infrastructure can result in alterations to adjustment portfolio. Evaluating the real changes in energy consumption as well as strategies for lowering GHG releases are essential. Although the majority of combined valuations concentrate on either specific regions or whole planet, local discussions have commenced investigating collaborations (Kyriakopoulos & Sebos, 2023). A lifecycle perspective may be employed to assess the marginal impacts of utilizing biogenic resources for CO^2 emission reduction, including indirect land use change. Micro-generation and alternative methods of enhancing electrical access can increase energy efficiency in households, mitigate adverse health effects associated with open-fire cooking, and improve study circumstances through superior illumination.

Conclusion

A person or organization can achieve climate neutrality by reducing their emissions to zero, which is a requirement that applies to both individuals and large businesses. It will take a number of years for many organizations to make the necessary technical investments and modify their behaviors in order to achieve climate neutrality, which means that they will not be able to rapidly cut their emissions to zero. In addition to the reductions they may have the option of investing in offsets, which are reductions of emissions that occur outside of their firm. This is an option for those who do not wish to wait. This choice contributes to the acceleration of the climate action across the planet for minimizing GHG emissions and imposes a "carbon price" on their own organization, which provides a financial inducement to them for reducing the emissions more quickly for their own organization. Moreover both mitigation as well as adaptation strategies are essential, as despite substantial mitigation measures, the climate will persist in its evolution in the forthcoming years, requiring adaptation. Here is a need to notice that offsets are not a long-term solution, nor do they take the place of the necessity for you or your organization to reach zero as quickly as possible.

The implications of this chapter are indeed fruitful but the dependence on secondary sources limits the practical significance of the research. Moreover, a few mitigation strategies are discussed in implications to combat the climatic variations across different sector, hence there is a need to examine the effectiveness of various approaches for mitigation and adaptation for developing sustainable habits. It also identified that scarce research is conducted in areas such as climate justice, ecosystem-based techniques for mitigating the CC, and the technological possibility of achieving net-zero emissions warranting future exploration. We also suggest that for addressing the CC successfully, there must be considerable research public involvement and advocacy.

References

Abbass, K., Qasim, M. Z., Song, H., Murshed, M., Mahmood, H., & Younis, I. (2022). A review of the global climate change impacts, adaptation, and sustainable mitigation measures. *Environmental Science and Pollution Research, 29*, 42539–42559. https://doi.org/10.1007/s11356-022-20232-5

Balsalobre-Lorente, D., Driha, O. M., Bekun, F. V., Sinha, A., & Adedoyin, F. F. (2020). Consequences of COVID-19 on the social isolation of the Chinese economy: Accounting for the role of reduction in carbon emissions. *Air Quality, Atmosphere & Health, 13*(12), 1439–1451. https://doi.org/10.1007/s11869-020-00951-3

Balsalobre-Lorente, D., Ibáñez-Luzón, L., Usman, M., & Shahbaz, M. (2022). The environmental Kuznets curve, based on the economic complexity, and the pollution haven hypothesis in PIIGS countries. *Renewable Energy, 185*, 1441–1455. https://doi.org/10.1016/j.renene.2022.01.104

Bataille, C., et al. (2018). A review of technology and policy deep decarbonization pathway options for making energy-intensive industry production consistent with the Paris Agreement. *Journal of Cleaner Production, 187*, 960–973. https://doi.org/10.1016/j.jclepro.2018.03.107

Bustreo, C., et al. (2019). How fusion power can contribute to a fully decarbonized European power mix after 2050. *Fusion Engineering and Design, 146*, 2189–2193. https://doi.org/10.1016/j.fusengdes.2019.03.150

Centre for Research on the Epidemiology of Disasters (CRED). (2019). *Natural disasters 2018*. CRED. https://emdat.be/sites/default/files/adsr_2018.pdf

Fahey, D. W., Doherty, S. J., Hibbard, K. A., Romanou, A., & Taylor, P. C. (2017). Physical drivers of climate change. In Wuebbles, D. J., Fahey, D. W., Hibbard, K. A., Dokken, D. J., Stewart, B. C., & Maycock, T. K., (Eds.), *Climate science special report: Fourth national climate assessment, I* (p. 79). U.S. Global Change Research Program. https://doi.org/10.7930/J0513WCR

Fawzy, S., Osman, A. I., Doran, J., & Rooney, D. W. (2020). Strategies for mitigation of climate change: A review. *Environmental Chemistry Letters, 18*, 2069–2094.

Feliciano, D., Recha, J., Ambaw, G., MacSween, K., Solomon, D., & Wollenberg, E. (2022). Assessment of agricultural emissions, climate change mitigation, and adaptation practices in Ethiopia. *Climate Policy, 22*(3), 1–18. https://doi.org/10.1080/14693062.2022.2075184

Fuss, S., et al. (2018). Negative emissions—part 2: Costs, potentials, and side effects. *Environmental Research Letters, 13*(6), 063002. https://doi.org/10.1088/1748-9326/aabf9f

Gerlach, T. (2011). Volcanic versus anthropogenic carbon dioxide. *Eos, Transactions American Geophysical Union, 92*(24), 201–202.

Goglio, P., et al. (2020). Advances and challenges of life cycle assessment (LCA) of greenhouse gas removal technologies to fight climate change. *Journal of Cleaner Production, 244*, 118896. https://doi.org/10.1016/j.jclepro.2019.118896

Hagen, M., & Azevedo, A. (2023). Influence of volcanic activity on weather and climate changes. *Atmospheric and Climate Sciences, 13*(2), 138–158.

Hayhoe, K., Wuebbles, D. J., Easterling, D. R., Fahey, D. W., Doherty, S., Kossin, J., Sweet, W., Vose, R., & Wehner, M. (2018). Our changing climate. In Reidmiller, D. R., Avery, C. W.,

Easterling, D. R., Kunkel, K. E., Lewis, K. L. M., Maycock, T. K., & Stewart, B. C. (Eds.), *Impacts, risks, and adaptation in the United States: Fourth national climate assessment, volume II* (p. 76). U.S. Global Change Research Program. https://doi.org/10.7930/NCA4.2018

Huang, W., Gao, Q.-X., Cao, G.-L., Ma, Z.-Y., Zhang, W.-D., & Chao, Q.-C. (2016). Effect of urban symbiosis development in China on GHG emissions reduction. *Advances in Climate Change Research, 7*(4), 247–253. https://doi.org/10.1016/j.accre.2016.10.002

IPCC. (2013a). *Climate change 2013: The physical science basis. Working Group I contribution to the fifth assessment report of the Intergovernmental Panel on Climate Change* (Stocker, T. F., Qin, D., Plattner, G.-K., Tignor, M., Allen, S. K., Boschung, J., Nauels, A., Xia, Y., Bex, V., & Midgley, P. M. Eds.). Cambridge University Press.

IPCC. (2013b). *IPCC factsheet: What is the IPCC?* https://www.ipcc.ch/site/assets/uploads/2018/02/FS_what_ipcc.pdf

Izaguirre, C., Losada, I., Camus, P., Vigh, J., & Stenek, V. (2021). Climate change risk to global port operations. *Nature Climate Change, 11*(1), 14–20. https://doi.org/10.1038/s41558-020-00946-x

Klein, R. J., Huq, S., Denton, F., Downing, T. E., Richels, R. G., Robinson, J. B., & Toth, F. L. (2007). Inter-relationships between adaptation and mitigation. In Parry, M., Canziani, O., Palutikof, J., van der Linden, P., & Hanson, C. (Eds.), *Climate change 2007: Impacts, adaptation, and vulnerability. Contribution of Working Group II to the Fourth Assessment Report of the Intergovernmental Panel on Climate Change* (pp. 745–777). Cambridge University Press.

Kumar, K., Singh, R., Sharma, A., & Lehri, B. P. S. (2024). Climate change adaptation strategies for achieving a net-zero economy. In R. Singh & D. Crowther (Eds.), *Transition Towards a Sustainable Future: Approaches to Global Sustainability, Markets, and Governance* (pp. 151–178). Springer. https://doi.org/10.1007/978-981-97-5756-5_8

Kyriakopoulos, G. L., & Sebos, I. (2023). Enhancing climate neutrality and resilience through coordinated climate action: Review of the synergies between mitigation and adaptation actions. *Climate, 11*(5), 105. https://doi.org/10.3390/cli11050105

Lawrence, M. G., et al. (2018). Evaluating climate geoengineering proposals in the context of the Paris Agreement temperature goals. *Nature Communications, 9*, 3734. https://doi.org/10.1038/s41467-018-05938-3

Leppänen, S., Saikkonen, L., & Ollikainen, M. (2014). Impact of climate change on cereal grain production in Russia: Mimeo.

Liang, S., Wang, D., He, T., & Yu, Y. (2019). Remote sensing of earth's energy budget: Synthesis and review. *International Journal of Digital Earth, 12*(7), 737–780.

Lipczynska-Kochany, E. (2018). Effect of climate change on humic substances and associated impacts on the quality of surface water and groundwater: A review. *Science of the Total Environment, 640*, 1548–1565. https://doi.org/10.1016/j.scitotenv.2018.05.277

Liu, Q., Xie, M., Wu, R., Xue, Q., Chen, B., Li, Z., & Li, X. (2021). From expanding areas to stable areas: Identification, classification and determinants of multiple frequency urban heat islands. *Ecological Indicators, 130*, 108046.

Locatelli, B. (2011). Synergies between adaptation and mitigation in a nutshell. In *Climate Change and Forests in the Congo Basin: Synergies between Adaptation and Mitigation* (pp. 11–20). CIFOR.

Lockley, A., et al. (2019). Geoengineering and the blockchain: Coordinating carbon dioxide removal and solar radiation management to tackle future emissions. *Frontiers in Engineering Management, 6*, 38–51. https://doi.org/10.1007/s42524-019-0010-y

Murshed, M., Abbass, K., & Rashid, S. (2021). Modelling renewable energy adoption across South Asian economies: Empirical evidence from Bangladesh, India, Pakistan, and Sri Lanka. *International Journal of Finance & Economics, 26*(4), 5425–5450. https://doi.org/10.1002/ijfe.2041

Murshed, M., Nurmakhanova, M., Al-Tal, R., Mahmood, H., Elheddad, M., & Ahmed, R. (2022). Can intra-regional trade, renewable energy use, foreign direct investments, and economic growth reduce ecological footprints in South Asia? *Energy Sources, Part b: Economics, Planning, and Policy.* https://doi.org/10.1080/15567249.2022.2038730

National Academy of Sciences. (2020). *Climate change: Evidence and causes: Update 2020.* The National Academies Press. https://doi.org/10.17226/25733

National Oceanic and Atmospheric Administration. (2021). *Trends in atmospheric methane.* https://esrl.noaa.gov/gmd/ccgg/trends_ch4

Nwokolo, S. C., Singh, R., Khan, S., & Kumar, A. (2024). Threats to the rapidity of sustainability transitions posed by technological changes. In Nwokolo, S. C., Singh, R., Khan, S., & Kumar, A. (Eds.), *Technological Pathways for Africa's Net-Zero Economy* (pp. 39–75). Elsevier. https://doi.org/10.1016/B978-0-443-31486-5.00002-1

Nwokolo, S. C., Singh, R., Khan, S., Kumar, A., & Luthra, S. (2023b). Remedies to the challenges of renewable energy deployment in Africa. In *Africa's Path to Net-Zero: CSR, Sustainability, Ethics & Governance* (pp. 45–68). Springer. https://doi.org/10.1007/978-3-031-44514-9_3

Nwokolo, S. C., Singh, R., Khan, S., Kumar, A., & Luthra, S. (2023a). Africa's awakening to climate action. In *Africa's Path to Net-Zero: CSR, Sustainability, Ethics & Governance* (pp. 120–135). Springer. https://doi.org/10.1007/978-3-031-44514-9_9

Raihan, A. (2023). A review of the global climate change impacts, adaptation strategies, and mitigation options in the socio-economic and environmental sectors. *Journal of Environmental Science and Economics, 2*(3), 36–58. https://doi.org/10.1016/j.jese.2023.02.002

Ricke, K. L., Millar, R. J., & MacMartin, D. G. (2017). Constraints on global temperature target overshoot. *Scientfic Reports, 7*, 14743.

Ripple, W. J., Wolf, C., Gregg, J. W., Rockström, J., Newsome, T. M., Law, B. E., Marques, L., Lenton, T. M., Xu, C., Hug, S., Simons, L., & King, S. D. A. (2023). The 2023 state of the climate report: Entering uncharted territory. *BioScience, 73*(12), 841–850.

Sahu, T. (2023). Climate change causes, impacts, action plan, and human health in India. *Sustainability, Agri, Food and Environmental Research, 11*, 000–000.

Schuurmans, C. (2021). The world heat budget: Expected changes. *Climate Change*, 1–15. CRC Press. https://doi.org/10.1201/9780367819535

Shahbaz, M., Balsalobre-Lorente, D., & Sinha, A. (2019). Foreign direct investment–CO2 emissions nexus in Middle East and North African countries: Importance of biomass energy consumption. *Journal of Cleaner Production, 217*, 603–614. https://doi.org/10.1016/j.jclepro.2019.01.223

Sharma, A. K., Sharma, M., Sharma, A. K., & Sharma, M. (2023). Mapping the impact of environmental pollutants on human health and environment: A systematic review and meta-analysis. *Journal of Geochemical Exploration*, 107325.

Sharma, R., Sinha, A., & Kautish, P. (2020). Examining the impacts of economic and demographic aspects on the ecological footprint in South and Southeast Asian countries. *Environmental Science and Pollution Research, 27*(29), 36970–36982. https://doi.org/10.1007/s11356-020-09707-4

Shi, Y., & Li, Y. (2024). Impacts of ocean acidification on physiology and ecology of marine invertebrates: A comprehensive review. *Aquatic Ecology, 58*(2), 207–226.

Sovacool, B. K., Griffiths, S., Kim, J., & Bazilian, M. (2021). Climate change and industrial F-gases: A critical and systematic review of developments, sociotechnical systems, and policy options for reducing synthetic greenhouse gas emissions. *Renewable and Sustainable Energy Reviews, 141*, 110759. https://doi.org/10.1016/j.rser.2021.110759

Venegas, R. M., Acevedo, J., & Treml, E. A. (2023). Three decades of ocean warming impacts on marine ecosystems: A review and perspective. *Deep Sea Research Part II: Topical Studies in Oceanography, 212*, 105318.

Victor, N., et al. (2018). The U.S. power sector decarbonization: Investigating technology options with MARKAL nine-region model. *Energy Economics, 73*, 410–425. https://doi.org/10.1016/j.eneco.2018.03.021

Wei, Y. M., Chen, K., Kang, J. N., Chen, W., Wang, X. Y., & Zhang, X. (2022). Policy and management of carbon peaking and carbon neutrality: A literature review. *Engineering, 14*, 52–63.

World Meteorological Organization. (2023). *The global climate 2011–2020: A decade of acceleration*. WMO. https://wmo.int/publication-series/global-climate-2011-2020-decade-of-acceleration

Wuebbles, D. J., Fahey, D. W., Hibbard, K. A., DeAngelo, B., Doherty, S., Hayhoe, K., Horton, R., Kossin, J. P., Taylor, P. C., Waple, A. M., & Weaver, C. P. (2017). Executive summary. In *Climate science special report: Fourth national climate assessment, volume I* (pp. 12–34). U.S. Global Change Research Program. https://doi.org/10.7930/J0DJ5CTG

Yan, J., et al. (2019). Negative emission technologies (NETs). *Applied Energy, 255*, 113749. https://doi.org/10.1016/j.apenergy.2019.113749

Yue, X.-L., & Gao, Q.-X. (2018). Contributions of natural systems and human activity to greenhouse gas emissions. *Advances in Climate Change Research, 9*, 243–252. https://doi.org/10.1016/j.accre.2018.12.003

Chapter 18
Darkside of Climate Neutrality and Smart Eco Innovations in Developed and Developing Countries

Amit Kumar Singh, Vikas Garg, and Pooja Kaushik

Abstract In this research we consider all the issues in climate neutrality as well as the use of smart eco innovations as critical way of solving problems associated with global climate change. It specifies climate neutrality as emission balancing that is achieved by carbon offsetting and carbon capture and storage. While the Paris Agreement calls for cooperation to restrain global warming, bringing about climate neutrality seems a mammoth task, especially for poor developing nations having little money and technological capacity. The chapter stresses on advent of eco innovations such as renewable energy systems, electric vehicles, and the circular economy practices to bring down emissions and promote sustainable development. Critical themes such as over reliance on developing technologies, environmental injustice associated with extracting resources, economic disparities and corporate green washing are addressed. It also examines the 'dark side' of climate neutrality in affluent countries, where reliance on the modern technology can obscure unsatisfactory systemic drivers of environmental degradation. The inability to stick to transition in developing countries exacerbates global equities since it is further hindered by financial restrictions and limited access to green technologies. It encourages global collaboration, equitable resource distribution, decentralised renewable systems, circular economies and inclusive technologies, to bridge these gaps. The result shows that as long as we want to realise net-zero emissions, we need focus on fair issues, ethical issues, as well as justice issues so that there is a fair, fair, fair, fair, fair and sustainable future path for all countries.

A. K. Singh
Department of Commerce, Delhi School of Economics, New Delhi 110007, India
e-mail: aksingh1@commerce.du.ac.in

V. Garg (✉)
Symbiosis Institute of Business Management, Noida, Symbiosis International (Deemed University), Pune, India
e-mail: vikasgargsir@gmail.com

P. Kaushik
Department of Management Studies, Galgotias College of Engineering and Technology, Greater Noida 201310, Uttar Pradesh, India
e-mail: poks3009@gmail.com; pooja.kaushik@galgotiacollege.edu

R. Singh and W. L. Filho (eds.), *Climate Neutrality Through Smart Eco-Innovation and Environmental Sustainability*, Climate Change Management,
https://doi.org/10.1007/978-3-031-83250-5_18

Keywords Climate neutrality · Eco-Innovations · Carbon capture and storage · Circular economy · Greenhouse gas emissions · Renewable energy systems

Introduction

Climate neutrality has been defined as the situation whereby emissions of greenhouse gases are balanced by the rate of the removal so that the net contribution to global warming is nil. This is also called as carbon neutral or net zero emission; this concept has been created much buzz as a principle strategy in combating climate change all across the world. Climate neutrality can be attained by controlling emission through cleaner technology, adopting more use of renewable energy and by option like carbon capture and carbon offset. With the worsening impact of climate change in the world, climate neutrality has now emerged as an environmental sustainable development goal as countries, businesses, and organizations strive to become climate neutral and reduce their Carbon Footprint rating.

The idea of climate niyogeneity refers not only to reducing greenhouse gas emissions from electricity generation or transport. It encompasses a larger framework, a way of dealing with the carbon sinks, the boosters of CO_2, which include forests, and oceans, among others (Eyime & Ushie, 2024). Efforts have been made on the international level to address both the causative factors as well as the adaptation measures in order to realize this goal. For instance, governments are increasingly putting more funds into the development of renewable energy, solar and wind power, encouragement of electric vehicles as well as green buildings to reduce the impact of carbon in cities. Both these strategies assist in progressing societies to the carbon neutral state in which emission accounted at zero or otherwise compensated by afforestation, reforestation or through carbon capture and storage (CCS) technologies (Nwokolo, 2023).

Climate Neutrality and Smart Eco Innovations: Definitions and Scope

Climate neutrality and the utilization of smart eco innovations considered as learning innovations that are oriented toward decreasing negative impact on the environment and toward initiating the process of sustainable development. They are important platforms that help the world in trying to achieve net zero targets (Joshi et al., 2024). It consists in technologies, systems and processes that allow the sectors in question to move from carbon intensive to low carbon circular and sustainable systems.

Climate Neutrality: Concept, Goals, and Mechanisms

Climate neutrality—or net zero emissions or carbon neutrality—means achieving an equal level of emissions of of greenhouse gases (GHGs) into the atmosphere and that absorbed. The goal is in striving to lower the temperature on the global stage and avoiding all the negative impacts of climate change. Climate neutrality or Net Zero, is an elaborate plan that entails a low use of carbon, accompanied by measures such as CCS, Carbon trading markets, transitioning to renewable energy among others (Falk, 2023).

The core mechanisms to achieve climate neutrality include:

1. **Carbon Reduction Technologies:** These include; Renewable energy systems; solar, wind, hydro electric energy, improved energy efficiency in industries, electrification of transport and heat (Gosling & Farley, 2024).
2. **Carbon Offsetting:** It involves compensating for emissions through projects like reforestation or renewable energy development, where the generated carbon savings are used to balance emissions produced elsewhere (Nwokolo, 2023).
3. **Carbon Capture and Storage (CCS):** This technology involves the removal of CO_2 emissions from industries and then the exclusion of these emissions from the atmosphere.
4. **Sustainable Agricultural Practices:** Farming sustains a major contribution to GHG emissions around the world. Several key strategies can be applied to reduce emissions from the sector—regenerative agriculture, practices like planting more trees, less reliance on fertilizers Ie, agroforestry (Nwokolo, 2023).

Literature Review

The Paris Agreement of 2015 is an international treaty to combat climate change that was signed by almost every country in the world. This agreement aimed at climate change below 2 °C and to try to achieve the level below 1.5 °C. Countries pledged to reach a state of zero carbon emissions in the second half of this century by signing the Paris Agreement, but some nations have set more ambitious goals—the European Union is planning to reach carbon neutrality by 2050. Such international efforts are indications that there is the realization that climate change is an international issue which requires individuals from all the world to duplicate. According to this arrangement, the developed countries have pledged to offer assisting funds to the developing countries in their efforts to fair and low carbon economies. This partnership is invaluable in making sure that the developing countries have the financial and technical backing for change and innovation for sustainable development, not having to pay for it through their necks in debt (Sachs & Kharas, 2024a, 2024b, 2024c).

The developing nations are worst affected by climate change, most of them do not have adequate resources that can be invested on renewable resources and green technologies. This have resulted into clamour for climate justice, That developed countries which have contributed immensely to emission of green house gases should bear more responsibility in terms of providing the necessary finance for emission reductions and the promotion of climate change adaptation in the developing world (Sachs, 2023).

Carbon offsetting is one of the most recognized instruments used to attain the state of climate neutrality. This enables companies organizations or even individuals to fund other projects that offset their carbon energy emissions in other projects. For example, offset programs usually cater for afforestation or social renewable energy project in the developing world. Although such programmes offer a flexible means of realising short-run emission reductions, it has attracted criticism because firms are not pressed into making concrete changes in emission-intensive practices (Barker et al., 2023).

It is very important to understand that climate-neutral transitions are not just about emissions slashing. It is also about the management of societal change towards entirely new economic paradigms that reflect sustainability. These comprise areas that would involve change from linear model which disposes off items to a cyclic model where such goods are recycled, mended and reused in order to cycle back and get the resources used to create them back. In detailed terms, the circular economy stands in stark difference to the traditional linear economy in the "take-make-dispose" model. In other words, the circular economy stands in support of the climate neutrality by encouraging recycling and upcycling while minimizing waste (Khan et al., 2024a). Climate mitigation plans involve changing emission through efficiency and cleaner technologies, adopting non-carbon and renewable energy sources and buying carbon credits. (Nwokolo, 2023).

It also includes the regulation of carbon reservoirs such as forests, and seas which are known to sequester large quantities of atmospheric CO_2. Sustainable development strategies both mitigation and adaptation are incorporated in global frameworks to reach this objective. For instance, governments worldwide are increasing the installation of renewable energy, such as solar and wind energy, encouraging the adoption of EVs, and enticing the development of green buildings to reduce urban carbon emissions. About such efforts' goals strive to bring societies on a pathway to becoming carbon neutral through outright reductions of emissions or otherwise through activities like afforestation and through technology such as CCS (Bhushan, 2024).

Global Context: International Usage of Both in Developed and Developing Country

The smart eco innovations have been equally adopted in the developed and the developing countries. Superiority in funding puts such nations in a position to deploy competent technologies, makes infrastructure, invulnerable and hyper modern technologies like electric vehicles, smart grids, and erudite systems of renewable power.

These countries are setting climate neutrality targets and defining the legislation necessary for eco innovations most often. For instance, the European Union has pledged to achieve zero carbon emissions by 2050 and this means that the EU will have to invest significantly in renewable energy, low carbon transport systems, and in greening of industries and buildings (Sachs, 2023).

In this respect, the issue comes down to the difficulties of using these innovations in developing countries. Though most of these countries are now exploring the use of renewable energy, they are hesitated by challenges including inadequate and unreliable infrastructure; lack of capital; and restricted access to technology. The developing nations have a chance to skip the traditional, carbon-fixed technologies directly implementing the more ecologically friendly ones. Emerging solutions such as distributed solar energy and affordable electric transport could provide these countries with a development model (Lee & Kumar, 2023a, 2023b, 2023c). India also futures solar energy; it intends to deliver gigawatts of solar power by 2030 (Sachs, 2023). Nevertheless, the realisation of these developments entails monetary and technological assistance from the developed countries through climate finance, technology sharing and cooperation.

Challenges and Opportunities of Smart Eco Innovations

Although there are definite prospects at Smart eco innovations, the process is far from perfect and filled with obstacles. One of the concerns is high cost per unit of research and development, and highly capital-intensive solutions. While costs of renewable energy technologies have been greatly reduced, fixed costs are still high; especially for developing countries with constrained budgets.

The Dark Side of Climate Neutrality, Particularly as It Applies to Developed Countries

However, the transition to climate neutrality to these countries has not come without black spots as developed countries embracing and adopting Technologies and policies in case of achieving climate neutrality. Several concerns have been proffered in these nations about the attainment of the net zero emission challenges such as-tech overreliance, environment. Horizontal, economic imparity, and corporate greening. These challenges demonstrate potential fallacies of technological determinism when one realizes that radical technology adoption involves more than solving technical problems that exclude societal, economic, and moral factors.

Technological Overdependence

Probably the most important consideration in developed nations is the fact that there is an overt dependency on top-notch technologies as the remedy to climate change. Sometimes CCS, EVs and other novel AI solutions are perceived as critical enablers for reaching Climate Neutrality. However, reliance on advanced technology essentially has some concerns.

First, some of those technologies are only emerging, and CCS and largescale energy storage solutions remain unproven and challenging. Moreover, the technology-generated solutions do not directly respond to the main drivers of environmental problems including overconsumption, environmentally destructive patterns of consumption and production, inequality and unfair distribution of resources in the world. For instance, despite the fact that EVs are environmentally friendly, manufacturing them, particularly processes like extraction of raw material for batteries like lithium and cobalt are highly destructive to the environment (Sorrell, 2023).

The heavily integrated AI and other sophisticated systems to ensure maximum energy utilization and minimum emissions are likely to have weaknesses, for instance, faulty system or misuse of data. Uncontrolled AI driven decisions, may often bias towards economical or corporate benefit leaving social and or environmental justice to the output of economic growth (Binns & Thompson, 2024a, 2024b, 2024c).

Environmental Injustice

Environmental injustice remains a pressing concern in developed countries' pursuit of climate neutrality. Although these nations are progressively adopting green technologies, the global shift toward sustainability often places significant burdens on developing countries, which host relocated carbon-intensive industries and provide raw materials essential for green technologies.

A notable example of environmental injustice is the relocation of carbon-intensive industries. As developed nations impose stricter environmental regulations, polluting industries—such as fossil fuel extraction, manufacturing, and heavy industry—often move to developing countries with more lenient environmental standards. This shift not only transfers environmental challenges to less affluent regions but also leads to serious health and social consequences for local communities. For example, the mining of rare earth elements and other essential materials for green technologies has been associated with environmental degradation and labour exploitation in countries like the Democratic Republic of Congo (Harrison, 2024).

Despite strict regulations governing e-waste recycling in many developed nations, large volumes of discarded electronics are sent to poorer regions, where they are often processed in unsafe, informal settings. This practice not only harms local environments but also deepens health disparities in these areas (Zhang & Liu, 2023).

Economic Inequality and Job Displacement

Other main worries include the economic impact in general that transition to climate neutrality is expected to bring on board, the most important factor being just how many jobs are to be lost in the old industries. The fossil fuel sector has been the lifeblood of most developed economies, and with the shift towards renewable energy, job losses are expected to be significant. Coal miners, oil rig workers, people in traditional automotive industries could lose their jobs when cleaner technologies replace fossil fuel-based systems. Green jobs are not an easy transition; sometimes it is also unfair. While the renewable energy sector is growing, many of these new jobs have different skill sets as well, which creates labour market mismatches (Sorrell, 2023).

Corporate Greenwashing

Another big challenge in reaching climate neutrality is corporate greenwashing. Greenwashing is when a company pretends to be responsible for the environment without changing business operations. On many occasions, the companies proudly claim themselves "carbon–neutral" or "green," based on limited actions they have taken or carbon offsets bought in place of actual emissions reductions. Greenwashing erodes the credibility of real world climate action by undermining both legitimacy and public confidence. Attention and resources funnel away from real sustainable businesses and projects. To exemplify, many multinationals have been discovered, which use imprecise environmental claims or false certification, in order to sell themselves as eco friendly, meanwhile engaging in actions that are ecologically harmful such as deforestation or fossil fuel exploitation (Harrison, 2024).

The Darkside of Climate Neutrality in Developing Countries

Climate initiatives are floating in a ridiculously large dangerous financial borrowing cycle. Although loans can provide the short term support green issues require, long term payments required from such loans can block economic development and constrain the ability of government to hire in priority areas including in areas of health, education and infrastructure. At the same time, these loans tend to have disadvantageous terms—often high interest rates and rigid repayment terms—add to the already hefty burden on national budget (Mishra, 2023a, 2023b).

Wealthy countries or international organisations do not provide financial grants to developing countries for renewable energy projects to meet their distinctive requirements. They may also be pressured externally to express themselves on technology that are out of line or unfit to embrace technology that are foreign businesses. Hence, poor countries may not be able to make autonomous and sustainable policies of economic development, failing thus to break out of the pattern of patterns of economic subordination (Sachs & Kharas, 2024a, 2024b, 2024c).

Technological Divide

The high cost and limited availability of green technology will constitute the key challenge. Renewable energy systems, electric automobiles, and energy efficient technology are unaffordable to many poor countries. Additionally, these systems are quite difficult to develop and manage for a lack of technical skills. For instance, many developing nations are challenged to establish the capability of running and executing smart grids as important segments of modern energy infrastructures (Lee & Kumar, 2023a, 2023b, 2023c). In addition, the transfer of technology from industrialised countries to poor countries is customarily circumscribed. What is good for wealthy nations to do with green technology is not necessarily what is good for the demands and local situations of developing countries. For example, solar power systems designed for the industrialized world can be too expensive, or impractical for people living in rural parts of Sub-Saharan Africa or South Asia with little or no infrastructure and grid connectivity. The mismatch complements or widens the technical gap, so that the developing nations do not get maximum advantage from global environmental progress (Chandra & Singh, 2023a, 2023b, 2023c).

Social and Cultural Resistance

In addition to financial and technological constraints, the integration of eco-innovations in underdeveloped nations is also tempered by social, and more often than not, cultural prejudice. It has to do with large scale green efforts like wind farms and solar parks as well as hydropower projects, all which can displace local residents or disrupt established industries in their path. Such initiatives may be viewed as a breach of the rights of indigenous communities, in particular, to land and resources, where they are not properly consulted, much less paid. Often they resist changing it because of the fear of losing ancestral lands, cultural erosion, disruptions to traditional livelihoods (Binns & Thompson, 2024a, 2024b, 2024c).

This highlights the importance of addressing cultural concerns to ensure equitable and inclusive adoption of sustainable practices. Furthermore, a lack of awareness and education regarding the advantages of green technologies significantly hampers their adoption in developing countries. In the absence of effective outreach and education, these technologies are often viewed as unfamiliar, irrelevant, or even disruptive to traditional ways of life (Graham & Patel, 2024a, 2024b).

The Role of Smart Eco Innovations in Mitigating the Darkside

Using Smart Eco Innovations as a Means of Addressing the Dark Side

Altogether, smart eco innovations bring beneficial solutions to some of the darker sides of the quest of climate neutrality, although in both developing and developed countries. Some of these technologies include decentralized renewable energy systems need to be integrated to provide adequate electric power for both industrial and residential purposes, and circular economy practices need to be employed as strategies to contest environmental, economic, and social challenges of attaining sustainable development. Furthermore, dramatic expansion of these innovations in a mutually beneficial manner globally is possible only through collective endeavours on the international level, especially for countries of the Global South.

Inclusive Technologies

On mitigating the adverse effects of climate neutrality, one of them is the use of inclusive technologies. Renewable energy technologies include but are not limited to solar microgrids, community based wind energy among others, there is potential to decentralize the renewable energy systems and enable communities to directly benefit from production without having to depend on central electricity grids. In the rural and other hard to reach areas as they are referred to in the developing nations, access to afford and sustained energy has also always been a sad myth. Approaches to decentralised energy solutions can make it affordable and economically feasible to provide electricity to these areas hence enhancing energy access and fighting poverty (Chandra & Singh, 2023a, 2023b, 2023c).

For instance solar microgrids have been successfully installed in several sub-Saharan nations for provision of electricity to local people/communities given a chance to generate electricity on their own. They remain relatively cheap, are simple to operate and do not require countless resources to expand. Such inclusive technologies not only serve to provide for human energy needs but also assisted in the development of the economy of those localities by providing employment in installation, maintenance and in the energy management systems. Second, it is safer since it can flexibly eliminate threats such as power loss or natural disasters frequently prevailing in many developing countries (Binns & Thompson, 2024a, 2024b, 2024c).

Apart from energy, other related technologies like cheap EVs, smart agriculture tools, and low-tech opportunities also outline additional prospects of closing the divide between the first and the third worlds. For instance, small scale electric cars and electric bicycles have been imported in rural areas to counteract traditional I.C engines automobiles and accept environment-friendly means of transport. Likewise, smart technologies like precision farming tools including smart irrigation systems, power by solar energy, can help to upraise the productivity of agriculture and improve the food security rate with less pollution (Graham & Patel, 2024a, 2024b).

Other important element of smart eco innovations is a circular economy practices that can address the negative impact of climate neutrality. The greater importance of the circular economy model is the importance of waste minimization, utilization or recycling of items for further use. This model is wholly different from the one that has long been called take make dispose, which is seen as one of the major causes of pollution and unfavourable impact on the environment. Because they optimize the utilization of resources, circular economy measures decrease the demand for new supplies and minimize emissions; moreover, new economic prospects in waste management and recycling industries (Nelson et al., 2024).

In the developing countries particularly the emerging economies, circular economy has great potential in generating employment and eliminating poverty. Recycling and waste management industries can generate employment opportunities for people especially for women and kids who carry out waste picking in the streets and sidewalks of developing countries. Awareness creation, use of incentives and rebates to enhance recycling rates as well as recognizing the informal sector to ensure its participation in recycling systems are good example of mechanisms that will ensure that they assist in the dual goal of environmental management and social policies (Chandra & Singh, 2023a, 2023b, 2023c). For example, in developing countries such as India and Brazil, informal recyclers employing independent collecting system also known as 'waste pickers' can benefit from training, by enhancing their market access and, thus recycling's stock value. The city of Amsterdam for example has been an advocate of circular economy and it has put strategies into place that ensure companies reduce their waste, retain materials for use and other designs that allow for recycling. In the same way, Rwanda and Kenya have implemented the circular economy for e-waste management and the recycling of e-waste and advancement green technology that generate the economy besides environmentally sustaining the disposal of electronics.

Global Collaboration

This is particularly important in ensuring that the fruits of smart eco innovations and the practice of the circular economy are felt by all nations, especially the developing ones, whose capacities may not easily support adoption of such solutions. As previous experiences have shown, successful international cooperation can help countries to acquire new green technologies, financial resources and external assistance to empower their countries to skip over not fit technologies and utilize the best modern technologies available.

Another important endeavour is the Green Climate Fund (GCF), which finances and project development and execution of climate change adaptation and mitigation in developing countries (Kumar et al., 2024). In areas such as renewable energy, sustainable soils and agricultures, and climate-liable structures, the GCF stays that developing countries can gain green technologies affordable without assuming high indebtedness or paired with any foreign investment (Mishra, 2023a, 2023b).

Ethical and Social Implications of Climate Neutrality and Eco Innovations

While climate neutrality and eco innovations are becoming the new trends in the world, apart from the technological perspective other ethical and social issues are also needed to be solved. Climate-neutrality in other words is not only about implementing low emission processes and the green economy, but more so about these changes being pro-justice, pro-equity alterations. This section discusses the ethics of pursuing climate objectives and social justice and environmental justice and the place of international relations in making certain that the benefits of eco-innovations are available for use across the global society.

Dual Consideration of Climate and Social Justice

Social Justice

Climate neutrality cannot be attained without social justice risking to deepen social disparities. For instance, the emphasis on green technologies such as electric vehicles and renewable energy systems and facilities may further bar rural or poor persons' needs from being met by the technologies or their means of embracing the technologies economically. Similarly, some climate adaptation policies or measures like carbon pricing or carbon taxes could harmed the poor since most of them depend on carbon intensive jobs for their living.

As a result, climate policies themselves must become fair and equitable. Climate policies should never be a tool that discriminates against certain people or communities during the journey towards clean energy. This ranges from financial assistance, technology investment and widely engaging in capacity development initiatives in order to make vulnerable groups competent to engage and benefit in the green economy (Sovacool, 2023).

Environmental Justice

Environmental justice is yet another ethical concern that is related to climate neutrality transition. It stresses that those who have suffered and continue to suffer since climate change and pollution are responsible should not be made to put up with it especially if they are or considered to be of lower "standing." When it comes to climate neutrality, there are environmental justice issues when an extensive eco-innovation or large-scale technologies and trends, including renewable energy systems, electric vehicles, and green construction, bring adverse impacts for a particular community.

For example, materials used in production of renewable energy technologies including lithium, cobalt and rare earths has been associated with social costs including environmental and human rights costs in various developing countries. These minerals, which are important to the production of batteries for electric cars and energy storage, are obtained from places that have negligible standards for the harms they cause to the environment and workers. Citizens of the local communities may end up witnessing polluted environments, migration and abuse of their resources as they try to meet the demands of transition needed for the wealthy countries' greener economy (Lepawsky, 2024). Responding to these environmental justice issues entails an integrated approach whereby citizens are put into consideration within decision making processes and their rights and or wants satisfied (McCauley & Heffron, 2023).

Policy Recommendations for the Transition Towards Climate Neutrality that Leaves no One Behind

There are many obstacles on the way to climatization, similarly as there are many questions concerning technological shifts but also social implications. In this section, the principles that can provide a fair and equitable approach to the delivery of climate neutrality are identified and discussed and include global solidarity, adaptation, creativity, and business responsibility. These recommendations are made with the view to make sure that both developed and develop countries can accrue the benefit of the transformation in a sustainable manner without causing injustice and closely linked with unsustainable development.

Global Cooperation and Equity

It, therefore, goes without saying that global cooperation is important in finding ways to combat climate change. The attempts toward climate neutrality should therefore be coordinated and with the developed countries offering capital, expertise and know-how to the Developing countries (Bocken et al., 2023). This support should extend to financing of renewable energy, as well as extending the supply of clean technologies and technical cooperation with developing nations to enable them to adopt low carbon development paths.

That is why the proposals for the modification the key climate change policy include the design of a new Global Climate Finance Mechanism, which guarantees fair delivery of the funds (Foley et al., 2023). The current model of climate finance is characterized by developed nations providing grants or concessional loans, which complicates things for developing nations particularly when they are pressed with debt (Jenkins et al., 2023). A better financing regime apparently would be to lock in long-term funding in sustainable activities in the developing world without adding more burden. The Green Climate Fund (GCF) is already an excellent advancement towards this concept but more should be done to increase efficiency and accessibility of the fund (Buchner et al., 2023).

Apart from money element, international technology transfer ought to be other main component of global partnership. This transfer should not be limited to mere sale of technologies; it should also entail the provision of information and skills in that will facilitate the optimization of technologies for the deem environment that faces developing countries (Sullivan & Mander, 2023). This can assist these nations to skip old and polluting technologies and go to extra sustainable ones. Initial obligation proposed under the Paris Agreement to focus on technology transfer is a good beginning, but more commitment and assistance are called for in order to enhance the agreement's effectiveness (Anderson & Gouldson, 2023).

There Will Be Local Adaptation and Innovation

At the local level, policy advice should centre on the idea that neighbourhoods and cities should be ready and able to innovate and unlock the promise of climate neutral tech. Bottom up solutions are required hence coming up with method that enable citizens especially those in the developing nations to determine their needs andospel them the way forward. This comprises promotion of innovation at the regional level and backing of SMEs that can implement innovations in sustainable technologies fitting to geographical locations.

Based on the policies recommendations, decentralised renewable energy systems like solar microgrids and small scale wind farms should be implemented among others. These systems not only offer energy to those who have been locked out of the grid but also result in employment in installation, maintenance as well as management of the energy (United Nations, 2024). Proposing the bottom-up promotion of such solutions to embrace renewable energy technologies in newly connected farmland districts or remote areas, that has for the future transfixed the innovative stream of energy transition could be highly effective (Hansen et al., 2024).

Corporate Governance that Controls and Punishes Unscrupulous Organisations

Business is a significant actor in the world's journey to net zero emissions; however, there are increasing worries around greenwash, where an organisation presents itself as sustainable while taking little to no actions. Tougher guidelines regarding firms' environmental declarations are necessary in order to limit the misrepresentation of those companies' contributions to sustainability objectives. They must make corporate carbon disclosures and progress to climate neutrality mandatory for governments to achieve corporate climate accountability. These reports should be produced by other parties of the structure to increase independence and credibleness.

A similarly important suggestion is to call for rewarding company investments in green innovation through tax incentives, subsidies, and other kinds of funding support. It can make business shift away from hidebound industries that pollute our atmosphere and economy towards clean technologies and activities. Governments should also also use proper tools such as carbon pricing mechanisms that capture the proper cost of carbon and promote the use of low carbon products (Cameron & Horne, 2024).

Limitations

Although the study provides a thorough examination of climate neutrality and intelligent eco-innovations, some of its limitations may restrict the breadth and practicality of its conclusions. It may be less innovative or provide less insight into ground-breaking or regionally specific research because it mostly summarises the body of existing literature and makes references to well-known international agreements and technology. The focus of discussions is on national and international initiatives, frequently ignoring local or micro-level effects and behaviours. For instance, it makes reference to rural regions and decentralised solutions, but it omits in-depth case studies that would enhance the text. Advanced technologies like carbon capture and storage, smart grids, and electric vehicles are heavily emphasised, however they might not be practical or accessible in environments with limited resources, especially in poor countries. Although this chapter recognises issues such as environmental injustices and corporate greenwashing, it does not provide a thorough analysis or answers for systemic inequalities related to eco-innovations and climate policy. Although social justice and ethics are covered, the paper doesn't go into great detail about how policies could lessen conflicts between social, economic, and environmental goals. Indigenous knowledge systems and their potential contribution to climate neutrality have received little attention.

Conclusion: Towards a Sustainable and Equitable Future

Making this move towards climate neutrality is a revolutionary chance to combat the effects of climate change, while promoting sustained development. But as the article explains, it is also complicated, courtesy of disparities between wealthy and poor countries. They have abundant resources and high technology developed countries, but have the problem of excessive dependence on experimental solutions, environmental injustice and also many inequalities in their economic transition. However, poor nations have fiscal constraints, technological gaps, and external pressures to adopt less than ideal solutions, which are further exacerbated by environmental and social barriers to green technology deployment. Promising smart eco innovations, including renewable energy systems, smart grid, and circular economy practices, appear to be necessary to solutions to these challenges, fair growth and loca adaptation. Essential to the transition of all nations is the development of policies promoting global collaboration, equitable financing, and technology transfer. Additionally, efforts driven by a virtuous foundation of corporate accountability and the fair transition of disadvantaged communities can be used as a foundation of ethics for climate neutral activity. A net-zero world requires a collaborative global framework that will bring together economic, social and environmental issues in order to develop a sustainable and equitable future.

Future Scope of Research

There are several avenues of future research in the study of climate neutrality and smart eco-innovations, including problem solutions, maximisation of potential for sustainable development, focusing on the problem solution. Future research will consider how to optimize renewable energy technologies, their scalability, cost and especially in poor countries.

More study is needed to make carbon capture and storage (CCS) systems more efficient and cost effective. Through investigating the space between circular economy and smart eco innovation, we can gain valuable insight into the reduction of waste and resource efficiency as it occurs around the world. The issue of equitable climate financing strategies that will permit developing countries to switch to low carbon technology without additional financial cost is another important area for research. The research in cutting edge decentralized renewable energy sources like solar micro-grids that are specifically designed for rural off grid areas have substantial room to improve on economic empowerment as well as energy access. Research could also consider ethical implications and the long term efficacy of carbon offset programs and whether they're properly supplemented by direct emission reductions. To foster real progress towards sustainability goals and keep public trust, we must also examine how governments frame the fight against greenwashing and corporate responsibility.

References

Anderson, B., & Gouldson, A. (2023). The role of technology transfer in global climate governance. *Environmental Policy and Governance, 33*(4), 220235.

Barker, T., Wright, D., & Lawrence, S. (2023). Carbon offsetting and its role in global climate policy. *Environmental Economics and Policy Studies, 25*(1), 6782.

Bhushan, M., Singh, R. (2024). Talent Management Facilitates Net-Zero Transition Through Employee Green Behavior. In: Singh, R., Khan, S., Kumar, A., Luthra, S., Chokshi, H. (eds.) *Net Zero Economy, Corporate Social Responsibility and Sustainable Value Creation. CSR, Sustainability, Ethics & Governance.* Springer, Cham. https://doi.org/10.1007/978-3-031-55779-8_7

Binns, R., & Thompson, D. (2024a). Cultural resistance to sustainable technologies in developing countries: A case study approach. *Sustainability and Development Studies, 36*(1), 88101.

Binns, R., & Thompson, D. (2024b). Inclusive technologies for sustainable development: Empowering local communities. *Technology and Society Review, 42*(3), 135148.

Binns, R., & Thompson, D. (2024c). The role of AI in climate change solutions: Opportunities and challenges. *Environmental Ethics and Technology Review, 28*(2), 123139.

Bocken, N. M. P., Short, S. W., & Rana, P. (2023). A circular economy perspective on sustainable business models. *Business Strategy and the Environment, 32*(1), 1229.

Buchner, B. K., Watson, C., & Ellis, M. (2023). Climate finance and its role in fostering equitable transitions. *Global Environmental Change, 75*, 6374.

Cameron, A., & Horne, M. (2024). Corporate accountability in the transition to a green economy: A global review. *Journal of Business Ethics, 154*(3), 102118.

Chandra, R., & Singh, M. (2023a). Renewable energy and the technological divide: Challenges for developing nations. *Energy and Technology Review, 45*(3), 132145.

Chandra, R., & Singh, M. (2023b). Renewable energy innovations and their global impact. *Journal of Energy Systems, 48*(3), 232246.

Chandra, R., & Singh, M. (2023c). The potential of renewable energy and inclusive technologies in developing countries. *Global Energy and Development Journal, 15*(2), 92106.

Eyime, E. E., & Ushie, O. J. (2024). Climate Risks and Economic Consequences of Rising Global CO_2 Emissions in Aviation, Shipping, and Heavy-Duty Transport. *Trends in Renewable Energy, 11*(1), 84–121. UNFCCC. (2023). The Paris Agreement: A guide. *United Nations Framework Convention on Climate Change.* https://unfccc.int/

Falk, J. (2023). Carbon neutrality and netzero emissions: Achieving global goals. *Climate Policy Review, 22*(1), 1230.

Foley, J. A., Ramankutty, N., & Tilman, D. (2023). Solutions for a sustainable food system in a changing climate. *Nature Sustainability, 6*(1), 2234.

Global Investment and Development in Africa, https://link.springer.com/chapter/https://doi.org/10.1007/97-3-031-44514-9_2

Gosling, C., & Farley, R. (2024). Smart grids and the future of energy systems. *Renewable Energy & Sustainability, 29*(2), 5472.

Graham, T., & Patel, S. (2024a). Circular economy in practice: Opportunities for sustainable development. *Environmental Economics and Policy Studies, 33*(1), 210227.

Graham, T., & Patel, S. (2024b). Resource colonialism and the global south: The environmental and social impacts of mining. *Journal of Environmental Policy and Planning, 29*(2), 201217.

Hansen, J., Levitt, M., & Thompson, E. (2024). Decentralized renewable energy systems for local economic development. *Renewable and Sustainable Energy Reviews, 35*(2), 4058.

Harrison, J. (2024). Environmental injustice and climate neutrality: The global consequences of developed nations' green transitions. *Global Environmental Politics, 24*(1), 5872.

Jenkins, A., & Tsang, P. (2023). The ethics of resource extraction: Cobalt mining and environmental justice in the Democratic Republic of Congo. *Journal of Global Environmental Change, 19*(4), 450463.

Joshi, A., Singh, R., Kumar, V., Rani, S. (2024). Net zero strategy: exploring opportunities, confronting challenges for developing countries. In: Singh, R., Crowther, D. (eds.) Transition Towards a Sustainable Future. Approaches to Global Sustainability, Markets, and Governance. Springer, Singapore. https://doi.org/10.1007/978-981-97-5756-5_11

Khan, S., Singh, R., Alnahas, J., Abbate, S., & Centobelli, P. (2024a). Navigating the Smart Circular Economy: A framework for Manufacturing Firms. *Journal of Cleaner Production*, 144007. https://doi.org/10.1016/j.jclepro.2024.144007

Khan, S., Singh, R., Alnahas, J., Abbate, S., & Centobelli, P. (2024b). Navigating the smart circular economy: A framework for manufacturing firms. *Journal of Cleaner Production, 480*, 144007.

Kumar, K., Singh, R., Sharma, A., Lehri, B. P. S. (2024). Climate change adaptation strategies for achieving net-zero economy. In: Singh, R., Crowther, D. (eds.) *Transition Towards a Sustainable Future*. Approaches to Global Sustainability, Markets, and Governance. Springer, Singapore. https://doi.org/10.1007/978-981-97-5756-5_8

Lee, S., & Kumar, S. (2023a). Smart eco-innovations and sustainable development: A global perspective. *Sustainable Development Studies, 12*(3), 94112.

Lee, S., & Kumar, S. (2023b). Technological barriers to sustainable development in emerging economies. *International Journal of Environmental Technology, 40*(2), 7689.

Lee, S., & Kumar, S. (2023c). The role of circular economy in achieving sustainability goals. *Journal of Sustainable Business Practices, 21*(4), 5672.

Lepawsky, J. (2024). The ethics of resource extraction in the green transition. *Journal of Global Environmental Justice, 10*(2), 125142.

McCauley, D., & Heffron, R. (2023). The social implications of renewable energy transitions. *Energy Policy and Social Justice Review, 19*(1), 87104.

Mishra, V. (2023a). Climate finance and the debt trap: A critical analysis of green loans in developing countries. *Global Economics Review, 21*(1), 5973.

Mishra, V. (2023b). Green finance and international cooperation for developing countries. *Global Economic Review, 22*(1), 7892.

Nelson, D., Martinez, L., & Garcia, T. (2024). Circular economy and climate neutrality: Opportunities and challenges. *Environmental Management Review, 33*(4), 312327.

Nwokolo, S.C., Singh, R., Khan, S., Kumar, A., Luthra, S. (2023). Remedies to the challenges of renewable energy deployment in Africa. In: Africa's Path to Net-Zero. CSR, Sustainability, Ethics & Governance. Springer, Cham. https://doi.org/10.1007/978-3-031-44514-9_3

Nwokolo, S. C., Singh, R., Khan, S., Kumar, A., Luthra, S. (2023). Africa's Awakening to Climate Action. In: Africa's Path to Net-Zero. CSR, Sustainability, Ethics & Governance. Springer, Cham. https://doi.org/10.1007/978-3-031-44514-9_9

Nwokolo, S. C., Singh, R., Khan, S., Kumar, A., Luthra, S. (2023). Global investment and development in Africa. In: Africa's Path to Net-Zero. CSR, Sustainability, Ethics & Governance. Springer, Cham. https://doi.org/10.1007/978-3-031-44514-9_2

Nwokolo, S. C., Singh, R., Khan, S., Kumar, A., Luthra, S. (2023). Impacts of climate change in Africa. In: Africa's Path to Net-Zero. CSR, Sustainability, Ethics & Governance. Springer, Cham. https://doi.org/10.1007/978-3-031-44514-9_7

Nwokolo, S. C., Singh, R., Khan, S., & Kumar, A. (2024). Threats to the rapidity of sustainability transitions posed by technological changes. In *Elsevier eBooks* (pp. 39–75). https://doi.org/10.1016/b978-0-443-31486-5.00002-1

Singh, R., A. G. (2024). ESG Excellence in the Digital World. In V. K. Ernesto D.R. Santibanez Gonzalez, Fintech for ESG and the Circular Economy. Wiley.

Sachs, J. D. (2023). *The age of sustainable development*. Columbia University Press.

Sachs, J., & Kharas, H. (2024a). Global collaboration for sustainable development: The role of climate finance. *World Development Perspectives, 34*(2), 117130.

Sachs, J., & Kharas, H. (2024b). Global governance for climate justice: Addressing inequality in the green transition. *Global Policy, 15*(3), 98110.

Sachs, J., & Kharas, H. (2024c). The future of climate finance: Funding solutions for developing countries. *World Development Perspectives, 33*(1), 99112.

Sorrell, S. (2023). Technological dependencies and the green transition: Risks and opportunities. *Energy Policy Studies, 41*(5), 250266.

Sovacool, B. K. (2023). Climate justice and the politics of ecoinnovations. *Environmental Politics, 32*(4), 503520.

Sullivan, A., & Mander, M. (2023). Just transitions in a lowcarbon economy: A framework for social equity. *Environmental Politics, 30*(5), 679697.

United Nations. (2024). Sustainable development goals: Goal 13 Climate Action. https://www.un.org/sustainabledevelopment/climateaction/

Zhang, Y., & Liu, Z. (2023). E-waste and environmental injustice: The hidden side of the green revolution. *Journal of Environmental Science and Technology, 39*(3), 210224.

Chapter 19
Climate Finance and Natural Capital: Linking Global Initiatives with Local Impact

Donkor Nawaah, Prince Yeboah Asare, and Michael Ayikwei Quarshie

Abstract Climate finance is a critical pillar in global efforts to combat climate change, channelling essential capital toward projects that reduce greenhouse gas emissions and promote sustainable climate action. This chapter aims to provide general guidance on how the development of climatic finance can be assessed. The most important factor is that the valuation of nature to any financial system is crucial if any climatic goals should be attained. In addition, in the global context, the role of global movements is difficult to underestimate. Meanwhile, adaptation and application on the local level are crucial and can be illustrated in a myriad of cases. The chapter showcases the examples of different countries where the concept of climate finance is much more of a fact of life. These examples serve to illustrate both the advantages and disadvantages of the public–private partnerships and the role of the nature-based solutions, together with developing trends like impact investing.

Keywords Climatic goals · Financing climate · Global movement · Sustainability · Valuation of nature

D. Nawaah
Department of Finance and Accounting, Woxsen University, Telangana, India
e-mail: donkor.nawaah@woxsen.edu.in

P. Y. Asare (✉)
Department of Business and Social Sciences Education, University of Cape Coast, Cape Coast, Ghana
e-mail: prince.asare@ucc.edu.gh

M. A. Quarshie
Department of Entrepreneurship and Business Sciences, University of Energy and Natural Resources, Sunyani, Ghana
e-mail: michael.quarshie@uenr.edu.gh

R. Singh and W. L. Filho (eds.), *Climate Neutrality Through Smart Eco-Innovation and Environmental Sustainability*, Climate Change Management,
https://doi.org/10.1007/978-3-031-83250-5_19

Introduction

The October 2024 flooding crisis in Spain is a timely reminder and clear example of the desperate need for climate finance that substantially contributes to local resilience efforts and adaptation towards increasingly erratic weather patterns. Flash floods, which left a toll of several dead in Spain in October 2024, are linked to a vicious weather system where cold air and warm air meet creating dense rain clouds; a phenomenon thought to be becoming more regular partly due to climate change. This local phenomenon known as Desprésion Aislada en Niveles Altos (DANA) can develop outside the influence of polar or subtropical jet streams producing high impact weather events that overwhelm local infrastructure and disaster response capabilities (McKibben, 2012).

Climate finance, an important field that intertwines both economic and environmental science, concerns the place of capital expenditure towards efforts that reduce greenhouse gas emissions and encourage sustainable development (Iacobuţă et al., 2022). It has become a key component on the international agenda at the centre of many global initiatives aimed at addressing environmental issues (Tao et al., 2022; Nwokolo et al., 2025). Consequently, efforts geared towards environmental sustainability have resulted in a drastic expansion in the role of the financial sector concerning the improvement of the adverse effect of climatic change. It is critical to recognise the financial worth of nature, as traditional economic systems frequently ignore the importance of ecosystems and biodiversity. The notion of natural capital is gaining popularity since it emphasises the true integration of nature's worth into economic frameworks (Weisheng et al., 2013). It is required to synthesize the economic increase with the protection of the planet's bio-system.

Movements (e.g., Fridays for Future, Extinction Rebellion (XR) and Climate Justice Alliance (CJA)) focused on raising awareness about climate issues on a global scale have defined the directions of the formation and development of climate finance. These movements, emerging as an answer to the growing level of risks to the environment, started to influence policies and financial flows both on a national and international level. Nevertheless, success or failure of global large-scale incentives on climate crucially depends on their local framing, a necessary requirement for development and implementation of climate financing strategies that would be most effective in a local context (Kumar et al., 2024). Generally, this chapter provides information on the character of climate financing and natural valuation methods, the influence of the global movement for climate and local framings. It was achieved through examination of case study in several countries with respect to climate financing.

Valuing Nature in Financial Terms

One of the key challenges in climate finance is the integration of environmental values into financial systems (Nwokolo et al., 2023a, 2023b, 2023c, 2023d, 2023e). The existing traditional economic models have often predicated on infinite natural resources hence over-exploitation (Sarkar, 2022). Over-exploitation caused severe environmental degradation and loss of biodiversity. The traditional approach did not appreciate the value of nature resulting in related disasters and therefore replaced "natural capital concept". Natural capital is the stock of natural resources and ecosystems such as forests, water bodies, and minerals, which offer a wide range of services needed by humans and other organisms in the environment (Joshi et al., 2024). For instance, the ecosystem services offered by natural capital include carbon sequestration, water purification, soil fertility, pollination, among others. Natural capital accounting presents the understanding and imitation of natural resources and ecosystem services into economic budgets. Consequently, ecosystem services that nature offers are made visible in economic production and relations. The use of natural capital accounting makes natural resources and ecosystems visible in the economic processes and in humans' livelihoods. This approach ensures that people run the environment better. The principle of natural capital is now well-known worldwide; it has been applied in several businesses and the financial systems of most countries to ensure environmental safeness.

Germany is another country that has heavily invested in natural capital in a policy called Energiewende. Energiewende entails a comprehensive transition from using nuclear energy and increasing the country's renewable energy resources (Joas et al., 2016). Because of valuing its natural resources and ecosystems by accounting their value in financial planning in increase in Germany, the value of ecosystem services like carbon sequestration and biodiversity has made the country justify increased public and private investment in renewable energy Germany is a world leader in climate protection because of incorporating natural capital into financial planning. This approach to the natural capital valuation reflects the country's promotion of the green bond market—financial instruments that fund projects including different renewable energy developments, energy efficiency improvements, and sustainable water management which provide positive environmental impacts (Climate Bonds Initiative [CBI], 2020). In the recent decade, China has rapidly expanded its green bond market, which reflects the government's efforts to ensure that the entire financial system is consistent with the country's sustainability goals. Also, the market expansion is proof that the country is assigning a monetary value to the positive environmental changes, incentivizing both private and public investors to support climate-friendly developments.

However, despite creating and developing the strategies, the process of nature valuation has several constraints. Valuating the ecosystem services is inherently complex and requires highly-detailed data, advanced models, and profound ecological knowledge. The benefit of plant pollination or soil fertility needs to be estimated using a range of sophisticated methods designed to capture the market and non-market value of the services. For instance, carbon sequestration can be valued by measuring the cost of the impacts leading to climate change that are avoided thanks to enhanced carbon storage in forests and wetlands (TEEB, 2010). However, these and other valuation exercises should be conducted with awareness of their limitations, which implies the discussion of the uncertainties and assumptions of both the ecological and economic models. Another issue of great concern connected with valuing nature is the risk that it may reduce the significance of environmental values to financial values. Monetary valuing of ecosystem services can help decision-makers by emphasising the economic value of natural resources and supporting sustainable management and protection.. However, certain aspects of nature cannot be calculated. Even though accounting for natural capital can help integrate environmental values into financial systems, it should be combined with other approaches not attempting to reduce the value of nature to financial endeavors only. As Nwokolo et al., (2023a, 2023b, 2023c, 2023d, 2023e) note, nature's worth comprised more than just its utility to human beings. That is, the cultural or intrinsic value of a sacred natural site which is of particular importance to some local population can hardly be calculated. In the same way, the financial worth of a beautiful unspoiled section of land is quite different from its beauty. Therefore, while natural capital accounting methods should be taken into consideration, they should be further adjusted to integrate with a broader range of approaches capable of recognizing and preserving non-economic values of nature.

Contemporary policy measures and developments demonstrate that efforts to integrate natural capital into financial systems are not only a matter of theoretical discussion. Both governments and international organizations are aware of the need to account for environmental values in their economic decision making. For instance, the Natural Capital Coalition (2016) explains that the Natural Capital Protocol has been developed with the aim of establishing a framework by which businesses and other organizations can account for their impacts and dependencies on natural capital. This protocol is acknowledged in demonstrating how to account for and measure natural capital values while providing a standardized form of natural capital accounting across a broad range of sectors. In addition, progress in the field of technology is also expected to promote the ability to value nature. With the greater availability of easily accessed data and such means as remote sensing, geographic information systems, and big data are expected to contribute to our attempts to more accurately and effectively evaluate the worth of various ecosystem services and other aspects of nature.

Global Movements Driving Climate Finance

Climate action has been a point of interest on the international agenda. The impetus for driving the momentum was due to various grassroots movements and advocacies that have significantly influenced the climate finance agenda. These movements have created a sense of urgency among the public that climate change is a significant threat, requiring immediate action. Most of these groups have attributed international policy reforms and commitments made at various levels to the financial pledges. Therefore, the role of group action in mobilizing climate finance is a critical lever in steering the world's response to climate change. One of the biggest movements is Fridays for Future, employed by a Swedish youth working to create climate change awareness. In August 2018, Greta Thunberg started her solo strike in front of the Swedish Parliament to push for stronger action on climate change. Later, her action caused a sensitization of the youth and other few adults which turned out to become the most significant environmental movement worldwide (Thunberg, 2019). The group's actions have created a lot of effort leading to governments and financial institutions to declare enhanced commitment to climate. Extinction Rebellion is another crucial movement founded in the UK in response to the government's lack of action to combat climate change. Using civil disobedience short of violence, the group managed to have the crisis in question addressed, and their tactics included blocking public highways, and mass participation in various acts. The actions of the ER group forced the government to declare the climate emergency soon after mass blockages of streets, leading to the police arresting participants who engaged in any form of damage. The government target, as reflected in 2019, is a legally binding target to reach net-zero greenhouse gas emissions by 2050 (Committee on Climate Change, 2019).

One of the most significant forces advocating for a Green New Deal in the United States is the Sunrise Movement, an organization of young activists. This grassroots group has been successful in changing national discussions and convincing politicians to adopt this policy. Although the Green New Deal has not yet been accepted, perhaps, for implementing a fair transition to a low-carbon economy, substantial climate finance is required, and the advocacy of the Sunrise Movement has made a great contribution (Fisher, 2019). In India, the global movement also has an impact on the climate fund as the country's local movement has called for addressing climate change by prioritizing and expanding the access to clean energy technologies such as solar photovoltaic. The country was set to have 175 GW of renewable capacity by 2022 and 450 GW by 2030 (Dey et al., 2022). The investments in clean off-grid solutions were led by the grassroots movement asking for more consideration in rural areas where electricity access is a luxury. Also, the funding was aimed to address the climate fund need to be aligned with the people since 46% of the total capacity was a part of the State's tender in India (Dubash et al., 2018). Such endeavours had led the country's financial institutions to improve climate research and divide the investments between climate goals and people's well-being.

The global movement had led to climate change funds divestment in fossil fuel campaigns which started in U.S. Universities and was spread worldwide. Investors, especially universities and pension funds, have been withdrawing their funds out of the fossil energy that is leading to the climate fund. These institutions have been reinvesting their funds in renewable energy technologies to minimize risks accompanied by climate change leading to financial loss (Barradale, (2010). One of the changes brought about by the countless global movements is an update of governmental and financial institution policies and financial commitments. There may be the need for climate finance to be made a priority considering that this legally binding budget and plan were accompanied by a commitment of £11.6 billion over five years to support international climate finance (Mc Guinness & Bullock, 2020). Currently, developing nations appear to be witnessing renewable energy advocacy to develop and create solar energy to power rural and urban homes. This is spearheaded by the public through grassroots advocacy. Taking these together, global movements have had a significant role in propelling climate finance. The financial flow for different countries has been prioritized, thanks to the relevant awareness by the global movements. Additionally, these movements have influenced policy changes within countries and financial institution reforms and as a result, they have ensured their redistributed priorities. In addition to that, owing to the effectiveness of the said movements, governments have been pressured to adopt decisive ambitious goals on climate and to back the said proclamations with allocated monetary budgets.

Role of Local Framing in Climate Finance

The global initiative of combating climate change is accompanied by substantial climate finance, drawn from both the public and the private sectors. Such finance is critical in supporting and facilitating various efforts and actions dealing with climate mitigation and adaption throughout the world. However, as the success of any local initiative depends on its applicability, both from the social, economic, and financial point of view, the local framing adapts these global strategies. Climate finance, adjusted to different social, economic, and environmental conceptions, allows the established actions and interventions to be not only possible but relevant. Thus a "localism" approach to climate change makes these global goals reachable by apprising region's difficulties and opportunities.

Local Context in Climate Finance

Climate change exerts its impact on different regions to a different extent due to varying geographical, economic development, and social organization. Therefore, the local implementation of measures of financing climate change is crucial as a general approach would appear vague. Local contexts and local framing involve applications and arrangements that address the needs and conditions of a specific region. Such an approach translates its implementations to where it is relevant and generally accounts for the socioeconomic conditions of the local community, cultural activities, which are often ignored in international strategies. Local knowledge is vital so that the financing in question is used for the best possible projects for a sustainable application in the future.

India may be the best example of such a practice that allows climate finance to be successful. Even though India is heavily populated, India faces issues of a low carbon transition on an unimaginably massive scale (Sovacool, 2019). International climate standards are rigged into India's policy on renewable energy. However, these standards have a local adjustment in the form of rural electrification, which must be a diverse one regarding the enormous population and conditions. One of the examples of a locally framed climate finance approach is the Indian government's Saubhagya scheme. With the high proportion of rural and remote regions in the country, the Indian government has recognized the difficulties of extending the national grid. Off-grid solar systems have been particularly heavily promoted, also considering the unsatisfactory or generally unreliable state of grid infrastructure in rural India in particular (Rajagopalan et al., 2022). Singh (2022) notes the success of Saubhagya and the role of local ownership in acceptance of these types of solutions. Accordingly, locally framed ownership and local benefit creation are the advantages of the approach. At the same time, South Africa provides another example of such an approach with a different direction being taken by the country. South Africa has developed a just transition framework in the face of its heavy reliance on coal being used for electricity generation and the associated risks (Mirzania et al., 2023). Coal mining and the coal energy industry are important employers and economic activities in certain regions of the country. The just transition framework is designed to ensure the social inclusivity of the coal use abandonment and the associated low-carbon transition by development of opportunities for new economic activities, investment in renewable energy projects tours which are not yet many and may serve as an opportunity for a substantial renewable energy investment. South Africa received 120 million euros from the European Union and 250 million dollars from the World Bank for projects which comply with the elements of a just transition framework in South Africa (Moyo, 2014). The example shows one possible way of ensuring local framing of the issue and therefore local ownership, namely, a focus on ensuring local benefits of climate and transition-related investments.

Ghana's climate-smart agriculture using climate finance as it is an example of a well-crafted strategy that contributes to achieving climate finance goals. Ghana is a developing country, which has addressed climate change through climate-smart rice agriculture. Mainly, climate finance efforts in Ghana are aimed at adaptation in the agricultural sector. Agriculture is an essential sector in Ghana's economy, and it provides a source of livelihood to millions. However, the sector is highly vulnerable due to the impacts associated with climate change, such as changes in the pattern of rainfall, high temperatures, and the extreme weather conditions (Arndt et al., 2015). Ghana's climate-smart agriculture strategy combines used climate finance with local knowledge and practices to other developed resilience measures. The reservoir's purpose is to build the adaptive capacity of small-scale farmers through sustainable farming systems, access to up-to-date climate information, and increased access of the population to climate risks. The strategy is funded by various international climate funds, for example, the Green Climate Fund (Amighini et al., 2022). Overall, Ghana's climate-smart agriculture is an example of successful climate finance because it demonstrates the local framing feature that benefits the specific needs of the intervention. The main difficulty of the local framing seems to be the need for adequate data and local expertise for assessing the actual needs of the region. Moreover, strong local institutions are critical for climate change execution and implementation.

Despite the challenges associated with local framing, it offers substantial opportunities. In particular, the provision of financial support is tailored to the actual needs of the relevant region and is therefore more effective and efficient (Singh, 2012). In addition, local framing helps to ensure that the targeted population perceives the financial support as an equitable opportunity or payback. It also helps to increase the level of commitment to a certain goal because it responds to the specific requirements of the given community, which is interested in the implementation of these programs. Overall, the local framing of climate finances is critical to achieving these global goals because it is the optimal and effective response to various challenges and opportunities. Case studies conducted in India, South Africa, and Ghana have shown that with a strong and developed infrastructure, obtaining and utilizing climate finance in such a manner is possible. As the role and significance of climate finance is only increasing, it becomes critical to adjust the given models to create an optimal and consistent transition to a low-carbon economy.

The Challenge of Insufficient Financial Resources

The most extraordinarily demanding issues in climate finance remain the deficit of financial resources, especially in developing countries. The Paris Agreement of 2015 reports that it aimed at collecting $100 billion annually by 2020 to guarantee climate financing to under-developed countries (UNFCCC as cited in Pickering et al., 2015). The necessary financial objectives have not currently been realized. According to the latest data, international climate finance flows have not yet reached the $100 billion marker, so the goals are not realizable now (UNFCCC, 2020 as cited in Sachs et al., 2022). It notes that this tendency has an adverse effect on the world community that cannot withstand the climate change crisis. At the same time, the pandemic of 2019 drastically complicates the situation because it demands substantial public funds. Governments pay more attention to rebuilding the economy and protecting the health of the people. COVID-19 results in financial support for world hygiene and human healthcare, diverting funds from climate (OECD, as cited in Buckle et al., 2020). It is essential to state that this factor affects economic tendencies; international funds that plan to finance climate are redirected to other goals. It is especially difficult for under-developed countries, as the issues of climate change affect them significantly more when fewer money resources are paid to ensure financing.

The complexity and accessibility of climate finance along with the total deficit of funding, plays a significant role. Receiving climate finance, international funds, contain a variety of issues, compliance standards, and conditions for their implementation. Such funds work on a particularly bureaucratic basis. They represent a range of different criteria and documentation that require thorough meetings. This mechanism significantly slows down the process of receiving funding and making it extremely difficult for under-developed countries (UNFCCC, as cited in Goldar et al., 2019). Such countries cannot always provide the necessary level of professional and technical skills to fulfill the desired conditions. The multiplicity of funding sources also complicates the efforts to mobilize resources. This is not just a reference to the various types of funds: in addition to multilateral climate funds, financing might come in the form of bilateral or private aid and investment. Most climate funds tend to have their own criteria, application procedures and result reporting requirements. This fragmentation of funding sources is inefficient and potentially results in overlapping areas of work.

Opportunities for Innovation in Climate Finance

The pessimistic assessment outlined at the beginning of the chapter notwithstanding, there are significant opportunities for innovation in climate finance. One new form of finance is blended finance. The idea is that using public funds and budget foresight enables climatic project implementation to be the lever for additional finance from the private sector. In India and other developing countries, blended finance is used to leverage private investments in renewable projects. Both public and private funds are used in the Indian Renewable Energy Development Agency, with the latter being offered the opportunity to invest without significant risks (CIF as cited in Mohamadi, 2021). This approach helps attract additional funding and mobilize greater private sector investment and innovation. In addition, microfinance may be used to improve the climate resiliency of the most vulnerable population. Microloans to be used for acquiring solar panels and cookstoves provide the poor people with an opportunity to recover or decrease costs and prepare them to adapt to the changing climatic conditions in advance. It helps them strengthen their economic situation and adapt to changes and invested money will also contribute to the achievement of broader climate objectives.

Public–Private Partnerships

Public–Private Partnerships (PPPs) also represent a key opportunity in the climate finance landscape. As demonstrated by the UNEP (as cited in Casady et al., 2024), PPPs can help fill the gap between the resources of public entities and expertise of the private sector, specifically in the areas of developing and managing large-scale projects. The strengths of PPPs lie in the possibility to pool resources across partners, share risks, and leverage the efficiency of the private sector in pursuing climate finance goals. In the context of renewable energy, UNEP (as cited in Casady et al., 2024) described projects of launching wind and solar energy operations through PPPs in Morocco and Kenya. In this scenario, public entities were able to offer support through policy frameworks and initial financing, while the private sector contributed through expertise and investment in both projects. As governments focus more on increasing the volume of climate finance, especially in the context of meeting the global climate targets, strengthening PPPs should be highly prioritised concerning resources and opportunities.

Mainstreaming Climate Finance into Traditional Financial Systems

The challenge or opportunity for the future of climate finance is its mainstreaming. To achieve the masses necessary to address the global climate needs, it is vital that it is incorporated into the mainstream financial system (Nwokolo et al., 2023a, 2023b, 2023c, 2023d, 2023e). This would mean that the decision-making of all financial stakeholders in all sectors should be influenced by consideration of climate investment as well as noninvestment. Furthermore, the flow of capital should be shifted from high-carbon undertakings towards sustainable, low-carbon gas. One of the key instruments for these processes is climate risk disclosure—disclosures issued by companies or financial organizations to report their climate-related risks and possibilities. By integrating this requirement into the regulatory environment, authorities can ensure that investors are able to adequately realize the risks of climate and appropriately redistribute their capital. Moreover, it is crucial that standards for thresholds and metrics for climate investments are adopted and implemented within the economy (Nwokolo et al., 2023a, 2023b, 2023c, 2023d, 2023e). These standards mean that the investment of climate finance follows the necessary threshold to qualify for significant climate goals and thus are impactful. At the same time, they need clear sets of metrics and standards, according to which recalibration might happen in case they are not met. Overall, the influential governments and financial stakeholders should focus on the development and implementation of standards.

Future Directions in Climate Finance

The future of climate finance is to be shaped by emerging trends and innovations aimed at aligning financial flows with the ambitions of the Paris Agreement. As the world community aspires to defeat climate change, it is essential to consider new and varied strategies and mechanisms that can be used for climate finance. Among those trends are nature-based solutions, impact investing, and advanced technologies. These novelties allow for promising not only to increase the scale of climate finance, but also to make it more sustainable, credible, and impactful.

Nature-Based Solutions: A Sustainable Trend in Climate Finance

Nature-based solutions are perceived as the most positive trend in, and sustainable approach to climate finance. NBs are defined as the sustainable management and restoration of ecosystems that provide us with a wide range of benefits, such as the supply of oxygen, water, and food, the pollination of crops, the control of climate and disease, and the maintenance of biodiversity (Seddon et al., 2020). These ecosystems are essential to address a wide range of environmental social, and, primarily, financial issues, such as carbon emissions. Thus, by conserving wildlife, it is feasible to prevent diseases and protect food resources. As a result of the increasing effects of climate change, there is a greater emphasis than ever on the need to adopt NBs. Nature-based solutions (Nbs) are increasingly cost-effective as compared to artificial, or engineered solutions that are employed in such cases (Nwokolo et al., 2023a, 2023b, 2023c, 2023d, 2023e). The reason is that the ecosystems are naturally designed to absorb carbon, cycle water, and protect from fires. Thus, an estimate applied by Haque et al. (2023) showed that millions of people who live near mangroves, small forests that are used for the "packaging" of storms, tsunami, and other water challenges, are protected from yearly hurricanes and torrents. Haque et al. point out the fact that mangrove plays a very significant role in preventing climate change since it acts as an effective sink of carbon. This type of ecosystem can sequester large amounts of carbon from the air as well as from surrounding bodies of water, such as seas and coastlines. The carbon-reducing capacity of mangrove is the result of the peculiarities of its structure, including a complex system of roots and dense, organic soil, which allows carbon to be trapped and stored for a relatively long period of time. In this way, mangroves are a vital component of the global system of carbon management. The fact that the mangrove is also capable of absorbing carbon from water makes it doubly useful, as it helps to maintain marine ecosystems. For this reason, it is important to reduce the destruction of this type of forest and search for ways to restore it.

The NbS integration into the climate finance is a significant change in the perception of climate finance and their focus towards more holistic and sustainable development. NbS allows us to restore and protect natural ecosystems and, at the same time, provide a diversification of other benefits, such as biodiversity conservation, improved livelihoods, and increased climate resilience for the local populations (Donatti et al., 2022). As a response to increasing awareness of the benefits of NbS, they are now increasingly present in national climate strategies and emerging international climate finance architecture, which contributes to the emphasis on NbS as a critical part of future climate finance. There are multiple examples of such NbS, for instance, forest and mangrove restoration projects. Kinol et al. (2023) argue that one of the most prominent trends affecting the trajectory of development for climate finance is the effective integration of Nature-based Solutions into it. The development of climate finance has increased the focus on natural ecosystems as an alternative to purely technological or infrastructural interventions aimed at addressing the impact of climate change. As the perspective of climate finance progresses, it is expected to gradually transform, focusing increasingly on natural ecosystems. The fundamental benefits of Nbs in the context of climate finance are associated with the far-reaching benefits of deep investment in the sector. Furthermore, Nbs offer several advantages besides ecological balance, including biodiversity preservation, increased resilience, and efficient resource management, making the approach central to future climate adaptation and mitigation. In addition, the consideration of such solutions in the context of climate finance is beneficial to global sustainability goals, ensuring that economic activities driven by investments are aimed at preserving the nature and human satisfaction.

Impact Investing

Impact investing is also one of the most powerful trends that help to shape the future of climate finance (Alaoui et al., 2024). Barber et al. (2021) describes impact investments as investments made with the intention to generate positive, measurable social and environmental impacts alongside financial returns. Impact investing became more popular with climate change since it could serve as a source of private finance for projects contributing to climate protection or adaptation. The most evident benefit of impact investing is its ability to motivate private finance to invest in projects focused on the transition to the low-carbon economy and social development. The result of the emergence of this new approach to the climate finance was the creation of multiple innovative financial instruments, such as green bonds and sustainability-linked loans. Green bonds are a new type of debt securities that are issued specifically to support projects that have a positive impact on the environment (CBI, 2020). These projects include renewable energy projects, energy efficiency, and pollution prevention projects. Green bonds became increasingly popular, and nowadays, both the government and private corporations issue such bonds to fund climate and contribute to the environmental sustainability of the earth. The accelerated growth of the green bonds market is an indicator of a broader trend of the growth of investor interest in financial instruments that yield financial returns and at the same time ensure a beneficial impact on the environment. Sustainability-linked loans are another important example of how investing impacts is shaping the future of climate finance (Nwokolo et al., 2023a, 2023b, 2023c, 2023d, 2023e). This type of loan provides financial incentives to the borrowers for reaching pre-agreed sustainability goals, such as reduction of greenhouse gas emissions or improvement of energy performance. This type of measurement allows encouraging the company to adopt or enhance their sustainable performance and, therefore, facilitates the achievement of global climate ambitions.

The Role of Technology in the Climate Financing

The space of climate finance is a rapidly evolving one, and as more developments come to the fore, technology will be inscribed in this dynamic landscape (Nwokolo et al., 2025). Adopting the technology of blockchain - one of the most important technological processes in this sphere, allows the tracing and recording of financial transfers. This is important because there are many stakeholders, and there may be problems with either corruption or misallocation of funds. The financial flows of the climate projects should be reflected in the respective classification decently and thus recoded and tracked. The OECD had earlier supported such solutions, noting the high potential of blockchain in measuring climate finance (Schletz et al., 2020). Other emerging technologies for climate assessment such as data analytics, and artificial intelligence tools could offer similar improvements in quantifying current climate risks and assessing these investments (Kamyab et al., 2023). This means that using these technologies may boost the accuracy of the environmental finance given that these technologies offer insights on some of the outcomes associated with interventions in other areas and can have predictive capabilities on how the overall climate would change with the intervention that is being planned. Digital platforms are also increasingly important tools for the dissemination and mobilization of climate finance. For example, climate crowdfunding platforms enable people and organizations to directly fund climate projects in a wide range of projects (Flórez-Parra et al., 2020). Thus, this opens the possibility of bringing climate finance into more hands than just those of a few stakeholders, allowing for the democratization of climate finance. Besides, these platforms would be able to pool together many small individual investments and thus in aggregate make it easier to fund the large-scale projects that we need to raise large amounts of money.

Conclusion

Innovations that align financial flows with the targets of the Paris Agreement will transform climate finance. Nature-based solutions, impact investing, and blockchain and AI technologies will define climate finance in the future by creating an effective, quality-based and far-reaching system. These trends will require the incorporation of climate finance into the main financial systems, resilience building through targeted investments, and the premise that financial flows facilitate the transition to a sustainable future with a limited carbon content. Opportunities to use these trends will allow overcoming climate finance challenges and contributing to a more resilient and sustainable world.

References

Alaoui, A. A., & Azdimousa, H. (2024). The future of impact investing: Strategies for scaling global impact. *International Journal of Economic Studies and Management (IJESM), 4*(2), 288–298. https://doi.org/10.5281/zenodo.10907693

Arndt, C., Asante, F., & Thurlow, J. (2015). Implications of climate change for Ghana's economy. *Sustainability, 7*(6), 72147231. https://doi.org/10.3390/su7067214

Barber, B. M., Morse, A., & Yasuda, A. (2021). Impact investing. *Journal of Financial Economics, 139*(1), 162–185. https://doi.org/10.1016/j.jfineco.2020.07.008

Barradale, M. J. (2010). Impact of public policy uncertainty on renewable energy investment: Wind power and the production tax credit. *Energy Policy, 38*(12), 7698–7709. https://doi.org/10.1016/j.enpol.2010.08.021

Buckle, S., Ellis, J., Jaber, A. A., Rocha, M., Anderson, B., & Bjersér, P. (2020). *Addressing the COVID-19 and climate crises*: Potential economic recovery pathways and their implications for climate change mitigation. *Ndcs and Broader Socio-Economic Goals*. https://doi.org/10.1787/2227779X

Casady, C. B., Cepparulo, A., & Giuriato, L. (2024). Public-Private Partnerships for Low-Carbon, Climate-Resilient Infrastructure: Insights from the Literature. *Journal of Cleaner Production*, 143338. https://doi.org/10.1016/j.jclepro.2024.143338

Climate Bonds Initiative. (2020). Green bond market summary Q3 2020. *Clim. Bonds, 11*, 1–2.

Committee on Climate Change. (2019). *Reducing UK emissions*. 2019 progress report to parliament.https://www.theccc.org.uk/publication/reducing-uk-emissions-2019-progress-report-to-parliament/.

Dey, S., Sreenivasulu, A., Veerendra, G. T. N., Rao, K. V., & Babu, P. A. (2022). Renewable energy present status and future potentials in India: An overview. *Innovation and Green Development, 1*(1), 100006. https://doi.org/10.1016/j.igd.2022.100006

Donatti, C. I., Andrade, A., Cohen-Shacham, E., Fedele, G., Hou-Jones, X., & Robyn, B. (2022). Ensuring that nature-based solutions for climate mitigation address multiple global challenges. *One Earth, 5*(5), 493–504. https://doi.org/10.1016/j.oneear.2022.04.010

Dubash, N. K., Khosla, R., Kelkar, U., & Lele, S. (2018). India and climate change: Evolving ideas and increasing policy engagement. *Annual Review of Environment and Resources, 43*(1), 395–424. https://doi.org/10.1146/annurev-environ-102017-025809

Fisher, D. R. (2019). The broader importance of #FridaysForFuture. *Nature Climate Change, 9*(6), 430431. https://doi.org/10.1038/s415580190484y

Flórez-Parra, J. M., Rubio Martín, G., & Rapallo Serrano, C. (2020). Corporate social responsibility and crowdfunding: The experience of the colectual platform in empowering economic and sustainable projects. *Sustainability, 12*(13), 5251. https://doi.org/10.3390/su12135251

Goldar, A., Sharma, S., Sawant, V., & Jain, S. (2019). *Climate change & technology transfer*: Barriers, technologies and mechanisms (No.382). Working Paper.https://hdl.handle.net/10419/203716.

Haque, E. U., Hayat, A., Afzaal, S., Asim, M., Hanif, S., & Awan, T. (2023). Disaster Hazards and Vulnerabilities in Agriculture: Role of Food Technologist. In Disaster Risk Reduction in Agriculture (pp. 407–428). Singapore: Springer Nature Singapore. https://doi.org/10.1007/978-981-99-1763-1_19.

Iacobuţă, G. I., Brandi, C., Dzebo, A., & Duron, S. D. E. (2022). Aligning climate and sustainable development finance through an SDG lens. The role of development assistance in implementing the Paris Agreement. *Global Environmental Change, 74*, 102509. https://doi.org/10.1016/j.gloenvcha.2022.102509.

Joas, F., Pahle, M., Flachsland, C., & Joas, A. (2016). Which goals are driving the Energiewende? Making sense of the German Energy Transformation. *Energy Policy, 95*, 42–51. https://doi.org/10.1016/j.enpol.2016.04.003

Joshi, A., Singh, R., Kumar, V., & Rani, S. (2024). Net Zero Strategy: Exploring Opportunities, Confronting Challenges for Developing Countries. *In Transition Towards a Sustainable Future* (pp. 233–252). Springer, Singapore. https://doi.org/10.1007/978-981-97-5756-5_11.

Kamyab, H., Khademi, T., Chelliapan, S., SaberiKamarposhti, M., Rezania, S., Yusuf, M., & Ahn, Y. (2023). *The latest innovative avenues for the utilization of artificial Intelligence and big data analytics in water resource management*. Results in Engineering, 101566.https://doi.org/10.1016/j.rineng.2023.101566.

Kinol, A. D., Arango-Quiroga, J., & Kuhl, L. (2023). *Opportunities for nature-based solutions to contribute to climate-resilient development pathways*. Current opinion in environmental sustainability, *62*, 101297.https://doi.org/10.1016/j.cosust.2023.101297

Kumar, K., Singh, R., Sharma, A., & Lehri, B. P. S. (2024). Climate Change Adaptation Strategies for Achieving Net-Zero Economy. *In Transition Towards a Sustainable Future* (pp. 163–183). Springer, Singapore. https://doi.org/10.1007/978-981-97-5756-5_8.

Mc Guinness, S., & Bullock, C. (2020). *Mobilising Finance for Biodiversity: A policy and institutional review of finance arrangements for biodiversity conservation in Ireland*. University College Dublin. https://doi.org/10.13140/RG.2.2.12378.90565.

McKibben, B. (2012). *The global warming reader: A century of writing about climate change.* Penguin.
Mirzania, P., Gordon, J. A., Balta-Ozkan, N., Sayan, R. C., & Marais, L. (2023). Barriers to powering past coal: Implications for a just energy transition in South Africa. *Energy Research & Social Science, 101*, 103122. https://doi.org/10.1016/j.erss.2023.103122
Mohamadi, F. (2021). Introduction to project finance in renewable energy infrastructure: Including public-private investments and non-mature markets. *Springer Nature.* https://doi.org/10.1007/978-3-030-68740-3
Moyo, T. (2014). Green economy/growth policies and their implementation in the context of the renewable energy sector: The case of Mozambique, South Africa and Zimbabwe. *International Journal of African Renaissance Studies-Multi-, Inter-and Transdisciplinary, 9*(2), 39–60. https://doi.org/10.1080/18186874.2014.987954
Natural Capital Coalition. (2016). Natural capital protocol. *Natural Capital Coalition.* https://doi.org/10.2139/ssrn.3073448
Nwokolo, S.C., Singh, R., Khan, S., & Kumar, A. (2025). Threats to the rapidity of sustainability transitions posed by technological changes. Technological Pathways for Africa's Net-Zero Economy. Elsevier, 39–75. https://doi.org/10.1016/B978-0-443-31486-5.00002-1.
Nwokolo, S. C., Singh, R., Khan, S., Kumar, A., & Luthra, S. (2023). Remedies to the Challenges of Renewable Energy Deployment in Africa. In Africa's Path to Net-Zero: Exploring Scenarios for a Sustainable Energy Transition (pp. 59–74). Cham: *Springer Nature Switzerland.* https://doi.org/10.1007/978-3-031-44514-9_3.
Nwokolo, S. C., Singh, R., Khan, S., Kumar, A., & Luthra, S. (2023). Africa's Awakening to Climate Action. In Africa's Path to Net-Zero: Exploring Scenarios for a Sustainable Energy Transition (pp. 299–310). Cham: *Springer Nature Switzerland.* https://doi.org/10.1007/978-3-031-44514-9_9.
Nwokolo, S. C., Singh, R., Khan, S., Kumar, A., & Luthra, S. (2023c). Africa's Path to Net-Zero. *CSR, Sustainability, Ethics & Governance.* https://doi.org/10.1016/B978-0-443-31486-5.00002-1
Nwokolo, S. C., Singh, R., Khan, S., Kumar, A., & Luthra, S. (2023). Global Investment and Development in Africa. In Africa's Path to Net-Zero: Exploring Scenarios for a Sustainable Energy Transition (pp. 15–58). Cham: *Springer Nature Switzerland.* https://doi.org/10.1007/978-3-031-44514-9_2.
Nwokolo, S. C., Singh, R., Khan, S., Kumar, A., & Luthra, S. (2023). Impacts of Climate Change in Africa. In Africa's Path to Net-Zero: Exploring Scenarios for a Sustainable Energy Transition (pp. 237–262). Cham: *Springer Nature Switzerland.* https://doi.org/10.1007/978-3-031-44514-9_7.
Pickering, J., Skovgaard, J., Kim, S., Roberts, J. T., Rossati, D., Stadelmann, M., & Reich, H. (2015). Acting on climate finance pledges: Inter-agency dynamics and relationships with aid in contributor states. *World Development, 68*, 149–162. https://doi.org/10.1016/j.worlddev.2014.10.033
Rajagopalan, S., & Breetz, H. L. (2022). Niches, narratives, and national policy: How India developed off-grid solar for rural electrification. *Environmental Innovation and Societal Transitions, 43*, 41–54. https://doi.org/10.1016/j.eist.2022.02.004
Sachs, J. D., Flanagan, O., Sorondo, M. S., Vendley, W., Annett, A., & Thorson, J. (Eds.). (2022). *Ethics in action for sustainable development.* Columbia University Press. https://doi.org/10.7312/sach20286.
Sarkar, A. (2022). Minimalonomics: A novel economic model to address environmental sustainability and earth's carrying capacity. *Journal of Cleaner Production, 371*, 133663. https://doi.org/10.1016/j.jclepro.2022.133663
Schletz, M., Nassiry, D., & Lee, M. K. (2020). *Blockchain and tokenized securities: The potential for green finance* (No. 1079). ADBI Working Paper Series. https://hdl.handle.net/10419/238436.

Seddon, N., Chausson, A., Berry, P., Girardin, C. A. J., Smith, A., & Turner, B. (2020). Understanding the value and limits of nature based solutions to climate change and other global challenges. *Philosophical Transactions of the Royal Society B: Biological Sciences, 375*(1794), 20190120. https://doi.org/10.1098/rstb.2019.0120

Singh, M. K. (2022). *Conclusion: Overcoming barriers to decentralized energy systems in India*: In eradicating energy poverty. *Palgrave Macmillan, Singapore.* https://doi.org/10.1007/978-981-16-7073-2_5

Singh, D., Sant, G., & Chunekar, A. (2012). Improving energy efficiency in India: Need for a targeted and tailored strategy. *Wiley Interdisciplinary Reviews: Energy and Environment, 1*(3), 298–307. https://doi.org/10.1002/wene.45

Sovacool, B. K. (2019). Visions of energy futures: Imagining and innovating low-carbon transitions. *Routledge.* https://doi.org/10.4324/9780367135171

TEEB. (2010). The economics of ecosystems and biodiversity ecological and economic foundations. *Earthscan.* https://doi.org/10.4324/9780203885353

Thunberg, G. (2019). No One Is Too Small to Make a Difference. *Penguin Books.* https://doi.org/10.4324/9780203488593

Timilsina, G. R. (2021). Financing climate change adaptation: International initiatives. *Sustainability, 13*(12), 6515. https://doi.org/10.3390/su13126515

Weisheng, L., MM Liu, A., Hongdi, W., & Zhongbing, W. (2013). Procurement innovation for public construction projects: A study of agent-construction system and public-private partnership in China. *Engineering, Construction and Architectural Management, 20*(6), 543-562. https://doi.org/10.1108/ECAM-09-2011-0084

The manufacturer's authorised representative in the EU is Springer Nature Customer Service Centre GmbH, Europaplatz 3, 69115 Heidelberg, Germany. If you have any concerns regarding our products, please contact ProductSafety@springernature.com

Printed and bound by CPI Group (UK) Ltd, Croydon, CR0 4YY
01/10/2026
02225932-0010